# BUILDING GEOGRAPHIC LITERACY
## An Interactive Approach

Fourth Edition

Charles A. Stansfield, Jr.
Rowan University of New Jersey

**Prentice Hall**
Upper Saddle River, New Jersey 07458

*Library of Congress Cataloging-in-Publication Data*

Stansfield, Charles A.
Building geographic literacy: an interactive approach / Charles A. Stansfield, Jr. – 4th ed.
p. cm.
ISBN 0–13–062210–9
1. Geography. I. Title.
G128.S724 2002
910--dc21 2001035316

Executive Editor: Dan Kaveney
Assistant Editor: Amanda Griffith
Editorial Assistant: Margaret Ziegler
Production Editor/Composition: Lithokraft II
Executive Managing Editor: Kathleen Schiaparelli
Assistant Managing Editor: Beth Sturla
Marketing Manager: Christine Henry
Managing Editor, Audio/Video Assets: Grace Hazeldine
Art Editor: Adam Velthaus
Art Director: Jayne Conte
Cover Designer: Bruce Kenselaar
Cover Credit: © 2002, 1998, 1994 Alan R. Epstein
Manufacturing Manager: Trudy Pisciotti
Assistant Manufacturing Manager: Michael Bell
Vice President of Production and Manufacturing: David W. Riccardi

Printed in the United States of America
10 9 8 7 6 5 4 3

ISBN 0-13-062210-9

Pearson Education Ltd., *London*
Pearson Education Australia Pty, Limited, *Sydney*
Pearson Education Singapore, Pte. Ltd.
Pearson Education North Asia Ltd., *Hong Kong*
Pearson Education Canada, Ltd., *Toronto*
Pearson Educación de Mexico, S.A. de C.V.
Pearson Education—Japan, *Tokyo*
Pearson Education Malaysia, Pte. Ltd.

# Contents

# Preface

*Building Geographic Literacy* is a workbook. It will involve you in creating maps, applying geographic principles, and learning basic place geography. Multiple copies of blank outline maps are provided for you to complete by correctly locating and labeling placenames. You will need to refer to an atlas or to maps in a geography textbook to supply the place locations that you identify on the outline maps.

In this new edition, all demographic, economic and cultural statistics have been updated as of the most recent data available at publication. This edition contains eight new tables and an additional 60 questions. There are seven new maps on which students are asked to enter data and interpret resulting patterns. Extra copies of regional and world maps are provided in Appendix B for the convenience of students wishing to redo maps and to enable instructors to assign map exercises of their own choice.

This workbook can be used with any standard world atlas or geography textbook. The placenames selected for placement and labeling on the outline maps are those most likely to appear on textbook maps and are commonly cited in geography texts.

Although the spelling of placenames reflects standard usage, some books and atlases may use slightly different transliterations from foreign languages, particularly Chinese and Arabic. These minor variations are not significant and will not interfere with the utility of this workbook.

You will be asked to *apply* your newly acquired knowledge of placename geography, and thus reinforce that knowledge, by making thematic maps. Thematic maps communicate qualities or characteristics or places in addition to *providing* place information. These qualities may be physical, economic, cultural, or political factors that vary geographically, that is, across space. For example, you will be making maps of population growth rates and per capita gross national product for countries in various regions.

This workbook is *interactive,* that is, you are asked to write answers, label maps, and make maps in the workbook itself as you go along. You will be studying placename geography by making placename maps, and by constructing thematic maps that apply placename comprehension to other geographic questions and concepts.

The first chapter explains the importance of knowing placename geography. Chapter 2 reviews Earth's location grid, the lines of latitude and longitude, and examines time zones and the International Date Line. The place geography of the continents, major islands, and oceans and seas is outlined in Chapter 3, which includes a glossary of important land and water features.

Chapters 4 to 13 comprise 10 regional chapters, each studying a major subdivision of Earth's lands and nations. Each chapter begins with an introduction to the region, followed by a survey of its important physical geography features.

"Objectives and Study Hints" highlight the significance of the geography of countries and peoples of the region, followed by a review of the region's political, economic, and cultural geography and regional demographics—the size, make up, and growth rates of its population. "Check Up" questions then test your geographic understanding of the region. Each chapter then summarizes one or two important geoconcepts related to the places just studied. Some chapters include an optional "Special Challenge"—an exercise that focuses your attention on a somewhat more complex geographic project or problem.

That geography is a dynamic subject has been demonstrated amply by the many boundary changes since the first edition. Chapter 6, initially titled "The USSR: The Devolution of Empire," has been rewritten completely. Further boundary changes in Europe—the fragmentation of Yugoslavia and the split of Czechoslovakia—also have been incorporated in maps and tables. Eritrea's secession from Ethiopia has been noted as well. Canada's 2000 creation of a new territory, Nunavut, has been noted on all maps of Canada and North America. The pace of boundary changes may slow, but future changes are likely in several regions. A concluding section, "Regional Watchlist," comments briefly on potential boundary changes one should be alert to after this edition goes to press.

It is important to note that, although the 10 major world regions used here are widely accepted by geographers, there are no right or wrong systems of regionalization. Regions are identified and bounded by geographers to facilitate geographic study. A region has some degree of internal similarity, and areas beyond it are characterized by different qualities. There are many possible, logical, and useful regionalization schemes because there are many different geographic criteria for describing regions. This workbook can be used with a variety of textbook regional systems.

I would like to thank my colleague Edward Behm at Rowan University for his many perceptive comments on the first edition. Professor W. Frank Ainsley, University of North Carolina at Wilmington; Richard Dixon, Southwest Texas State University; and Michael P. Peterson, University of Nebraska at Omaha, contributed many useful comments on the third edition. I appreciate their diligent and perceptive criticisms.

This fourth edition clearly has benefitted from the superbly professional and efficient efforts of Dan Kaveney, Executive Editor. Once again, thanks to Paul Corey for his enthusiastic support of the original concept. I'd like to thank Megan Hill for her thorough copyediting, and the staff of Lithokraft II for their careful formatting. Mapquest prepared the new maps required by this edition, and redrafted many other base maps.

Charles A. Stansfield, Jr.
Rowan University of New Jersey

# 1

# The Significance of Place

## WHY STUDY PLACE GEOGRAPHY?

As a patient regains consciousness, recovering from the effects of anesthesia, two questions are asked to determine his or her level of awareness: "Do you know what day it is?" "Do you know where you are?" Successfully answering these questions on the level required ("It's Tuesday and I'm in the hospital in Denver") indicates regained consciousness. Successfully answering those same basic questions on a broader philosophical level indicates a generally educated person. Do you truly understand where you are in time and place?

History deals with time contexts; geography deals with *spatial contexts.* History explains the time sequence of events—what happened when, and why. Geography focuses on understanding the significance of location: Where is it? What is it like? Why is it there? Understanding the fundamental facts of geography—placename geography—provides a key to world understanding. People live in a time and a place, and events happen in the context of a time and a place. It is necessary for us to see our own time and place in relation to past times and other places.

Where we live has a history, and we begin to truly know a place when we know something about how it came to be. If we can relate the history of a particular place to the history of the state, the nation, and the world, then we understand the historical context of that place. Every place has links with its own local past, with other places near and far, and with broader historical events and trends as well.

Where we live also has a geography. Any particular place on Earth has a web of interconnections with other places. No point on the planet can exist in isolation, either in the historical or geographical sense. The character of any place is shaped by its location relative to other places, in addition to its local surroundings. Placename geography provides a basic framework for understanding the world around us.

## SOME BASIC QUESTIONS IN GEOGRAPHY

The literal meaning of geography is "description of the Earth." It is an ancient science, for people have always been interested in what other places are like and how they relate to home. Description in geography is a good beginning, but only the first step. Modern geography starts with accurate description of location (Where is it?) and various qualities of location (What is it like?). Placename geography is to more advanced geographic studies as learning to read is to studying literature; one cannot interpret the symbolism in poetry without first knowing how to read, and one cannot understand the importance of London's *relative* location without knowing London's *exact* location.

Thinking geographically requires a good knowledge of placename geography. Geographic thinking then uses this basic knowledge to help analyze the significance of location; to explain the physical, cultural, economic, or political qualities of that place; and to understand linkages between that place and other places. If the first pair of related questions in geography are "where is it, and what is it like?" then the next organizational question is "why there?" Geography seeks explanations for the characteristics and interactions of places and patterns. This third question could be paraphrased as "so what?" (relevance). In other words, now that exact location is known, and the various qualities of that location have been described, and an explanation of *why* that pattern or place exists where it does and is like it is, the *relevance* of that place to other people in other places can be examined. We can now look for *links* and try to understand the reasons for those links and their impact.

Place geography is essential information. Developing a general knowledge of placename geography gives you a useful competency. As in learning a foreign language or a computer language, however, you must *use* this knowledge to keep it current and to reward your effort in learning it. Applying placename geography to the challenge of understanding current events not only refreshes your place geography, but also justifies its study. The ultimate goal of mastering placename geography is to better comprehend what we see and hear happening around the world. When we can perceive *connections*—connections between what happens in one place and what happens in other places—then we have developed a deeper understanding of events, causes, and consequences.

This workbook will help you to develop your knowledge of placename geography. It also demonstrates the relevance of placename geography, providing opportunities to apply this knowledge and appreciate its usefulness to an educated person. For these reasons, you will be asked not only to complete placename maps, but also to interpret *thematic maps* (maps that organize and display other types of geographic information) and to construct some thematic maps yourself.

Geographers look at locations on two scales—site and situation. *Site* refers to the exact location. It is a very specific description of the location's characteristics in terms of physical geography (climate, topography, etc.) and human geography (population density, language, culture, etc.). *Situation* is a broader definition of location. Situation is relative location, relative to a larger region, a country, or the world.

For example, the *site* of early New York City was at the southern tip of Manhattan Island, an elongated rocky island surrounded by the Hudson River and the channels known as the East and Harlem rivers. But the *situation* of New York City is at the inner end of New York Bay on the naturally deep Hudson River, which is navigable by oceangoing ships to Albany. The situation of New York City is at the seaward end of a great natural routeway up the Hudson Valley and westward through the Mohawk Valley to the Lake Ontario Plain, then southwestward along the Lake Erie Plain to the vast interior lowland at the continent's heart. The situation, or relative location, of New York City is that of a superb natural harbor at the Atlantic coast terminus of a great natural highway cutting through the Appalachian Mountain system to the highly productive midwest of America.

## FIVE THEMES IN GEOGRAPHY

At some point early in their study of geography, students confront the realization that geography seems to be concerned with a great variety of topics. A course in world regional geography, for example, will introduce and apply data and ideas from geology, glaciology, climate, oceanography, pedology (study of soils), natural vegetation, and ecology. Historical studies, economics, political science, and anthropology likewise will supply information and insights to aid in comprehending world geography. These many disciplines, whose viewpoints and data can be vital to an understanding of geography, add up to an impressive, perhaps even daunting, list. There are moments when geographic studies seem to be encyclopedic, requiring at least some knowledge of the physical, biological, and social sciences.

Students of geography need guidance through the mountains of potentially useful information. That guidance is supplied by focusing on five fundamental themes in geography. These are location, place, relationships within places, movement, and regions.

### Location

As already noted, location, or place geography, is the obvious starting point in any geographic study. Location can be described at two levels—absolute location and relative location. Absolute location can be identified precisely using the Earth's grid system of degrees of latitude and longitude, described in detail in Chapter 2. Degrees of latitude and longitude are degrees of a circle, which contains 360° of arc. As the sphere of the Earth has an average circumference of 24, 900 miles, a degree of *latitude* is one-360th of 24,900 miles, or 69 miles (actually, the earth is *spheroid,* or spherelike, rather than a perfect sphere; 24,900 miles is the average circumference of a very slightly pear-shaped spheroid). Degrees of *longitude,* unlike average degrees of latitude, can be generalized in length only at the Equator as 69 miles. Because the distance between meridians of longitude steadily decreases toward the poles, degrees of longitude are not constant in miles. Each degree of latitude or longitude can be subdivided into minutes and seconds. Minutes and seconds of latitude or longitude have the same relationship to degrees of a circle as minutes and seconds have to hours. Each degree of a circle is subdivided into 60 minutes, and each minute contains 60 seconds. A map grid partial notation of 30° 17′ 37″N, for example, would read 30 degrees, 17 minutes, 37 seconds of latitude North. One minute of latitude is about 1.15 miles in length; one second of latitude is about 100 feet. The actual distance between longitudinal degrees diminishes as the meridians converge at the Poles. With the possibility of determining location down to one-sixtieth of one-sixtieth of one degree, this is a reasonably precise and reliable position indicator as positions of the globe's land and water features have not changed appreciably in recorded history.

Determining location on the Earth's grid with precision is only the beginning of wisdom. Questions of relative location carry geographic inquiry to a level beyond absolute location. Relative location is a key concern in understanding interdependence at geographic scales ranging from local through regional and national to global. When one looks at the relative locations of the Persian Gulf oilfields and the major oil markets in Europe, Japan, and North America, their relative locations give new significance to locations like the Straits of Hormuz, the Suez Canal, and the Straits of Gibralter. How would the discovery of major new oilfields in th South China Sea, the Arctic Basin, or Australia affect the significance of relative locations around the world?

## Place

Every place on Earth has a unique geographic complex of physical and cultural characteristics. Physical characteristics include landforms, water features, climate, soils, natural vegetation, and animal life. Cultural characteristics include population density and settlement patterns, race and ethnicity, language and religion, levels of economic development, architecture and material culture, and philosophies of economic and political organization. Consider the concept of "home" or "homeland." Home and homeland are intensely emotional concepts; around the world, people can find themselves fighting for, perhaps dying for, their homeland. This passionate devotion to a place reflects the unique blend of physical and cultural characteristics that personify places.

With very few exceptions (Antarctica, the remote Arctic, etc.), each place on Earth has been shaped, modified, and affected by human actions. Whereas the human/cultural imprints on the landscape are at their most obvious in urban areas, the countryside rarely is solely the product of physical geographic processes. In rural environments, people have displaced natural vegetation and animal life with their domesticated crops and animals, pasture grasses, and commercially desirable species of trees. People have cut down forests, drained swamps, cut and bored through mountains, and otherwise shaped and modified the physical world.

## Relationships within Places

Humans continuously interact with their environments. The decisions that people make concerning land use, natural resource exploitation, and settlement patterns are influenced by those people's cultural values, past experience, and goals as well as by their level of economic development and available technologies. People of different cultures *perceive* their environment and its potentials differently according to their cultural backgrounds. It is as through we all see the world through tinted glasses—the "tint" represents our cultural background, our values, our experiences as a people and as an individual, and our general understanding of humans' relationships with the physical world. Consider, for example, the varying interpretations of land use and resources utilization in the lower Mississippi Valley and Delta. Native-American perceptions of these swampy backwaters, "Bayou Country," of the natural levees formed by periodic floods along riverbanks, and of the better-drained uplands surely were different than those perceptions of the European and African Americans who introduced commercial agriculture into the area. Sugarcane fields and pasturelands both were created from drained and diked swamplands. Much of the "Bayou" backwater country was used more for hunting and fishing than for agriculture. If, however, this part of the world had been "invaded" by people from southern China and Southeast Asia rather than primarily Western Europeans and West Africans, would people of these different cultures have valued the swampy Bayous as potential rice paddies? In the 1930s, the discovery of oil and natural gas fields in Louisiana and neighboring Texas introduced another element in the relationships there between humans and environments. People continually reevalate a place's advantages and disadvantages for human settlement.

## Movement

"Movement" refers to spatial interactions. These are economic, cultural, and political interactions that may take place within a very local area, such as commuting patterns in a metropolitan area, or at any scale up to global, as in international trade. Visit any highway-oriented business "strip" in the United States or Canada.

Reading the labels or packaging materials in a typical discount clothing, appliances, household product and variety store will reveal the scope and importance of international trade. Televisions assembled in Mexico, VCRs from Korea, cameras from Japan, shoes from Brazil, and kitchen knives from Germany and France are displayed along with clothing from Malaysia, India, China, Sri Lanka—the list goes on and on. Supermarket shelves may hold some surprises for those who haven't paid close attention to labels. Fresh fruit from Chile and Israel, fresh tomatoes from the Netherlands and Mexico, canned fish from Norway and Canada, even spaghetti noodles from Turkey and Italy demonstrate the complex spatial interactions in our interdependent world.

### Regions

The last of the five fundamental themes in geography is that of regions. With the exception of three introductory chapters, this workbook is organized by world regions, and the nature of regions is developed in depth in a geoconcept later in this chapter. Regions are important tools in the study of geography because they enable geographers to focus on a convenient and manageable subdivision of the world.

## PLACING EVENTS IN THE CONTEXT OF TIME AND PLACE

What's happening? We're all interested in current events, both out of natural curiosity and the practical sense that these events may directly affect us and our future. *What* is happening can be better understood if we know *where* it is happening (see Figure 1–1).

Consider the following major recent and contemporary events:

- The Soviet Union has broken up into 15 republics; the largest of those republics, Russia, could further fragment if the Chechen Revolt succeeds.
- Religious fundamentalism is sweeping through the Muslim world from Indonesia to Morocco, from (former Soviet) central Asia to Arabia.
- The European Economic Community (EC) has become the European Union (EU), aspiring to a political union to match its economic union.
- The Pacific Rim countries are becoming a major force in world trade, led by enormously successful Japan, and followed now by South Korea, Taiwan and Singapore.
- China's experiments with economic freedoms have led to less successful attempts to achieve political freedom.
- Iraq precipitated a world crisis by invading Kuwait; Saudi Arabia faces an uncertain future.
- India's state of Kashmir is seeking greater autonomy, perhaps a prelude to joining Pakistan.

Each of these mega-events and trends must be evaluated in light of the history and culture of the nations and peoples involved. In each case, the geography of the trend or the event is also highly significant.

These geographic facts help explain what is happening, and where. Geographic relationships help us to understand *linkages.* The Russian Federated Republic is the largest country in the world. It dominated the other 14 republics of the former USSR. In most of the 14 smaller republics, the ethnic groups for which they are named form the majority of the population.

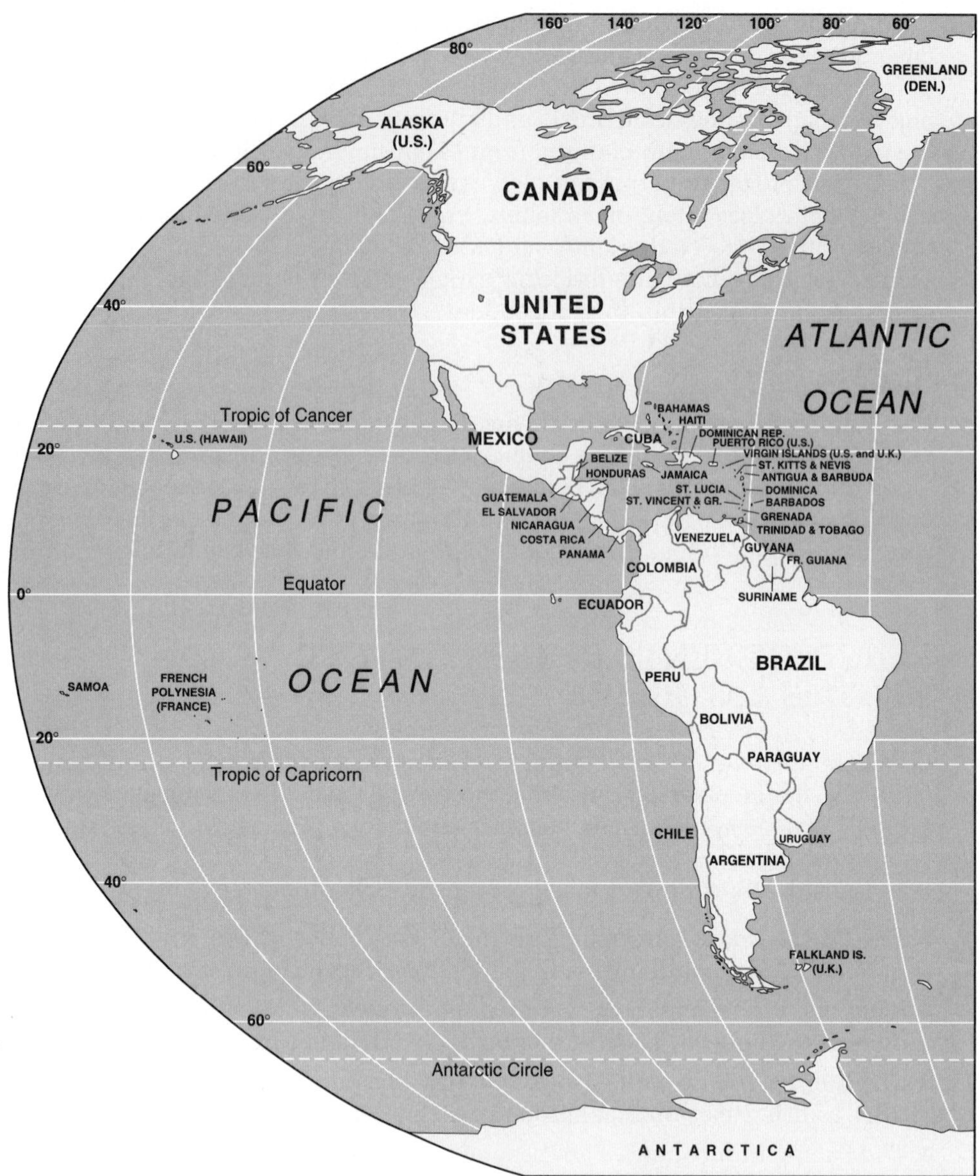


**FIGURE 1–1**
This political map reveals a highly compartmentalized world. The numerous political entities range in size from Russia to minute but significant countries such as Singapore, Malta, and Grenada. The names of these political entities evoke images of different environments, peoples, cultures, and levels of well-being. However, the political boundaries that segregate Earth's 6.1 billion inhabitants do not clearly reveal the underlying geographic complexities of our world. (Technical note: Political boundaries come and go, and this map reflects the recent union of East and West Germany and the troubled merger of North and South Yemen; the newly independent states (former USSR); the fragmentation of Yugoslavia; the split of the Czech Republic and Slovakia; and the independence of Eritrea from Ethiopia. At this writing, the sovereignty status of Western Sahara is unresolved; it is currently administered by Morocco.

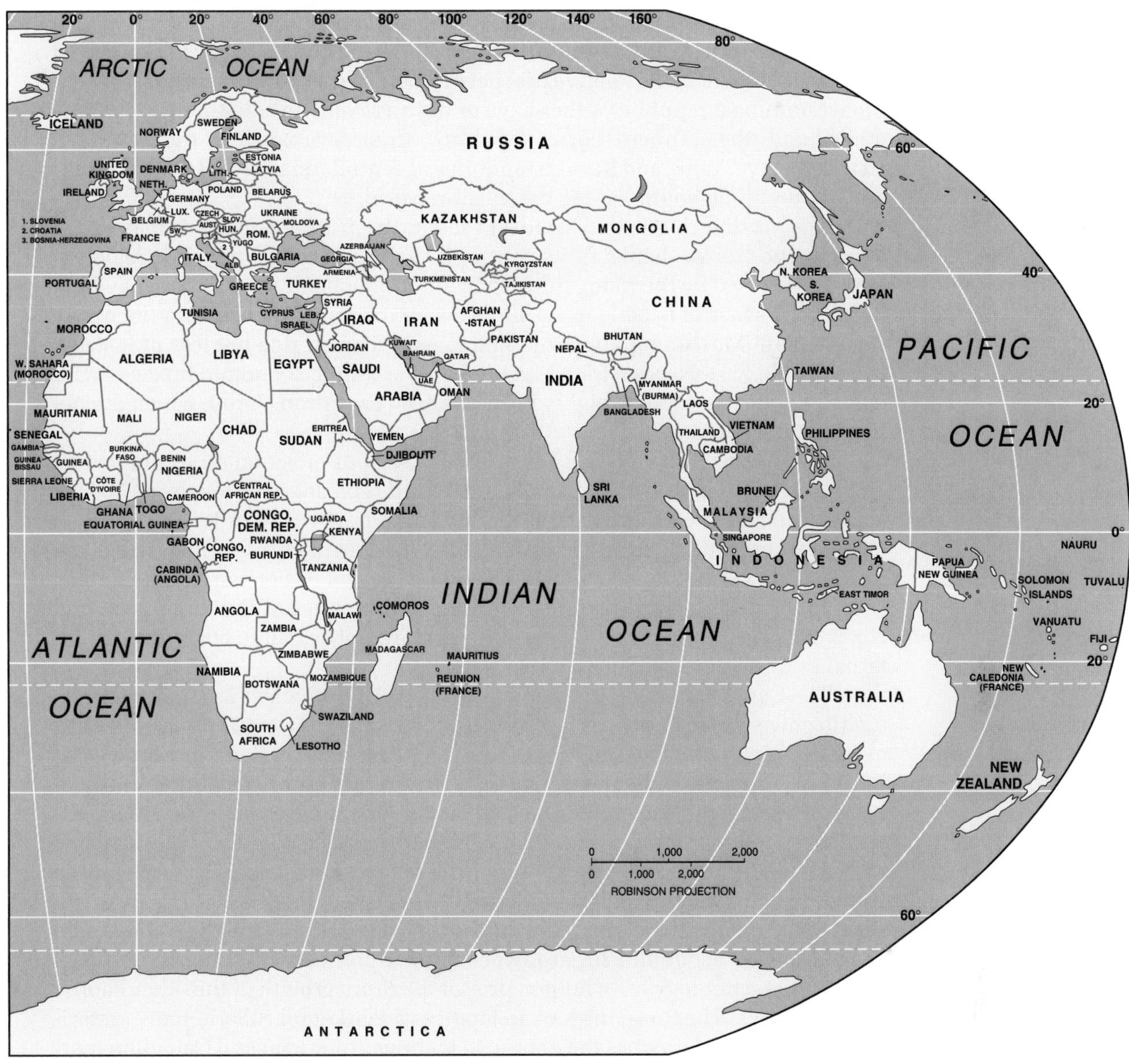
ARCTIC OCEAN
RUSSIA
KAZAKHSTAN
MONGOLIA
CHINA
INDIA
PACIFIC OCEAN
INDIAN OCEAN
ATLANTIC OCEAN
AUSTRALIA
ANTARCTICA
ROBINSON PROJECTION

The "western" republics, such as Lithuania, consider themselves more "European" than the ethnic Russians in both location and cultural identification, and they wish to share in the relative prosperity of Europe, especially the EU. Clearly, the westernmost republics either have or desire strong links to the West. On the other hand, the southern tier of republics, from Azerbaijan to the Turkmen, Uzbek, Kyrgyz, Tajik, and Kazakh republics of central Asia, have strong religious and cultural links southward to the Muslim world.

The Muslim-dominated Middle East–North Africa region has a great deal of clout in the world, thanks to its control of half the world's oil. This region is strongly affected by the rising tide of Muslim fundamentalism, which may help deepen the sense of Islamic unity and a desire to establish links with the newly independent Muslim republics of Central Asia. The oil-rich but less industrialized Middle East–North Africa Muslim belt has strong economic linkages with Japan and the prosperous, highly industrialized European Union, which is a net importer of energy.

Iraq's invasion of oil-rich Kuwait had a serious effect on all oil-importing countries, both industrialized and Third-World. The defeat of Iraq by a U.S.-led coalition may have brought about the birth of a new world order; it definitely changed the political geography of the Middle East. In today's fast-industralizing, increasingly interdependent world, the oil of Iraq and its Persian Gulf neighbors links them closely with the high-technology economies of Europe, Japan, and North America. Saudi Arabia's absolute monarchy has had to contend with internal as well as external threats.

The European Union is second only to the United States as an overall economic power. The EU's success reflects the economic power of very large political units as compared to the relatively small internal national economies of individual European countries. The EU's rising star contrasts sharply with economic stagnation in eastern Europe and Russia; as a huge economic unit, the EU will be an important trade partner for all the other world regions.

The Pacific Rim of Asia is one of the fastest-industrializing parts of the world, which is remarkable considering its relatively modest natural-resource base. Rapidly industrializing Taiwan, South Korea, Hong Kong, and Singapore are led by the economic superpower Japan. American and Japanese investments and technology have helped propel the rapid growth of the other Pacific Rim countries. The Russians look to Japan as a market for Siberian raw materials, and Japan looks across the Pacific to its huge American and Canadian markets. Much of Japan's oil comes from the Middle East, another interregional linkage.

China, relatively rich in natural resources but among the world's poorest nations in economic output per person, is experimenting with Western-inspired economic reforms in an attempt to develop faster. A possible economic partnership among the Pacific Rim countries and China could become a fourth enormous economic bloc comparable to the EU, the United States, Canada, and Mexico, and Russia (currently less productive, but potentially a huge economic power).

India, which in its cultural and ethnic variety is more comparable to the continent of Europe rather than to a single country, may face strong disruptive tensions. Such a destabilization of South Asia would create problems for the world community.

Increasingly, Americans and Canadians live, work, compete, buy, and sell in a global economy in which the United States no longer clearly predominates but rather is now first among many major players.

# GEOCONCEPT

## Regions

A region is an area of Earth that is distinctive in some respect. Regions must possess some degree of uniformity in their character or in their degree of internal interaction. Regions are the basis for subdividing larger geographic units, such as the whole Earth or a continent.

Everyone is familiar with regions. As children grow up, they learn to recognize their home neighborhood, a region in miniature. One's neighborhood has some degree of internal unity or similarity, some unique blend of physical, cultural, and land-use characteristics. Perhaps the neighborhood is made up largely of one ethnic or racial group. Most residents may share a similar income or lifestyle. Each neighborhood has boundaries—edges where people recognize a transition to another neighborhood or distinctively different area. Common types of urban or suburban neighborhood boundaries include very busy streets or highways, industrial districts, or parks.

On the blank page provided (Figure 1–2), draw a map of the region you know best—your home neighborhood. Think first about how large an area (in city blocks or miles) is involved. Perhaps your neighborhood is a small town and you'll find it easy to determine boundaries. If your neighborhood is in a larger city or suburb, it may have no official boundaries or designation, but it probably will have a commonly recognized name. What are the boundaries of your neighborhood? They may be ill-defined, but don't worry too much about precision at this point. Start drawing—get into it. It might help to sketch the approximate boundaries first, such as abrupt changes in land use—from housing to factories, from farms to forest, from row houses to apartments. Anything that restricts movement across it, such as an interstate highway, river, or railroad, may be a good choice for a boundary. Where is the center of your neighborhood? Is it a shopping center, school, park, or post office? Draw a map of your neighborhood and label everything you think is important.

Regions are especially useful concepts because they provide a reasonable way to break large areas into smaller ones for ease of study and to help you remember geographic information. It is easier to learn placename geography country by country, or continent by continent, than it is to tackle the entire globe as one unit! Regions are to geography what eras or periods are to history.

As mentioned in the Preface, this workbook's division of the world into 10 regions is not the only possible or "correct" regionalizing scheme. The 10 regions used here emphasize cultural and historical characteristics rather than physical factors.

# WORLD REGIONS

Make a map showing the 10 major world regions as used in this workbook. Use the blank outline map of the world (Figure 1–3), which shows national boundaries, and refer to a labeled world political map such as Figure 1–1. Use pen or pencil to make heavy, wide lines along regional boundaries to emphasize them over the national boundaries. Use a different color for the regional boundaries.

The United States and Canada together form the first region. Draw a regional boundary line along the U.S.–Mexican border, and continue it as a dashed line through the Gulf of Mexico, between Florida and Cuba, and between Florida and

**FIGURE 1–2**
A map of your neighborhood.

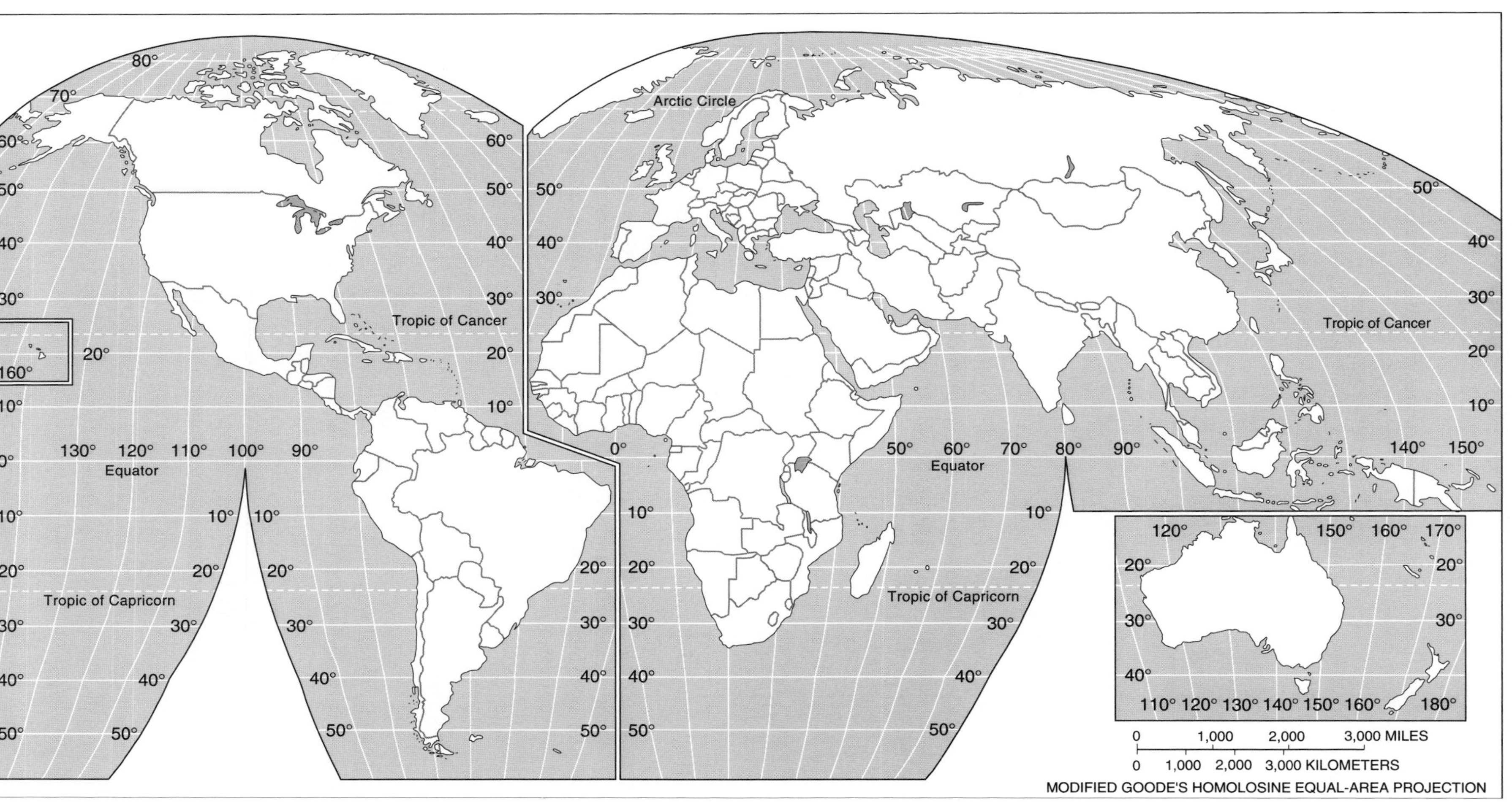


**FIGURE 1–3**
The 10 major world regions used in this workbook.

the Bahamas. Draw a dashed line between Alaska and Siberia, extending that line due north to the North Pole. The Canadian archipelago, or group of islands north of mainland Canada, is part of the U.S.–Canada region. Greenland, the world's largest island, traditionally is considered part of the "new world" rather than the "old world," or Eastern Hemisphere. Greenland is a largely self-governing territory of Denmark; its Inuit (Eskimo) name is Kalaallit Nunaat. Draw a dashed line through the Denmark Strait between Greenland and Iceland and around the eastern edge of Greenland and to the North Pole. (Iceland is assigned to Europe.) Label the region "U.S. and Canada."

Europe is another region. Draw a regional boundary line along the western boundary of Russia, and of Turkey, which controls a tiny piece of the European continent. Draw a dashed line from the Turkey–Greece boundary through the Mediterranean Sea to the southern tip of Spain (be sure that this water boundary lies south of the islands of Sicily, Sardinia, and Crete). Draw a dashed line between the islands of Iceland and Greenland. Label this region "Europe."

The western boundary of Russia, Transcaucasia, and Central Asia with the Europe region has been established. Now continue the boundary of Russia, Transcaucasia, and Central Asia along the southern boundaries of "Transcaucasia"—Georgia, Armenia and Azerbaijan, then along the southern boundaries of Turkmenistan, Uzbekistan and Tajikistan. Follow the eastern boundaries of Tajikistan, Krgyzstan and Kazakhstan and the Russian (Siberian) border in Asia to the Pacific. Label this region "Russia, Transcaucasia, and Central Asia."

Draw a dashed water boundary line north of Australia, separating it from the Indonesian islands and Papua New Guinea. Label Australia, New Zealand, and the Pacific Islands as such.

Draw a regional boundary line along China's national boundaries with Vietnam, Laos, Myanmar (Burma), India, Bhutan, Nepal, Pakistan, and Afghanistan to the border of the newly independent states, which is already outlined. Draw a dashed-line water boundary between China and the Philippines and between Japan and Russia. Label the region thus enclosed as "East Asia."

Draw a regional boundary along the western boundaries of Afghanistan and Pakistan with Iran. Draw a similar boundary along the eastern boundaries of India and Bangladesh with Myanmar (Burma). The boundary of this region with East Asia to the north is already defined. Label this new region "South Asia."

You have already bounded the southeastern corner of the Asian mainland—Myanmar (Burma), Thailand, Malaysia, Laos, Cambodia, and Vietnam, together with the island groups of Indonesia, the Philippines, and Papua New Guinea. Label this region "Southeast Asia."

Draw a dashed-line boundary midway through the Gulf of Aden (south of the Arabian peninsula), then northwestward midway through the Red Sea to the Sudan–Ethiopia boundary. Now draw a solid regional boundary line along the Sudan–Ethiopia line, then between Sudan and its southern neighbors (Kenya, Uganda, Zaire), then between Sudan and the Central African Republic (CAR). Continue the regional boundary between the Central African Republic and Chad. Continue the boundary westward to the Atlantic along the southern edges of Chad, Niger, Mali, and Mauritania. Label the continent of Africa south of this boundary as "Africa south of the Sahara."

The next region, Middle East–North Africa, you have already bounded by establishing the regional boundaries of its neighbors—Africa south of the Sahara, South Asia, the newly independent states, and Europe. Label this new region "Middle East–North Africa."

Finally, Latin America is all of the Americas south of Mexico's border with the United States, which you already have drawn in. Label this last region "Latin America."

The continent of Antarctica is omitted from the list of regions because it is the only continent without a permanent population and the only one not effectively incorporated into the political world of states and territorial administration.

## SOME POPULATION CHARACTERISTICS OF WORLD REGIONS

The focus of geography is on the often complicated interactions of people, their cultural traits, economic and political systems, and their technologies with the physical environment. In a sense, all geography is cultural or human geography. Among the many geographic criteria for identifying world regions are the demographic (population studies) characteristics of these regions.

Table 1–1 lists the regional shares of Earth's human population for 2000 and a projection to the year 2025. One obvious grouping of world regions is based on relative degree of economic development, as noted on Table 1–1.

**TABLE 1–1** Regional Shares of World Population (in millions)

| Region | 2000 | % of Total | Projection to 2025 | % of Projected Total |
|---|---|---|---|---|
| World | 6067 | — | 7810 | — |
| Developed World | 1184 | 19.5 | 1236 | 15.8 |
| Less Developed World | 4883 | 80.5 | 6575 | 84.2 |
| U.S. & Canada | 306 | 5.0 | 374 | 4.8 |
| Europe | 583 | 9.6 | 577 | 7.4 |
| Russia, Transcaucasia, Central Asia | 218 | 3.6 | 223 | 2.8 |
| Australia, New Zealand, Pacific Islands | 31 | 0.5 | 39 | 0.5 |
| East Asia | 1493 | 24.6 | 1669 | 21.4 |
| South Asia | 1352 | 22.3 | 1879 | 24.0 |
| Southeast Asia | 528 | 8.7 | 717 | 9.2 |
| Africa South | 566 | 9.3 | 899 | 11.5 |
| Middle East – North Africa | 473 | 7.8 | 731 | 9.3 |
| Latin America | 518 | 8.5 | 703 | 9.0 |

**1–1** Currently, the "developed world" (U.S. and Canada, most of Europe, Japan, Australia and New Zealand, and Russia) contains about what proportion of the globe's population? (Circle the correct answer.): 10%, 20%, 50%, 75%.

**1–2** By the year 2025, the developed world's share of total population will have (Circle the correct answer.): increased; decreased; remained constant.

**1–3** Which world region contains the smallest proportion of population, both current and projected? ______________________________

**1–4** Which region currently has the largest share of people? ____________
Which region is expected to have the largest share by 2025?

______________________________

**1–5** Which of these regions is *not* projected to experience a relative decline in its share of world population by 2025? (Circle the correct answer.)

a. Europe

b. Russia, Transcaucasia and Central Asia

c. U.S. and Canada

d. Africa south of the Sahara

On the map of world regions which you have recently constructed, Figure 1–3, rank-order these regions in terms of their current share of world population. For example, East Asia now contains the largest share. Label that region with a large "1," and continue through all the regions.

## GETTING STARTED

On a blank world outline map (Figure 1–4), label as many countries and physical features as you can. Label the continents and major islands in as much detail as you can. Label the oceans. Put a star or other symbol on the map to show your present location. When you have completed your world map from memory, set it aside. You will want to compare this first effort with what you know by the time you finish this workbook!

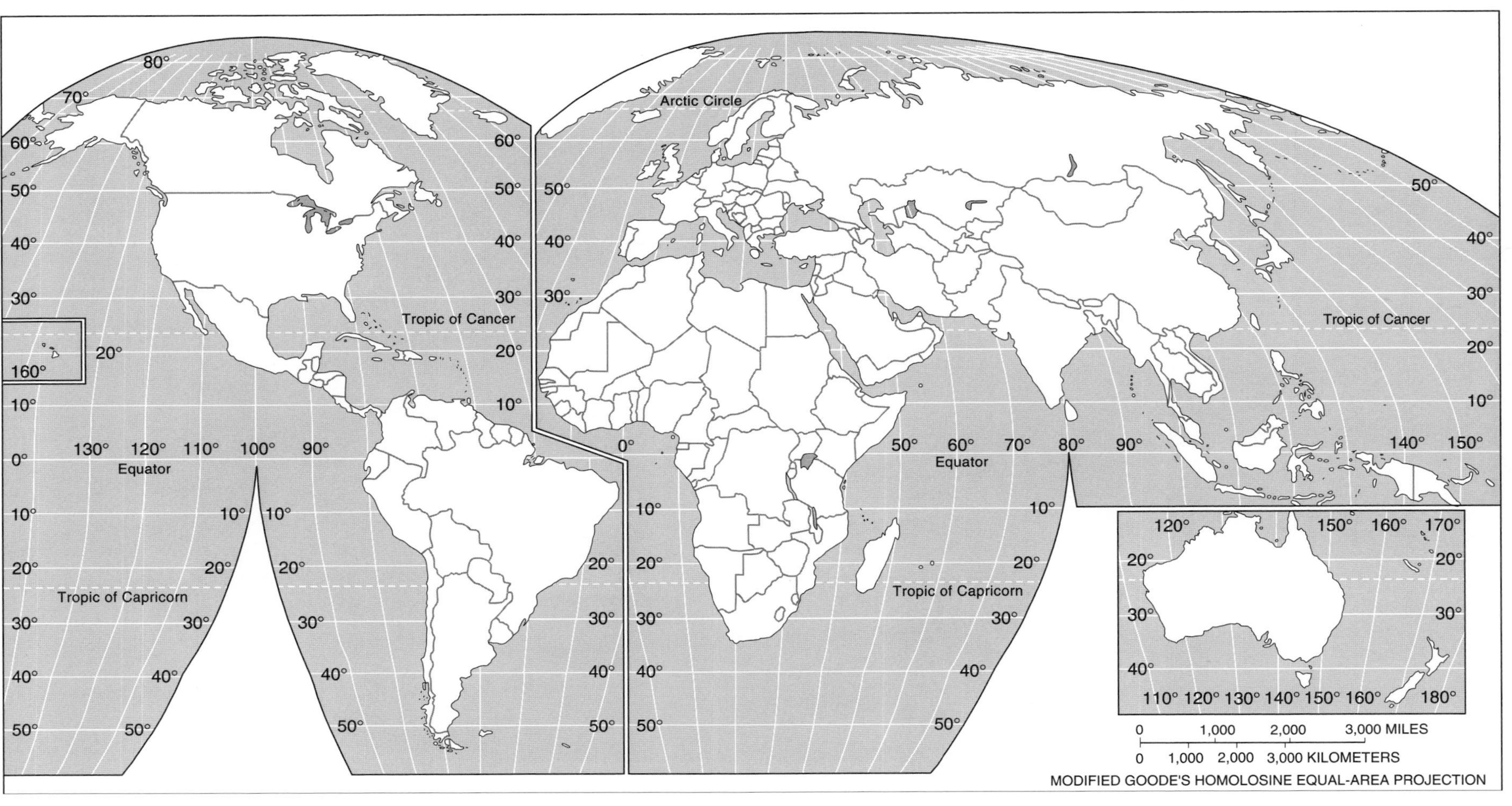


**FIGURE 1–4**
What I knew about placename geography when I started this workbook.

# 2

# The Frame of Reference: Earth's Location Grid, Time Zones, and the International Date Line

## LATITUDE AND LONGITUDE

Because Earth's surface is vast and complex, we need a uniform system for identifying locations, especially locations *relative* to other places. On a sphere like the Earth, directions like "north" or "west" have no meaning unless everyone agrees on a starting point to which they can refer. After all, if you head "north" and continue far enough in a straight line, you eventually will cross the North Pole and find yourself heading south! Maintaining a course in the constant direction "east" eventually will bring you back to the starting point; in fact, you would come back to your starting point from the west!

Earth rotates about an axis, an imaginary line connecting the poles of rotation. At the poles themselves, theoretically there is no motion of rotation at all because the rest of the surface rotates around the poles. These two poles, north and south, provide natural reference points. They make it possible to determine the Equator, the line connecting all points on the globe that are exactly "equally distant from " or halfway between the North and South Poles (Figure 2–1). The Equator divides Earth into two half-spheres, or hemispheres, a Northern Hemisphere and a Southern Hemisphere. We use the Equator as the starting point in identifying locations on a north–south basis. Any point *not* on the Equator must be either north of it or south of it. On Earth, the farthest north you can get from the Equator is one-quarter of a circle. One-quarter of a 360° circle is 90°. So, the North Pole is 90° north of the Equator and the South Pole is 90° south of the Equator. Because the poles are points, not lines, we need no further location information other than "90° N" or "90° S" to state the precise location of either pole. (Continuing in a straight line from the Equator for more than one-quarter of a circle would carry you beyond the North Pole, and thus you would be heading southward.)

But what if the location of a place is halfway between the Equator and the North Pole? The place would be at 45° N latitude. However, 45° N latitude is more than one place; it is a line connecting *all* the points around the globe that are halfway between the Equator and the North Pole. For example, Salem, Oregon, is close to 45° N. So are Bordeaux, France; Torino, Italy; and Harbin, China. Obviously, knowing that a place is "45° N" is not enough information to pinpoint its location! It is like being told that your seat in a huge stadium was in the fifth row. *Where* in the fifth row would be your next question!

So, how can we specify an east–west location? The answer is to establish a reference grid. A *grid* is a system of straight lines crossing one another at right angles. When the lines are labeled, a precise location can be determined. For example, if a classroom has five rows of eight seats each, a student can be assigned the

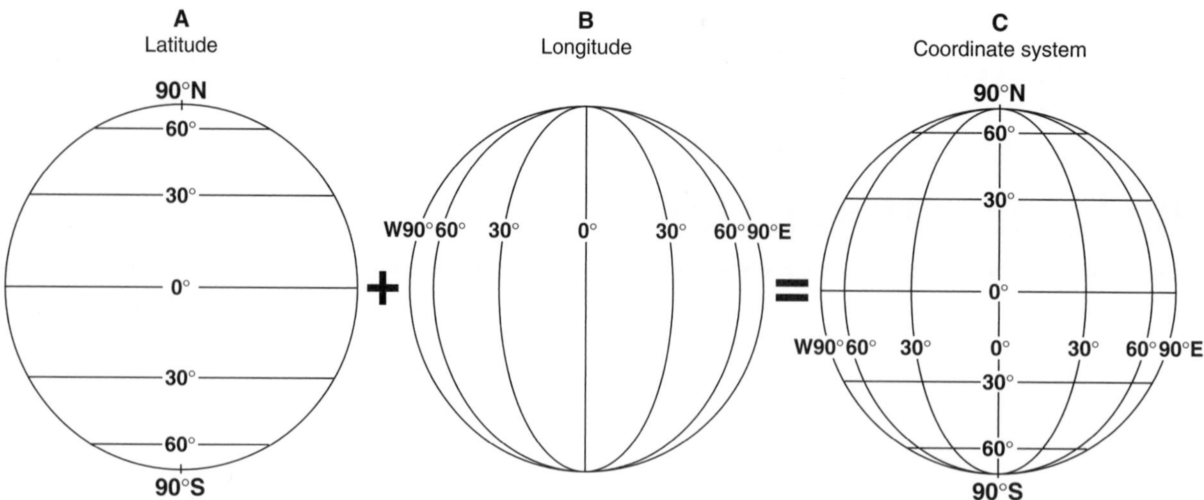


**FIGURE 2–1**
Longitude and latitude. The lines of latitude (A) measure distances north and south of the Equator. The lines of longitude (B) measure distances east and west of the Prime Meridian to 180° east–west. When latitude and longitude are combined into a coordinate system (C), the location of any place on Earth can be determined.

second seat in the third row. The numbering of rows usually is obvious—numbered with reference to the front of a classroom, a stage, or a playing field. But the numbering of seats on the rows may be less obvious. Is it the second seat in from the windows or the second seat in from the aisle or a wall? The student must know where the "second seat" is being measured *from. One always needs a starting line or point of reference.*

Earth's grid system has the same problem that our example classroom grid does—whereas distance north or south of the Equator (a reference line) is easy to measure, what reference can we use for determining east–west relative locations? There is no "natural" reference point or line from which to start measuring east and west. East or west *of what*?

An arbitrary choice of starting point must be made. Just as the Equator is the baseline, or zero line, for measuring latitude, people must agree on an east–west baseline, or zero line, from which to measure longitude.

Lines of latitude run east–west, connecting all points located the same degree of a circle north or south of the Equator. They are called *parallels* because they are parallel to each other and to the Equator. Look at a globe and you can see that the parallels of latitude never meet, or get closer to one another. Although the two poles are points, not lines, they too indicate distance north or south of the Equator.

Lines running straight over the globe's surface connecting the North and South Poles are called *meridians.* Look at Figure 2–1(B), and you can see that an infinite number of meridians could be drawn connecting the poles. No particular meridian is by nature more important than all the others, so the nations of Earth got together in 1884 and decided to use the Greenwich Meridian as the 0° reference meridian, or Prime Meridian. Greenwich is a suburb of London and the location of the Royal Astronomical Observatory. The observatory played an important early role in determining accurate east–west distances because an observatory can measure precise time through observation of the Sun and stars.

People have always measured time by the Sun. When the Sun was at its highest point above the horizon on its daily transit across the sky, then it was noon, the middle of the daylight hours. Half of a 24-hour day before or after noon was midnight, the middle of the hours of darkness. As Earth rotates on its axis, at the same time revolving around the Sun, at any moment half of Earth is always in sunlight, and half is always in darkness. At any one point on Earth, a period of daylight always is followed by a period of darkness. On an *annual* basis, every place on Earth spends half the time in daylight and half the time in the dark (half the time facing the Sun; half the—time facing away from the Sun).

## CHECK UP

**2–1** The Equator is a line connecting all points that are halfway (equally) between which two points on Earth's grid? ____________________

________________________________________

**2–2** The Equator is the zero line, or base line, from which we measure distances ________________ or ________________ (compass directions).

**2–3** How many degrees of a full circle can you go northward or southward from the Equator before you would, if you continued in a straight line, begin to approach the Equator again? ____________________

________________________________________

**2–4** In measuring distances east–west on Earth's location grid, the meridian (line of longitude) that serves as the baseline (zero line) most commonly is the meridian that passes directly over ________________ (city).

**2–5** How many degrees of a full circle can you travel eastward or westward from the Zero (Prime) Meridian before heading back toward that Prime Meridian? ____________________

*Coordinates* are exact locations on a map grid where a parallel crosses a meridian. Remember that a *parallel,* or line of latitude, represents distance north or south of the Equator, and a *meridian,* or line of longitude, represents distance east or west of the Zero (Prime) Meridian. As Figure 2–1(C) shows, latitude + longitude = coordinate system. Except for the North Pole and South Pole, which are at 90° latitude but have no longitude, all locations on Earth's grid must be defined by both a latitude and a longitude reading. By custom, latitude is always listed first; 30° N, 60° E, for example. The Equator, 0° latitude, does not need a further label; any other degree of latitude must be labeled north (N) or south (S).

Similarly, the Prime Meridian, or 0° longitude, does not need a label of east or west because it is neither—it is the line from which we begin measuring east or west. The meridian 180°, exactly halfway around the world from the Prime Meridian, does not carry an east (E) or west (W) designation either, because it is *both* 180° east of, and 180° west of, the Prime Meridian.

**2–6** What is the latitude, in degrees, of a line halfway between the Equator and the North Pole? (Be sure to label this latitude "N.") ____________________

**2–7** Measuring from the Equator, what is the latitude of a line that is two-thirds of the way from the Equator toward the South Pole? ____________________

**2–8** What is the longitude of a meridian halfway between the Prime Meridian and the 180° meridian, traveling eastward from the Prime Meridian?

________________________________________

Look at Figure 1–1, which shows the latitude–longitude grid with parallels and meridians labeled by degrees. For the following coordinates, circle the correct identity of the locations on the globe.

**2–9** 0°, 0°

a. over Central Australia
b. in Brazil
c. in the Atlantic Ocean south and west of Africa
d. at the South Pole

**2–10** 40° N, 75° W

a. over northern Alaska
b. near Philadelphia
c. in the Bahamas
d. near Moscow

**2–11** 51° N, 0°

a. over Antarctica
b. in central Siberia
c. near London
d. on the Equator in the central Pacific Ocean

**2–12** 33° N, 118° W

a. near Los Angeles
b. near Chicago
c. near Rio de Janiero, Brazil
d. near the North Pole

*Note:* You may need to *interpolate*, or measure between, the parallels of latitude and meridians of longitude labeled on maps. For example, 95° W is halfway between meridians labeled 90° W and 100° W; 51° N would be a tenth of the distance between 50° N and 60° N (starting from 50° N), and so on.

## GEOCONCEPTS

### Understanding Maps

Maps communicate information, as pictures or words do. Maps give us information about places, such as their exact location, various qualities or characteristics, and relative location—where are they in relation to other places, and how is this relative location significant? Maps *organize* information about places and give us insights into the importance of a location and of geographic patterns in countries, regions, and the world.

To communicate *effectively,* maps must communicate *selectively.* A single world map, for example, can't show the place location of all communities having over a few thousand population, names of all physical features like streams and mountains, population density, roads and railroads, membership in religious bodies, languages spoken, types of crops grown, and income levels. Nobody would be able to make any sense of so many layers of detailed information on one map.

A good map should select and organize information that we need and communicate it so that we can understand it. Most maps referred to in this workbook, and most maps that you will make, focus on place geography. Some *thematic* maps will be created or interpreted by you to help demonstrate the relevance of place geography: why it is useful to know where places are to understand what is happening there and what it means to us. Thematic maps are special-purpose maps that show the geographic patterns of some qualities of the physical, cultural, political, or economic environments in which we live.

To read maps well, you need to know a little about map scale and about the problem of map projections (the flat representation of Earth's round surface onto a sheet of paper). We'll look at scale first.

## Map Scale

A map is like a scale model—a faithful representation of the original, but produced at a smaller scale and showing only important details. Scale on a map shows the ratio of distance and area on Earth to the distance and area on the map (Figure 2–2). Scale usually is expressed as a proportional fraction (1/1000), a ratio (1:1,000,000), a line drawn on the map (bar scale), or a verbal statement ("one inch equals one mile").

A large-scale map shows a small area in a high degree of detail; a small-scale map shows a larger area but in less detail. A scale of 1:5000 means that one unit of measure on the map (inch, centimeter, or whatever) represents 5000 of those *same* units on Earth itself. A scale of 1:5000 is a large-scale map compared to a scale of 1:1,000,000, in which one unit of measure on the map represents a million of the same units on Earth. Scale is comparable to fractions, where 1/2 is a larger slice of pie than 1/10.

**2–13** Classify the following ratios as to whether they are large, medium, or small scale in relation to one another:

1:1000 ______________________

1:63,360 ______________________

1:100,000 ______________________

**2–14** What is the largest possible scale (without enlargement of reality)?

______________________

**2–15** If a map is reduced in scale, what happens to the amount of detail that can be legibly shown? ______________________

______________________

## Map Projections

A map is a representation of all or part of Earth's surface. Earth's surface is curved and cannot be represented on a flat surface, such as a map, without some form of distortion (Figure 2–3). However, some qualities of Earth's surface can be truly or approximately represented on a flat map. By choosing a map projection having the desired attributes, we can select which classes of information we wish to accurately portray. Of course, the globe has maximum accuracy in portraying true distance, true direction, true shape, and true area. Unfortunately, globes are much too cumbersome for most purposes.

No map projection can give us all four of the true measures just listed. In choosing a projection, we sacrifice at least one of these attributes of the globe to achieve at least one of the others. Here are characteristics of the meridian and parallel grids *on the globe* (but only variously true for flat maps, depending on the projection):

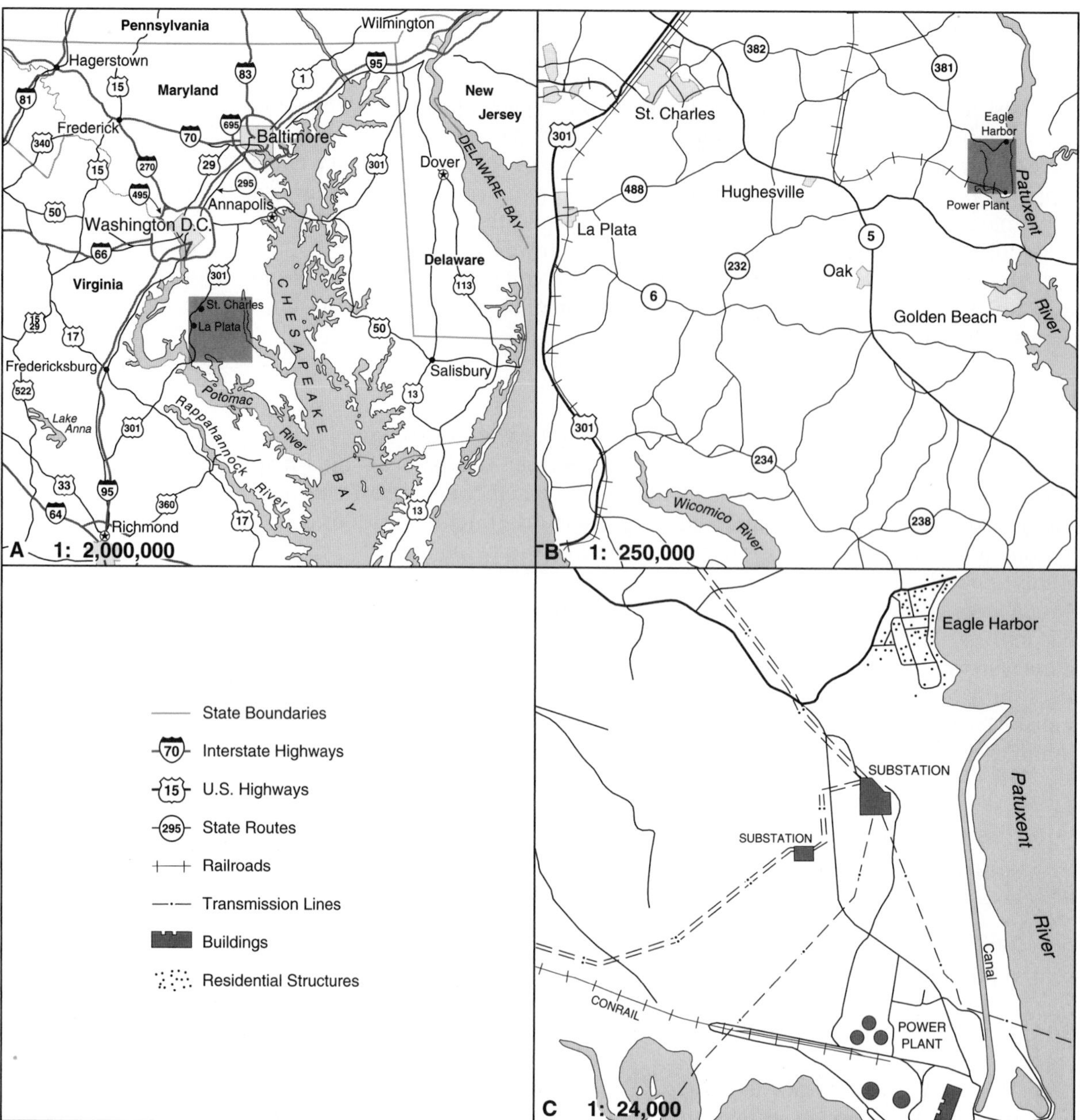


**FIGURE 2–2**
These three maps illustrate the concept of scale—the ratio of map size to Earth size. A portion of Maryland is shown at three progressively larger scales; each portion shows a smaller area in greater detail. Scale is chosen in accordance with the level of detail appropriate to the area studied or the purpose of the map.

**FIGURE 2–3 (*on facing page*)**
Projecting a round Earth onto flat paper. In the process of creating maps, cartographers have to reduce the actual size of the area depicted and increase the degree of generalization. Unavoidable distortion occurs when the three-dimensional Earth is depicted on a two-dimensional surface. (A) Process of converting Earth to a map. (B) Robinson projection, which attempts to reduce many negative aspects of distortion without interruption. (C) Goode's Homolosine projection, which maintains the property of equal area by utilizing interruption. (D) World scale projection used in this workbook, a modified Goode's Homolosine designed to maximize land area. Note that all of these projections follow the convention of placing north at the top, a European-oriented viewpoint that began with early Greek maps.

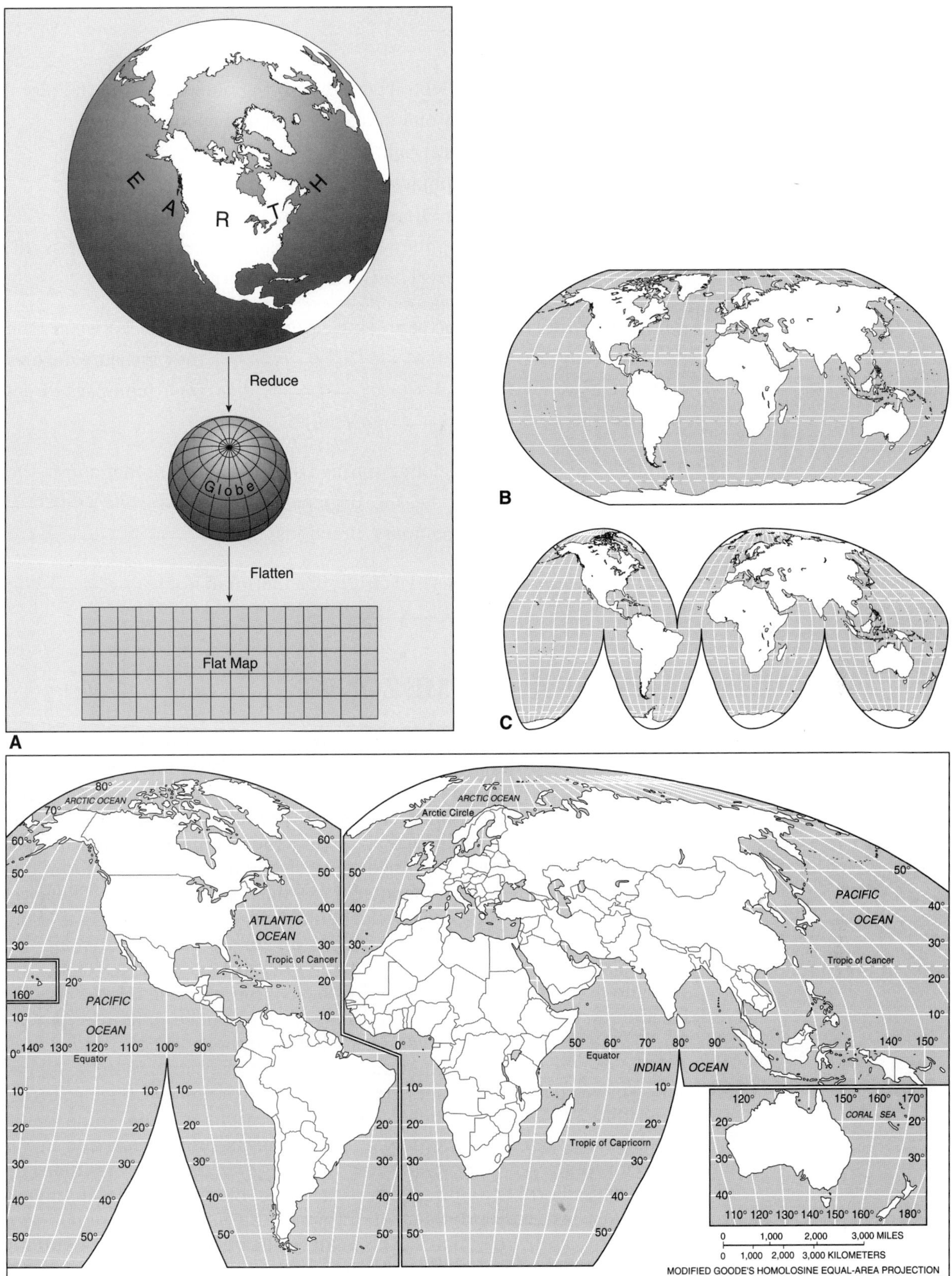
E A R T H
Reduce
Globe
Flatten
Flat Map
A
B
C
ARCTIC OCEAN
ARCTIC OCEAN
Arctic Circle
ATLANTIC OCEAN
PACIFIC OCEAN
PACIFIC OCEAN
INDIAN OCEAN
Tropic of Cancer
Tropic of Cancer
Equator
Equator
Tropic of Capricorn
CORAL SEA
0 1,000 2,000 3,000 MILES
0 1,000 2,000 3,000 KILOMETERS
MODIFIED GOODE'S HOMOLOSINE EQUAL-AREA PROJECTION
D

1. Meridians converge on the globe at both the North Pole and the South Pole.
2. Meridians are approximately one-half the length of the Equator.
3. Meridians intersect parallels at right angles.
4. Parallels decrease in length poleward.
5. Along any given parallel, the intervals between meridians are equal.
6. Along different parallels, the interval between meridians decreases poleward.
7. Along all meridians, the interval between parallels is constant.

**2–16** Meridians on the globe are lines running directly between the ________________ and the ________________; they measure degrees of a circle ________________ or ________________ (compass directions) of the ________________ meridian.

**2–17** Parallels are lines on the globe running parallel to one another and to the ________________; they measure degrees of a circle ________________ or ________________ (compass directions) of the ________________.

**2–18** The total length, in degrees of a circle, of a meridian is ________________; of a parallel is ________________.

## SOLAR TIME AND TIME ZONES

Earth rotates on its axis from west to east (Figure 2–4). Your home location moves out of the dark and into the sun-illuminated half of the globe once every day (unless you live close to either pole), and dawn is always in the eastern sky. Places east of your home always experience dawn before you do. Places farther west than your home see the dawn some time after you. On the rotating globe, it is *always* noon someplace and it is always midnight someplace. Of course, noon, midnight, and any other time are fleeting moments for any fixed point on Earth. The locations of noon or midnight keep changing, but these times, reflecting Earth–Sun relationships, are always present somewhere and are always present along the same meridian at any one moment.

Imagine that it is noon, measured by observation of the overhead Sun, at your location. At that precise moment, places to your east have already had noon, Sun time. Those places now are somewhere in the afternoon by solar time. Locations west of you are yet to see solar noon for that day; they still are in the morning hours. The word *meridian* refers to not only lines of longitude, but also to the highest position of the Sun in the sky, which is local noon. "A.M." time (as in 9 A.M.) abbreviates the Latin words *ante meridian,* meaning *before* the Sun's meridian (morning). "P.M." is *post meridian,* meaning *after* the Sun's meridian (afternoon). Because Earth is rotating from west to east, the noon meridian travels from east to west, just like the moment of sunrise and the moment of sunset. Every place on the same North Pole–South Pole connecting line (meridian of longitude) will experience noon, and all other solar times, at the same moment.

Thus, measuring time on Earth is controlled entirely by our planet's relationship with the Sun. This relationship also controls the "march of the seasons," as you can see in Figure 2–5. Study of this figure reveals some key facts:

1. Earth's axis is *not* perpendicular to its plane of orbit, but is tilted 23½°. This simple fact causes the change of seasons, because the axial tilt changes the intensity of solar rays received at any one latitude as Earth revolves around the Sun.

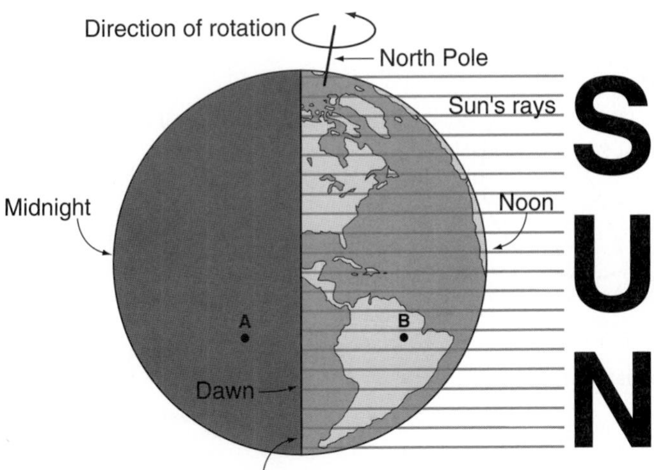


**FIGURE 2–4**
Solar time and Earth's rotation. Looking down on the North Pole, Earth rotates counterclockwise on its axis. The Sun always rises in the east and sets in the west. At A, it is night, but dawn is approaching. At B, it is morning and will be noon shortly. The calendar date is approximately June 21, as the North Pole is in 24-hour daylight.

2. The combination of the 23½° tilt and revolution around the Sun creates four key events every year:
   a. *Summer solstice,* when the Sun is directly overhead at the Tropic of Cancer (23½° N) and the solar rays strike the Northern Hemisphere more directly, creating summertime in that hemisphere.
   b. *Winter solstice,* when the opposite is true: the Sun is directly overhead at the Tropic of Capricorn (23½° S) and the solar rays strike the Southern Hemisphere more directly, causing summertime there, but winter in the Northern Hemisphere.
   c. *Spring and fall equinoxes,* when the Sun is overhead at the Equator and the solar rays strike the Northern and Southern Hemispheres equally, causing the transition seasons of spring and fall.

A knowledge of Earth–Sun relationships tells you that on one of the two equinoxes, March 21 or September 23 (the dates vary a day or so from year to year), the Sun's noon position will be directly overhead, at a 90° angle to the surface, *only* at the Equator. There can be only one point at a time at which the Sun is directly overhead. If the date is March 21 and the Sun is directly overhead, 90° above you, at noon, you must be on the Equator, 0° latitude. On the same date, if the Sun is just on the southern horizon at noon, or 0° above you, you must be at the North Pole, 90° N (Figure 2–5).

Because the axis of rotation is *not* at a perfect right angle to the plane of the orbit around the Sun, but is tilted 23½°, the overhead rays of the Sun are not always over the Equator. The overhead rays move up to the Tropic of Cancer, 23½° N, by the summer solstice, about June 21, and then, taking six months to do so, move to the Tropic of Capricorn, 23½° S, on December 22, the winter solstice. This introduces an additional calculation to determine latitude by noon Sun angle on dates other than an equinox.

Although people learned to measure latitude early using fairly simple instruments, longitude (east–west distances) was far more difficult to determine accurately. That problem was solved when people remembered that *time,* measured by the Sun, is different east to west at any one moment. It takes 24 hours for solar noon to return again to any particular place. The noon meridian travels through

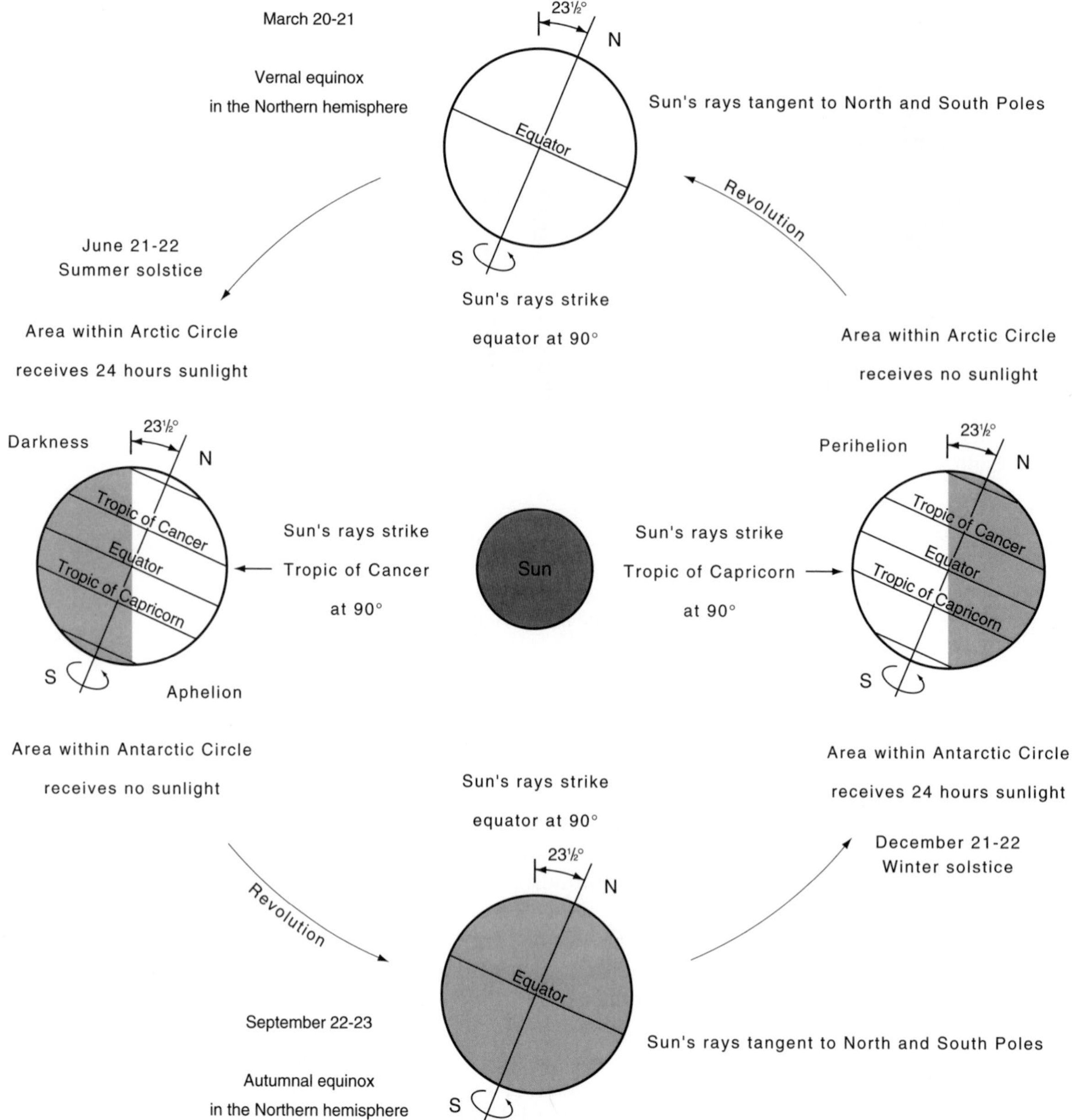


**FIGURE 2–5**
March of the seasons: A diagram of the positions of Earth in relation to the Sun during a single year. The sun strikes Earth at any one location at different angles during the course of a year.

360° of a circle in 24 hours, so it travels 15° of a circle (1/24th of 360°) in an hour. When traveling, if you carry a clock set to solar time at the longitude where your trip begins, and then, as you travel east or west, compare it with another clock you set to local solar time, you can subtract the time difference in hours, multiply by 15°, and thus determine how many degrees east or west you have traveled—your new longitude. Imagine that you have set an accurate clock to solar time at Greenwich, England. Then you travel eastward to Salerno, Italy. When the local solar time in Salerno is noon, your clock says that it is 11 A.M. at Greenwich. In another hour, the noon meridian will reach Greenwich. Therefore, you must be 15° of a circle east of Greenwich, or at longitude 15° E (Figure 2–6).

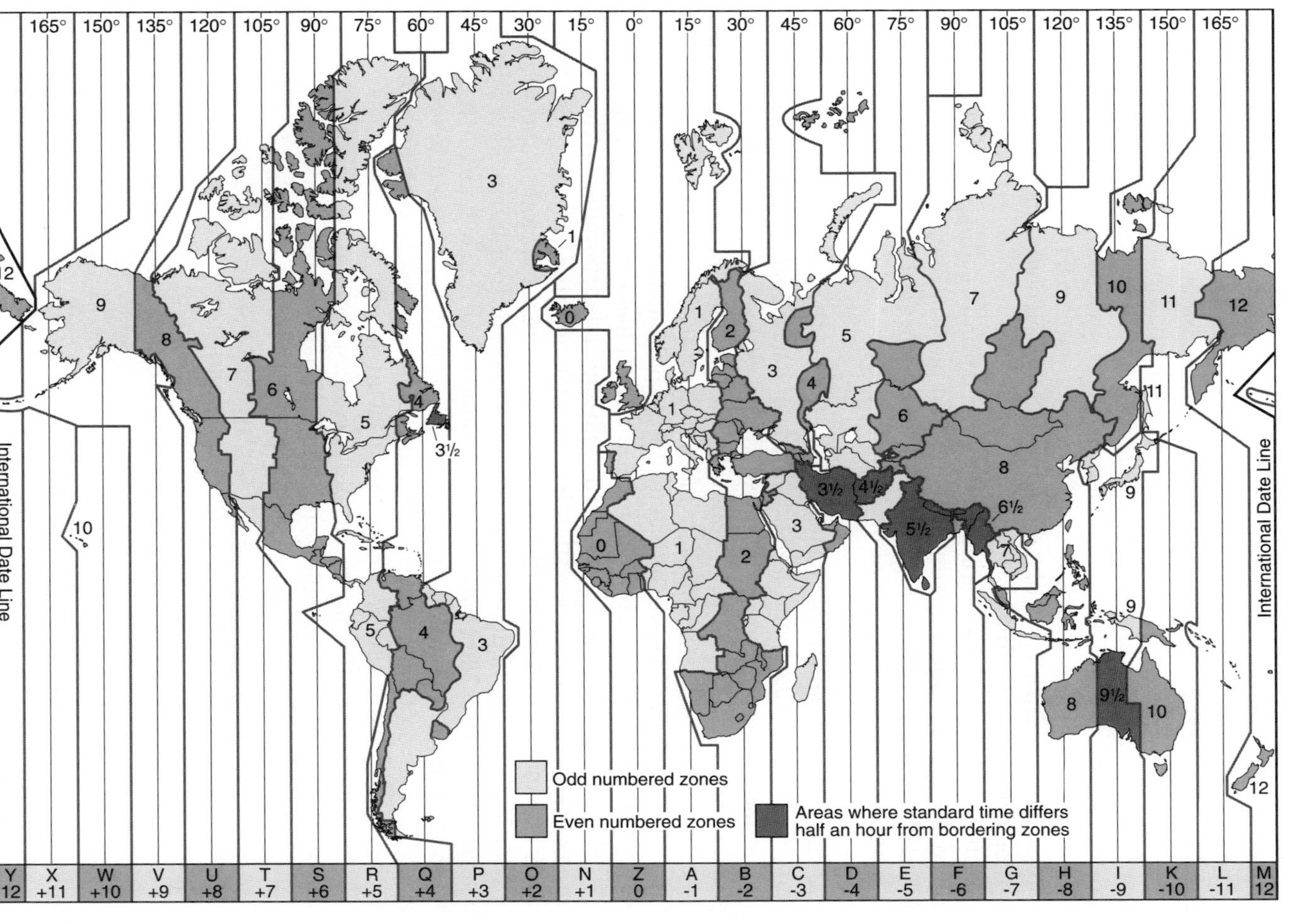


**FIGURE 2–6**
Standard time zones of the world. Each time zone covers approximately 15° of longitude. Most countries observe "standard time," under which the same time of day is observed within each zone for a distance of approximately 7½° on either side of the standard meridian (0°, 15°, 30°, 45°, 60°, etc.). (*Source*: Based on data from the U.S. Navy Oceanographic Office.)

With the development of super-accurate clocks late in the eighteenth century, it became possible for the first time to use time differences to measure longitude. Now people had accurate ways to determine *both* latitude and longitude. They needed only a calendar, a table of solar noon Sun angles at latitudes by calendar date, a sextant (an instrument that precisely measures the Sun's angle above the horizon), and a good clock. At last, people had the capability of measuring east–west *and* north–south distances and could exactly pinpoint any location on Earth's grid.

The Greenwich Meridian is, for most maps, the Prime Meridian, or 0° longitude, the base from which east–west distances are measured in degrees of a circle. French maps once used the Paris Meridian as their prime, and Russian maps still use the Moscow Meridian as the Russian prime, but to avoid confusion, most maps now use the Greenwich Meridian. Greenwich is the standard because the British had the first really accurate clock (chronometer) and thus had the first really accurate maps. They, of course, used their own observatory at Greenwich as their east–west baseline.

In describing an Earth-grid location, the latitude reading is given first, followed by the longitude. Philadelphia, for example, is at about 40° N, 75° W. (It is 40° of a circle north of the Equator and 75° of a circle west of Greenwich.) Except at the poles, where there are no meridians, both coordinates must be given to locate a point.

If you travel more than 180° (half a complete circle) away from Greenwich, you then are heading around the other side of Earth, back closer to Greenwich, and the longitude numbers decrease from 180° toward 0° at Greenwich. Just as the Equator doesn't need a north or south designation, the 180° meridian, or longitude line, is neither east nor west (in fact, it is both at the same time). The furthest east *or* west you can be from the Prime (Zero) Meridian is 180°.

That time changes as one moves east or west had little significance to people before the development of modern, high-speed transportation. When it took three weeks or more to sail the Atlantic from England to the United States, the few hours difference in solar time did not matter. London is five hours "ahead" (east) of New York; the noon meridian passes over London approximately five hours before it passes over New York. When it is noon by the Sun in London, it is about 7 A.M. local solar time in New York. In sailing-ship days, one couldn't expect to reach a distant destination at any particular time, anyhow. It was usually a guess as to how many days or even weeks would be involved.

But communication (by telegraph) and transportation became much faster, and by the late nineteenth century, the longitudinal time differences had become important. For example, a train might leave Philadelphia, 75° W, at 8 A.M. Monday. It traveled for 24 hours (elapsed time) before reaching St. Louis at 90° W. An accurate wristwatch on one of the passengers would read 8 A.M. on Tuesday morning when the train arrived in St. Louis. But clocks in St. Louis, set to local solar time, would say 7 A.M. The wristwatch was accurately reporting the time back in Philadelphia.

On the other hand, Pittsburgh is only 5° of longitude away from Philadelphia, and St. Louis is 15° away in east–west degrees of a circle. Local Sun time at Pittsburgh thus is 20 minutes behind Philadelphia's local Sun time. Should train schedules reflect time differences of less than an hour? What about the five minutes of difference in solar time between New York and Philadelphia? The problem was complicated further by development of long-distance telegraph and telephone service.

Standard time zones were the answer. Every 15° change in longitude produces a one-hour difference in time. So, every fifteenth meridian (0°, 15°, 30°, etc.) is the center of a zone that would span $7\frac{1}{2}°$ each side of that time zone's central meridian, halfway toward the next whole-hour meridian. As the map of actual time zones shows (Figure 2–6), the system was adopted but actual time-zone boundaries have been adjusted for practical reasons.

For example, strict adherence to the time zones' boundaries would split some metropolitan areas. Tampa, Florida, is located close to the halfway point between the 75° W meridian, used to determine Eastern Time in the United States, and the 90° W meridian, which is central to the Central Time Zone. Half the shoppers, commuters, schoolchildren, and workers in the metropolitan area can't be in a different time zone from the other half–imagine the confusion! Thus, when it is solar noon at 75° W, it is standard noon throughout the Eastern Time Zone. It is unimportant that it is not exactly noon solar time throughout the entire zone. Convenience and the ability to coordinate schedules are essential.

## THE INTERNATIONAL DATE LINE

High-speed transportation and virtually instantaneous communications highlight another problem that affected very few people before the modern era: determining what day it is. As long as people stayed in one place, today was Monday, tomorrow Tuesday, and so on. But today, people routinely fly from the United States to Asia. What day is it there? There has to be a line of longitude that everyone agrees upon, where we can declare that the "solar noon" has finished one complete circuit around the Earth, and we will say that Monday is over and Tuesday begins. After existing for one round trip, 24 hours, noon, Monday must "die" and be replaced by noon, Tuesday, and so on.

The International Date Line, where the day officially changes, has been assigned to 180°, exactly opposite the Greenwich Meridian. 180° lies mostly across the sparsely populated mid-Pacific Ocean (Figure 2–6). For convenience, people commonly redraw the line so that it "jogs" (varies) from 180° longitude over Siberia, and jogs again to avoid placing some of Alaska's Aleutian Islands on different days. It zigzags from 180° again to avoid causing similar problems for island groups that are in frequent touch with one another, just as time-zone boundaries are adjusted to avoid dividing metropolitan areas.

Thus, the new day begins on Earth at midnight at the International Date Line. During the next 24 hours, the new day advances westward around the world, finally covering the entire surface at the end of the 24 hours. For an instant, there is only one calendar day on Earth as the midnight meridian passes over the Date Line. The first minute of the next day, 12:01 A.M. Monday, now moves around Earth during the next 24 hours. It will disappear at the Date Line and will become 12:01 A.M. Tuesday.

Except right at midnight on the Date Line, two calendar days are present on Earth. The "newer" of the two always extends westward from the Date Line to the current position of the midnight meridian, with the older day extending the rest of the way around the world back to the Date Line. Thus, when you cross the International Date Line going from west to east, the day you enter is the "older one," the earlier date. For example, you could go from December 25 to December 24. When crossing the Date Line the other direction, moving from east to west, you enter the newer day, as going from December 24 on the eastern side of the Date Line to December 25 on the western side.

## CHECK UP

The diagram in Figure 2–7 presents meridians of longitude. The International Date Line is at longitude 180°. In the following check-up questions, you will use these facts:

a. 15° of longitude equals one hour of time.
b. All places *east* of a given point have *later* local solar times; all places *west* of a given point have *earlier* local solar times.
c. When crossing the International Date Line, going east, you lose a day; going west, you gain a day.

**2–19** From your knowledge of Earth–Sun relations, explain how you know that 15° of longitude = 1 hour of time: ______________________________

______________________________

**2–20** In Figure 2–7, what is the difference in hours of Sun time between A and F?

______________________________

**2–21** A plane leaves C at exactly noon, local solar time, on Tuesday, January 23, flies exactly seven hours, and lands at A. What are the day, date, and hour at A *when* the plane *lands*? (Circle the correct answer.)

a. 2 A.M., Wednesday, January 24
b. 7 P.M., Tuesday, January 23
c. 7 P.M., Wednesday, January 24
d. 2 P.M., Tuesday, January 23

**2–22** If it is Friday, 10:30 P.M. at A, what are the day and hour at C and at F?

(C) Day: ______________
Hour: ______________
(F) Day: ______________
Hour: ______________

**2–23** A ship bound westward for F from D comes to A on the afternoon of Wednesday, June 12. What will be the day and date on that ship the following day (24 hours later) after it crosses the International Date Line?

Day: ______________ Date: ______________

**2–24** When it is noon at Greenwich, England, what time is it where the Prime Meridian crosses the Arctic Circle? ______________ The Antarctic Circle? ______________ Explain: ______________

______________________________

**2–25** Under what circumstance is it the same calendar day everywhere on Earth?

______________________________

**2–26** Assume that Earth stopped revolving around the Sun, but remained rotating at its December 21 position. What would the length of daylight be at the Antarctic Circle on June 22? ______________ On September 22?

______________

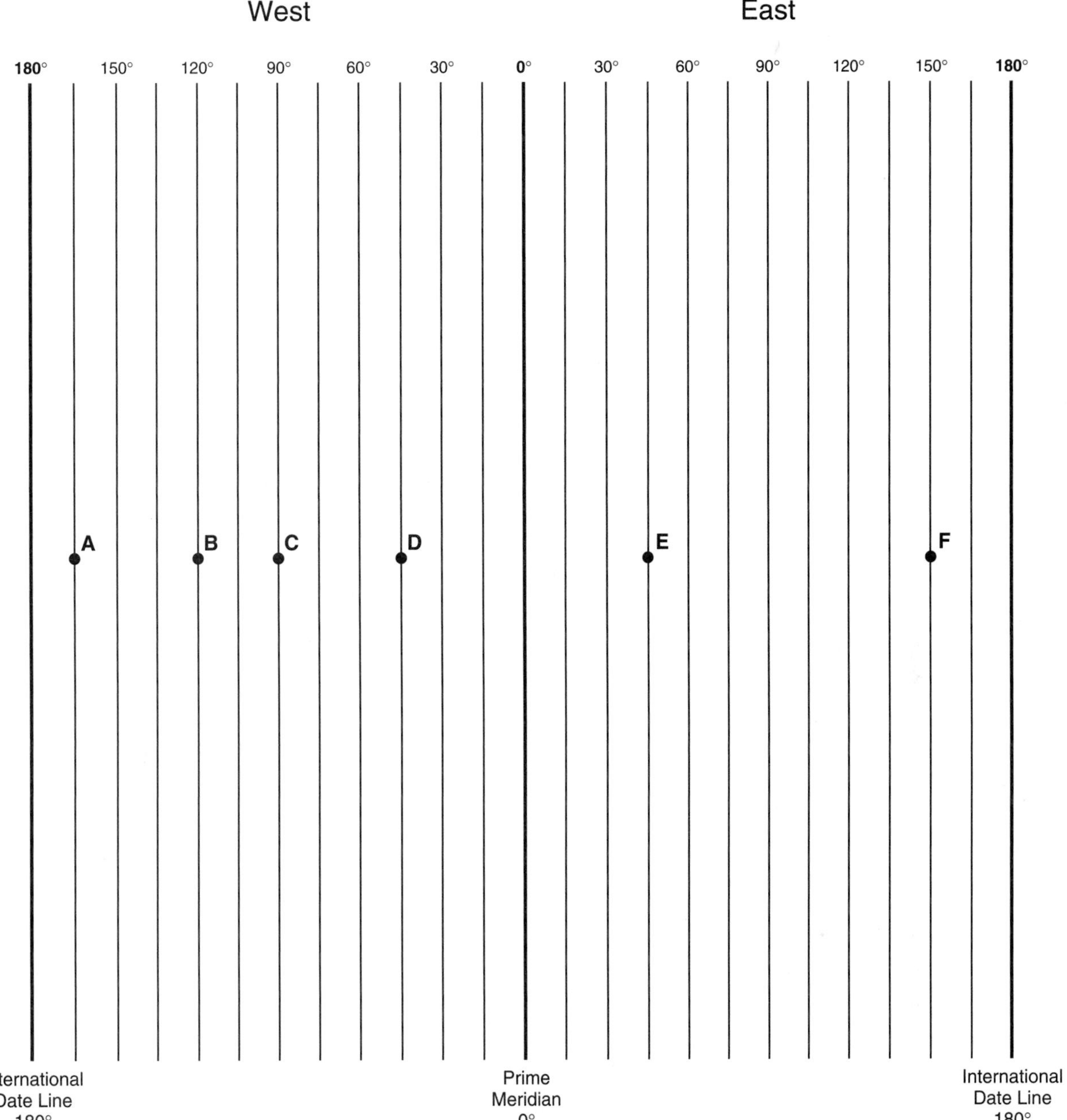


**FIGURE 2–7**
Meridians of longitude.

**2–27** Under the conditions in question 2–26, would there be any seasonal changes in the length of day and night at any latitude at any time during the year? ______________________________

**2–28** Are there errors in the following sets of coordinates? If so, circle them:

| | |
|---|---|
| 14° N, 70° W | 91° S, 134° W |
| 78° N, 192° E | 0°, 180° |
| 90° N | 0°, 0° |
| 23° N, 3° W | 65° E, 109° W |
| 89° N, 9° E | 180° |

# 3

# Lands and Seas

## INTRODUCTION

The basic place geography of planet Earth seems fairly simple and straightforward, with seven continents and four oceans. But is it that simple? Although Europe is a continent in historical and cultural tradition, isn't it really part of the great landmass of Eurasia? Also, it is obvious on a globe or world map that the vast body of saltwater that surrounds all the continents is one continuous "world-ocean." But people have chosen to give portions of this world-ocean dozens of different names, like Pacific Ocean and Baltic Sea.

Building a knowledge of place geography is challenge enough without attempting to justify these long-standing traditions or to dispute the basis of them. This workbook accepts the traditional continents and oceans and generally recognized placenames as they are. To do otherwise would decrease your ability to place historical and current events within a geographic frame and to recognize the geographic relationships among different places. To communicate, we must all use the same geographic vocabulary, both in *generic* placenames (seas, bays, straits, peninsulas, plateaus, and so on) and *specific* placenames.

## A QUESTION OF PERSPECTIVE

Physical features sometimes are known by different names to different peoples and cultures. It is the English Channel to the British, whereas to the French it is La Manch. Arabs refer to the Arabian Gulf rather than the Persian Gulf, although it is the identical water feature. A government may change a country's name or the spelling of the name, as when Burma became Myanmar. Often, names are a matter of perspective. The "Middle East" is "middle" from a Western European viewpoint, for example.

As noted, transliteration of Chinese and Arabic placenames can be confusing. In many cases, an alternative placename or spelling will be placed in parentheses after the more common name—for example, Beijing (Peking). Preferred names and spellings follow *Goode's World Atlas* as the placename authority.

Also, in some instances, the *de facto* (in fact) territory controlled by a country may differ from its *de jure* (legal) boundaries as recognized by other states. The boundaries of Israel are a case in point.

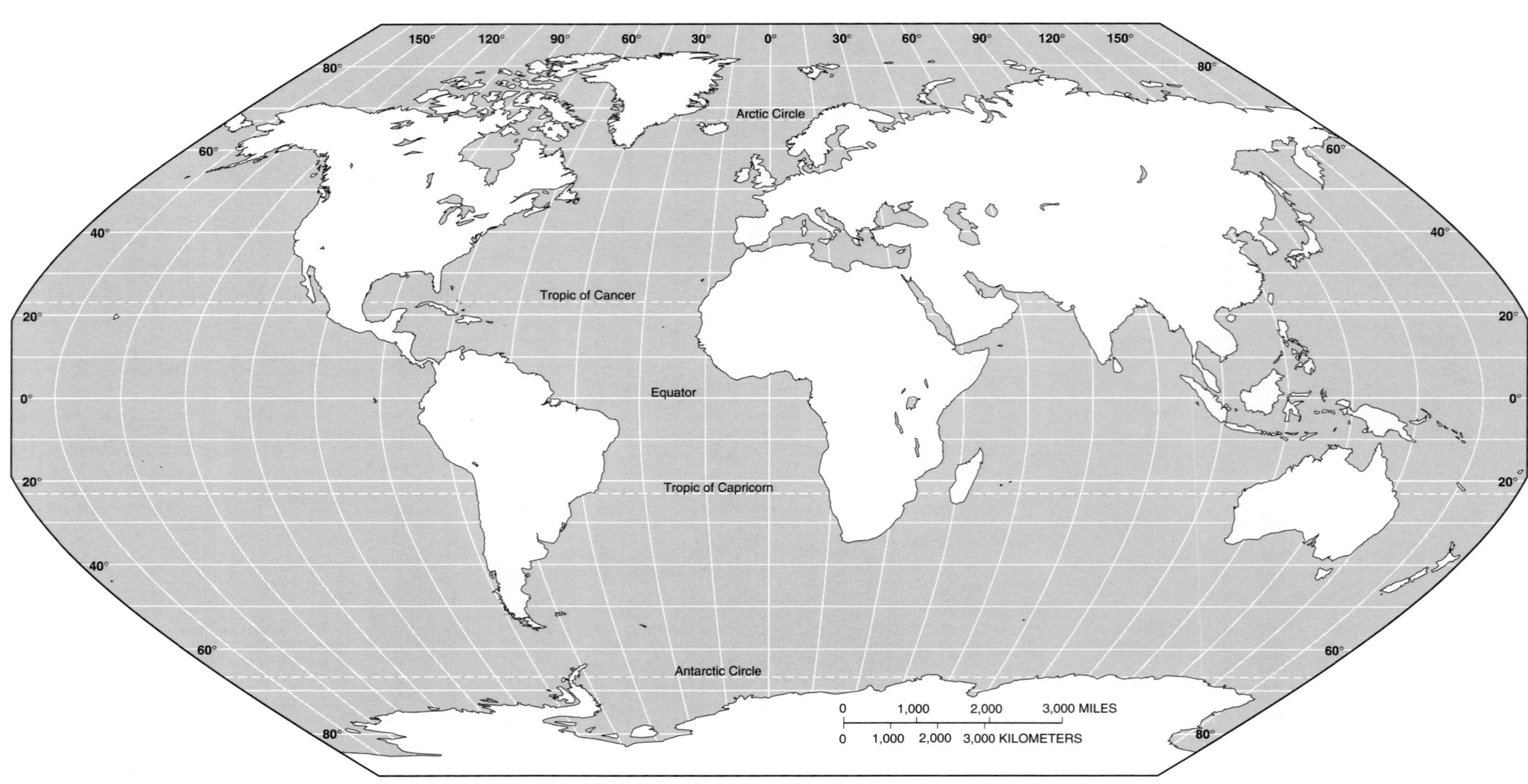


**FIGURE 3–1**
Continents and oceans.

On the outline map of the world in Figure 3–1, label the continents and oceans. Then, on the map, number the continents and oceans in order of their relative size, with number one the largest:

| Continent | Percent of Earth's land |
|---|---|
| Asia* | 29.7 |
| Africa | 20.4 |
| North America | 16.3 |
| South America | 12.0 |
| Antarctica | 8.9 |
| Europe* | 7.0 |
| Australia | 5.2 |

| Oceans | Percent of Earth's water area |
|---|---|
| Pacific | 46.0 |
| Atlantic | 23.9 |
| Indian | 20.3 |
| Arctic | 2.6 |

# GENERIC PLACENAMES

Before proceeding to the proper names of specific places, it is useful to review the "family names," or generic names, of places. For example, in the placename "Hudson Bay," "bay" is the generic name, and "Hudson" is the specific bay. If you are not sure what a bay looks like on a map, it is more difficult to identify any particular one, like Hudson Bay.

The following glossary describes each major generic term used in placenames and lists an example or two of each. Locate and label each example on the appropriate world, regional, or subregional map (the best choice of map is indicated for each example).

# MAP REFERENCES

Whenever you are asked to locate and label physical features (bays, straits, rivers, mountains, etc.), a map reference ordinarily will be provided. Customarily, the latitude reference will be listed first, followed by longitude. Latitude is always given as N (North) or S (South) of the Equator; longitude is identified as E (East) or W (West) of the Prime Meridian at Greenwich. In this workbook, most map references are rounded to the nearest full degree and are *representative* for linear or large features like rivers, bays, and the like. Map references are *not* given for the largest-scale physical features, oceans or continents. You will need to *interpolate* for many map references. This means that if a reference is given as 35° N, 17° W, and your map shows latitude and longitude lines at 10-degree intervals (10° N, 20° N, etc.), you'll need to estimate that 35° N is halfway between 30° N and 40° N, and that 17° W is seven-tenths of the distance from 10° W toward 20° W.

## Archipelago

Any set of closely grouped islands in any large water body. *Examples:* the islands of Indonesia: 5° S, 119° E (Southeast Asia map) (Figure 3–2); Japan: 40° N, 133° W (world map) (Figure 3–3).

*We will honor the traditional separation of the huge landmass of Eurasia into two continents.

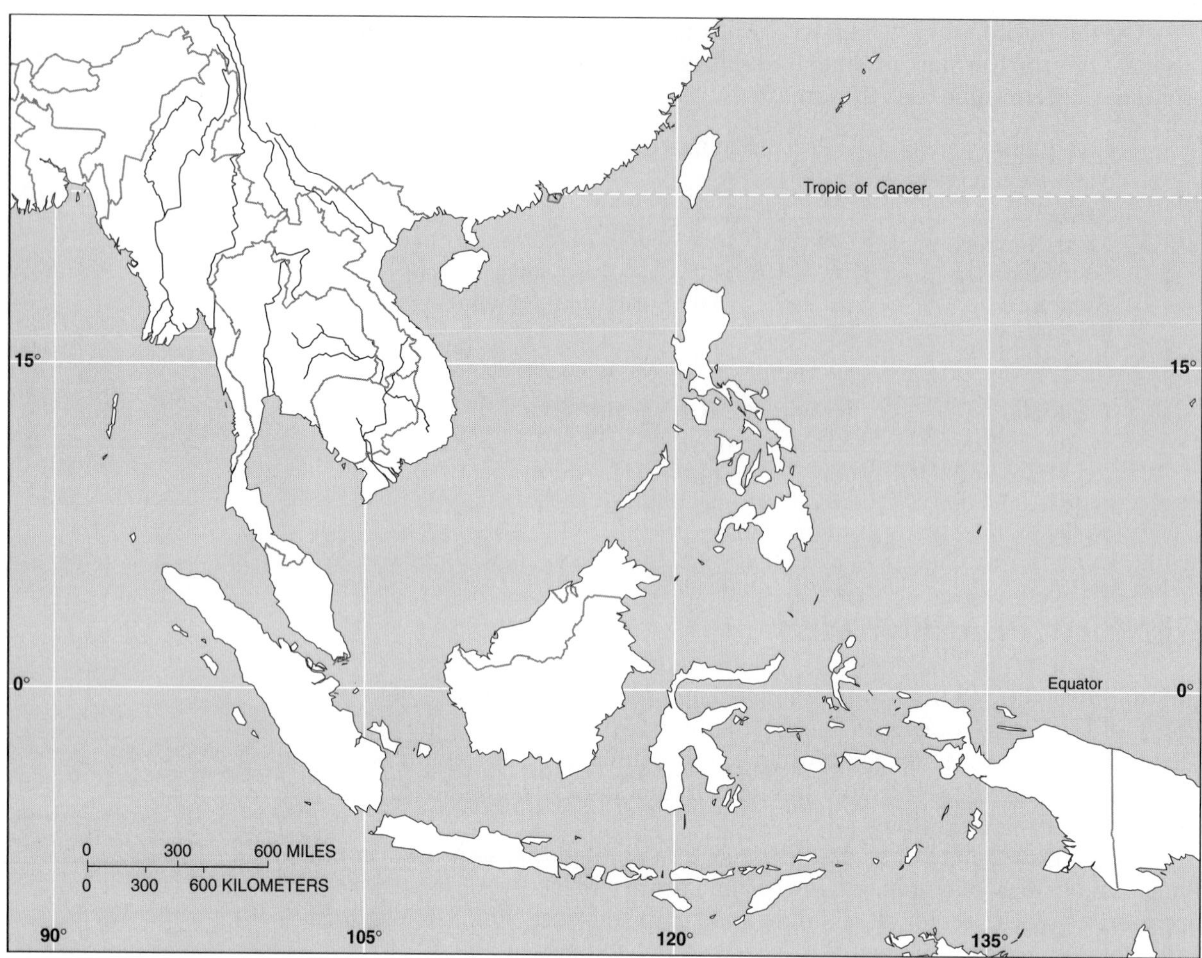


**FIGURE 3–2**
Southeast Asia map.

## Bay

A body of water partly enclosed by land. Compared to a gulf (see *Gulf*), a bay *generally* is smaller and more surrounded by land. The bay's mouth, on opening to the larger body of water, may be wider than a gulf's mouth. *Examples:* Bay of Biscay: 45° N, 4° W (Europe map) (Figure 3–4); Bay of Bengal: 17° N, 87° E (world map) (Figure 3–3).

## Cape

A point of land that extends into a lake, sea, or ocean. Some capes are the terminal points of continents, like southern Africa's Cape of Good Hope and Cape Agulhas. Others are smaller features of more local importance, like Cape Cod, Massachusetts. A cape may be formed of islands, or islands plus mainland, as at Cape Canaveral, Florida. *Examples:* Cape Hatteras 35° N, 75°W (North America map) (Figure 3–5); Cape Horn 55° S, 70° W (world map) (Figure 3–3).

## Channel

A wide strait (see *Strait*) of water between two landmasses that lie close to each other. *Example:* English Channel: 50° N, 1° W (Europe map).

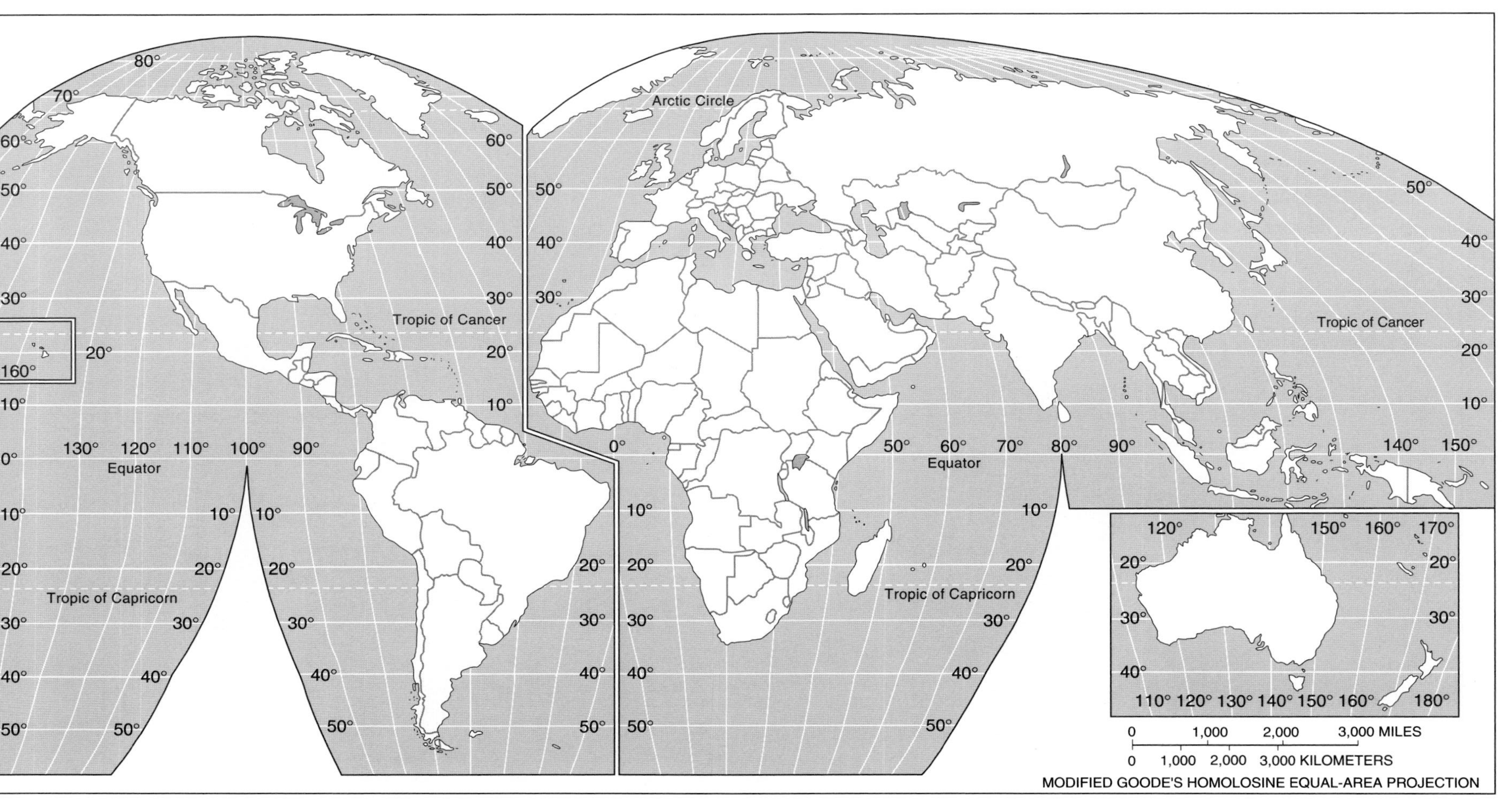


**FIGURE 3–3**
World map.

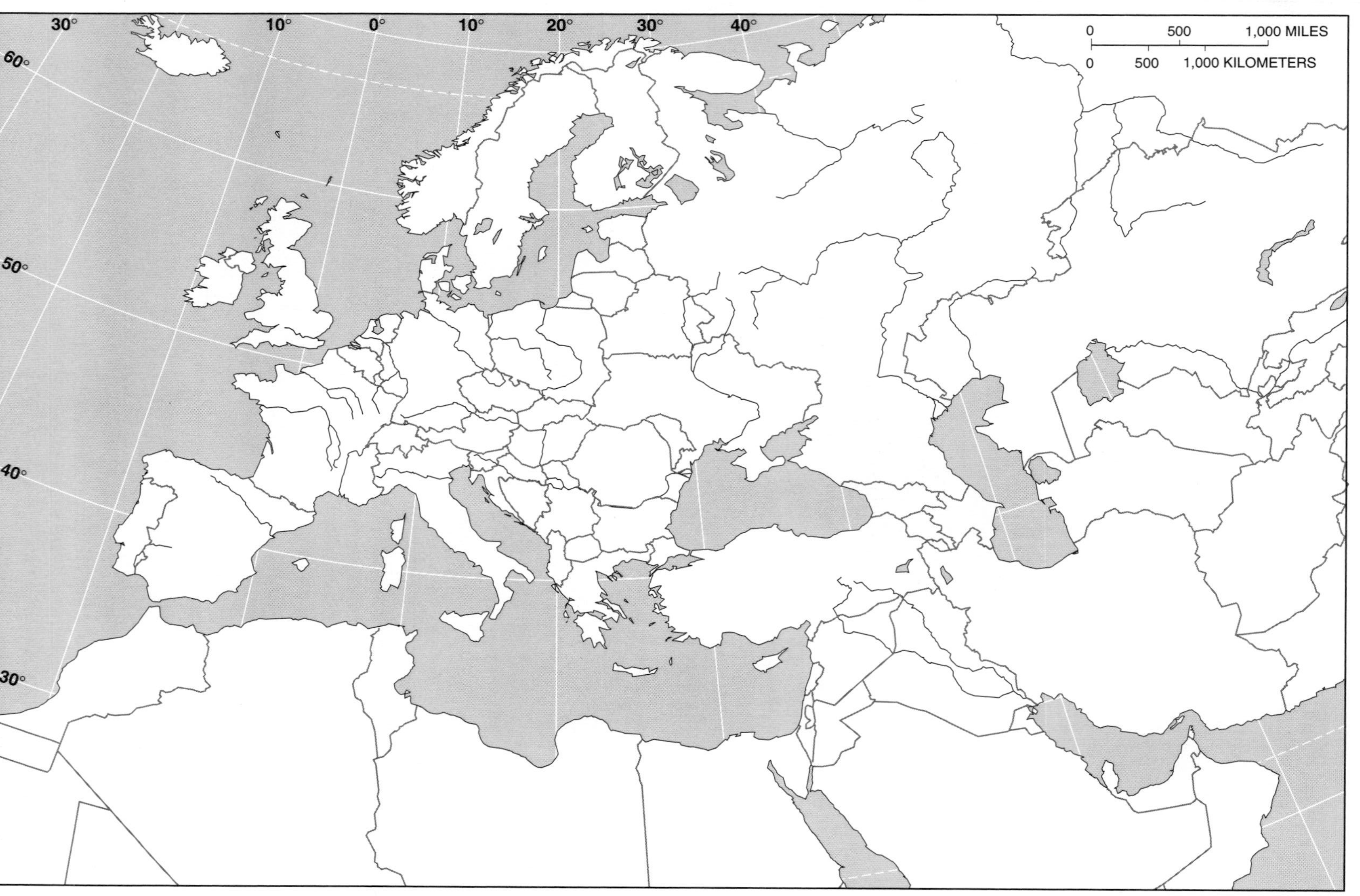


**FIGURE 3–4**
Europe map.

**FIGURE 3–5**
North America map.

**3–1** The English Channel narrows into which strait? ________________

## Delta

A low, flat plain formed at the mouth of a river by that river's deposit of sediments (sand, silt, clay). In addition to the fanlike, arched triangle shape at the mouth of the Nile River that resembles the Greek letter delta and gives this feature its name, deltas can have various shapes, and the name is applied to all such river-deposited features of whatever shape. *Examples:* Mississippi River Delta: 28° N, 90° W (North America map); Ganges River Delta: 22° N, 90° E (South Asia map) (Figure 3–6).

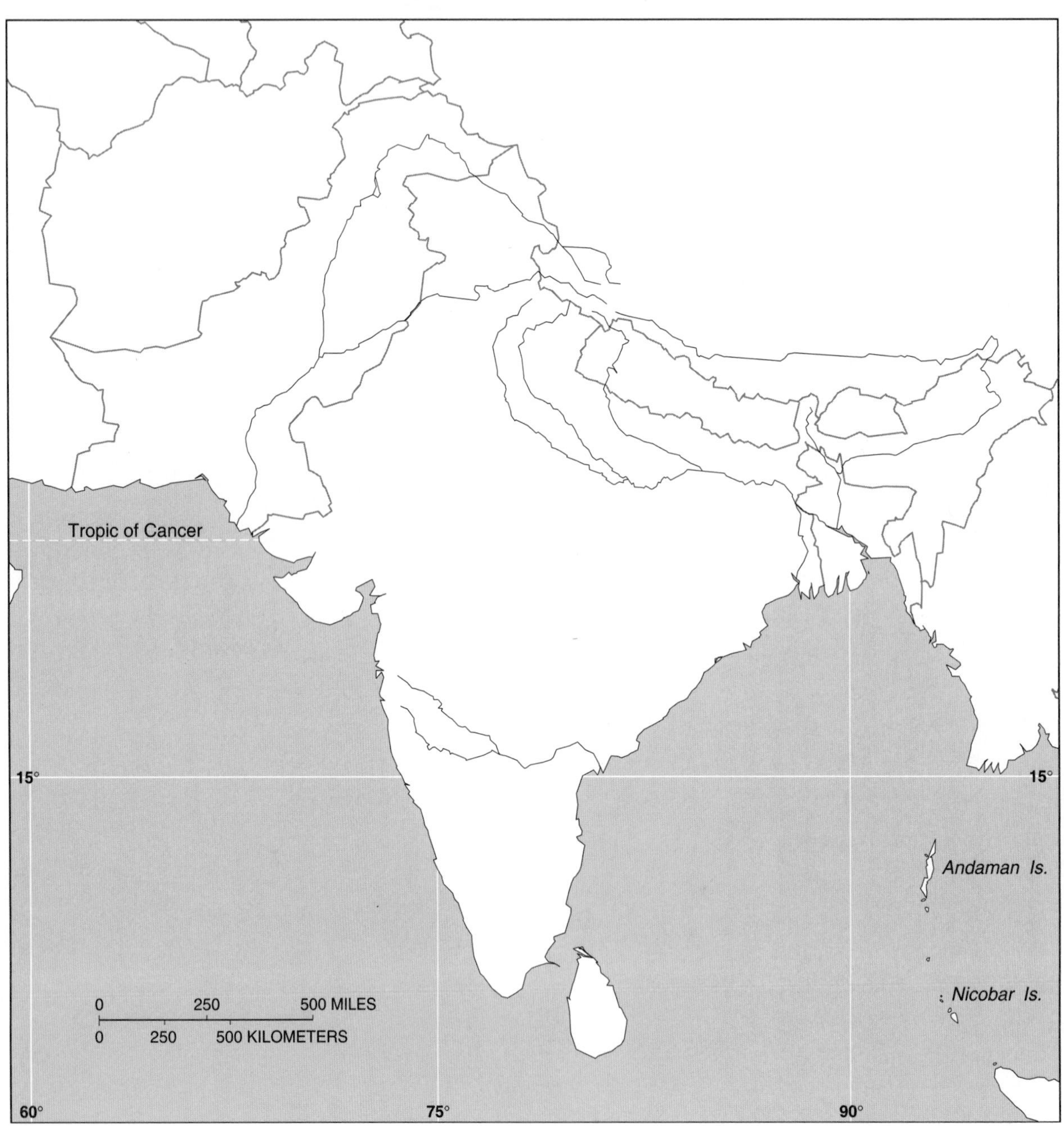


**FIGURE 3–6**
South Asia map.

**3–2** Considering its origins, is a delta a safe place during a flood?

______________________________

## Estuary

The widening mouth of a river into the sea where freshwater and saltwater mix and tidal effects are strong. Estuaries sometimes are drowned river valleys where the sea level has risen. *Examples:* Lower Thames River: 51° N, 0° (Europe map); Rio de la Plata: 35° S, 55° W (South America map) (Figure 3–7).

**3–3** Why are "drowned-valley"—type estuaries usually good harbors? ________

______________________________

## Fjord

A long, narrow, usually deep ocean inlet along a mountainous coast, formed by glacial erosion of river valleys. Coastlines with fjords (or fiords) are extremely irregular, with many narrow indentations. *Examples:* coast of Norway: 60° N, 5° E (Europe map); coast of Alaska: 56° N, 135° W (North America map).

**3–4** Is it accidental that coastlines with fjords usually are found in relatively cold-climate mountain coasts?

______________________________

## Gulf

An arm of the ocean that penetrates into a surrounding landmass. The Gulf of Mexico is the world's largest gulf. *Examples:* Persian Gulf: 27° N, 53° E (Middle East map) (Figure 3–8); Gulf of Alaska: 58° N, 45° W (North America map).

**3–5** Compare the Persian Gulf and Hudson Bay (56° N, 85° W) on the world map. Are gulfs always larger than bays? ________________

## Isthmus

A narrow strip of land connecting two larger bodies of land and separating two water bodies. *Examples:* Isthmus of Panama: 8°N, 80° W; Isthmus of Suez: 30° N, 33° E (world map).

**3–6** If you wished to shorten the route of oil tankers between the Persian Gulf and Japan, which isthmus would you cut by a canal, similar to the canals through the isthmuses of Panama and Suez? ________________

## Peninsula

A piece of land projecting into a body of water and surrounded on three sides by water. *Examples:* Alaskan peninsula: 65° N, 150° W (North America map); Scandinavian peninsula: 60° N, 12° E (world map).

**3–7** Which other U.S. state is composed mostly of a large peninsula?

______________________________

## Plain

A large area of relatively flat land. *Examples:* Great North European Plain: 53° N, 10° E (Europe map); Great Plains of the United States and Canada: 40° N, 100° W (North America map).

**FIGURE 3–7**
South America map.

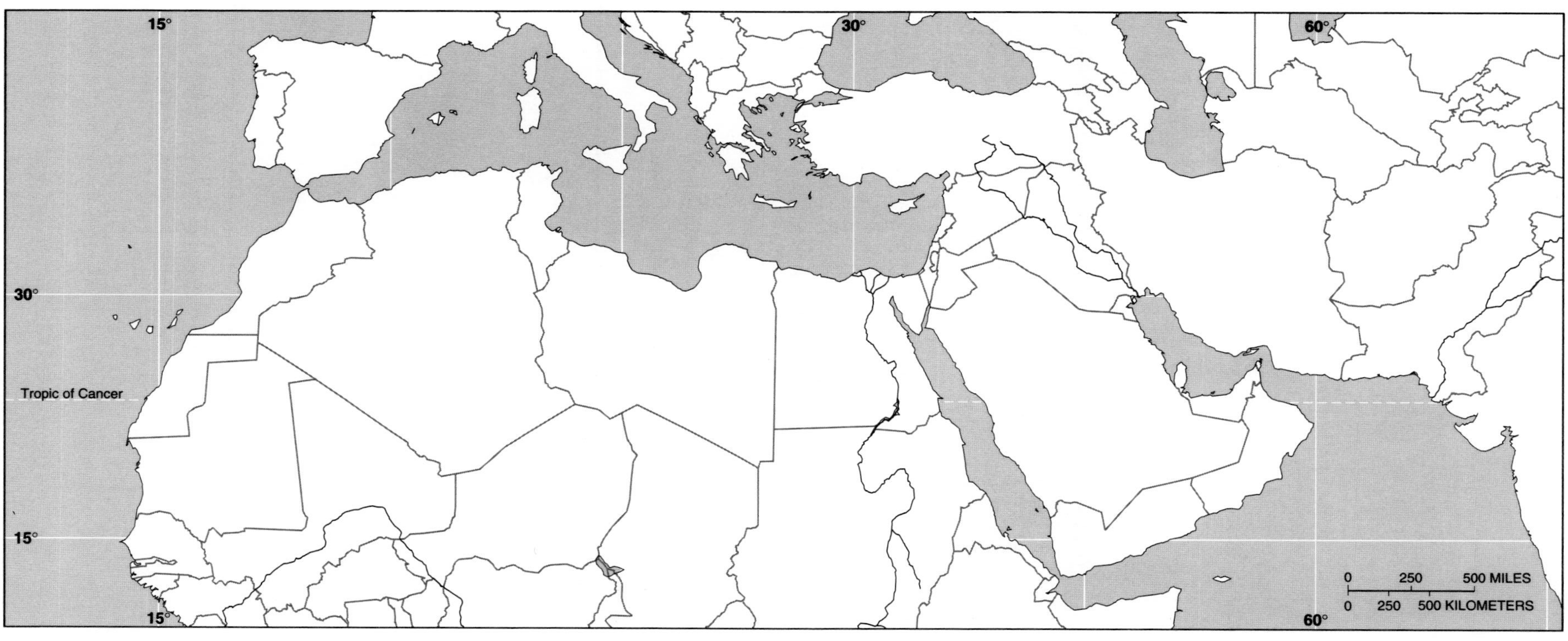


**FIGURE 3–8**
Middle East map.

**3–8** Which continent has the smallest percentage of its total area in plains? (Circle the correct answer.) Africa, Europe, North America.

### Strait

A narrow strip of water connecting two larger water bodies. *Examples:* Strait of Gibraltar: 36° N, 5°W (world map); Strait of Malacca, 3° N, 100° E (Southeast Asia map).

**3–9** Which country controls the strait connecting the Black Sea and the Mediterranean? ____________________

## MAJOR ISLANDS ASSOCIATED WITH CONTINENTS

Most islands and island groups (archipelagoes) are grouped with nearby continents or major world regions. Significant exceptions are New Zealand and the three major island groups of the Pacific Ocean—Polynesia, Melanesia, and Micronesia. These islands usually are grouped as "Oceania"—the islands of the ocean-dominated region.

### Pacific Basin

Locate and label these islands on the map of the Pacific basin in Figure 3–9:

Tasmania: 42° S, 148° E
New Zealand (North and South islands): 42° S, 174° E
Polynesian Triangle: Hawaii–New Zealand–Easter Island: 20° N, 157° W; 42° S, 174° E; 27° S,110°W
Melanesia: The islands east of New Guinea and north of New Zealand
Micronesia: The "tiny islands" (that's what the name means) north of Melanesia and east of the Philippines

### Europe

Locate and label these islands on a map of Europe (Figure 3–10):

| | | | |
|---|---|---|---|
| Great Britain: | 55° N, 3° W | Sardinia: | 40° N, 9° E |
| Ireland: | 53° N, 7° W | Crete: | 35° N, 25° E |
| Iceland: | 65° N, 19° W | Corsica: | 43° N, 8° E |
| Sicily: | 38° N, 14° E | | |

### North America

Locate and label these islands on the map of North America in Figure 3–11:

| | | | |
|---|---|---|---|
| Cuba: | 22° N, 80° W | Vancouver: | 50° N, 126° W |
| Hispaniola (Haiti and Dominican Republic): | 13° N, 72° W | Cape Breton: | 46° N, 61° W |
| Puerto Rico: | 13° N, 67° W | Ellesmere: | 80° N, 80° W |
| Jamaica: | 18° N, 77° W | Greenland: | 70° N, 40° W |
| Newfoundland: | 49° N, 56° W | Aleutians: | 52° N, 170° W |
| Victoria: | 70° N, 110° W | Prince Edward: | 46° N, 63° W |
| Baffin: | 65° N, 70° W | | |

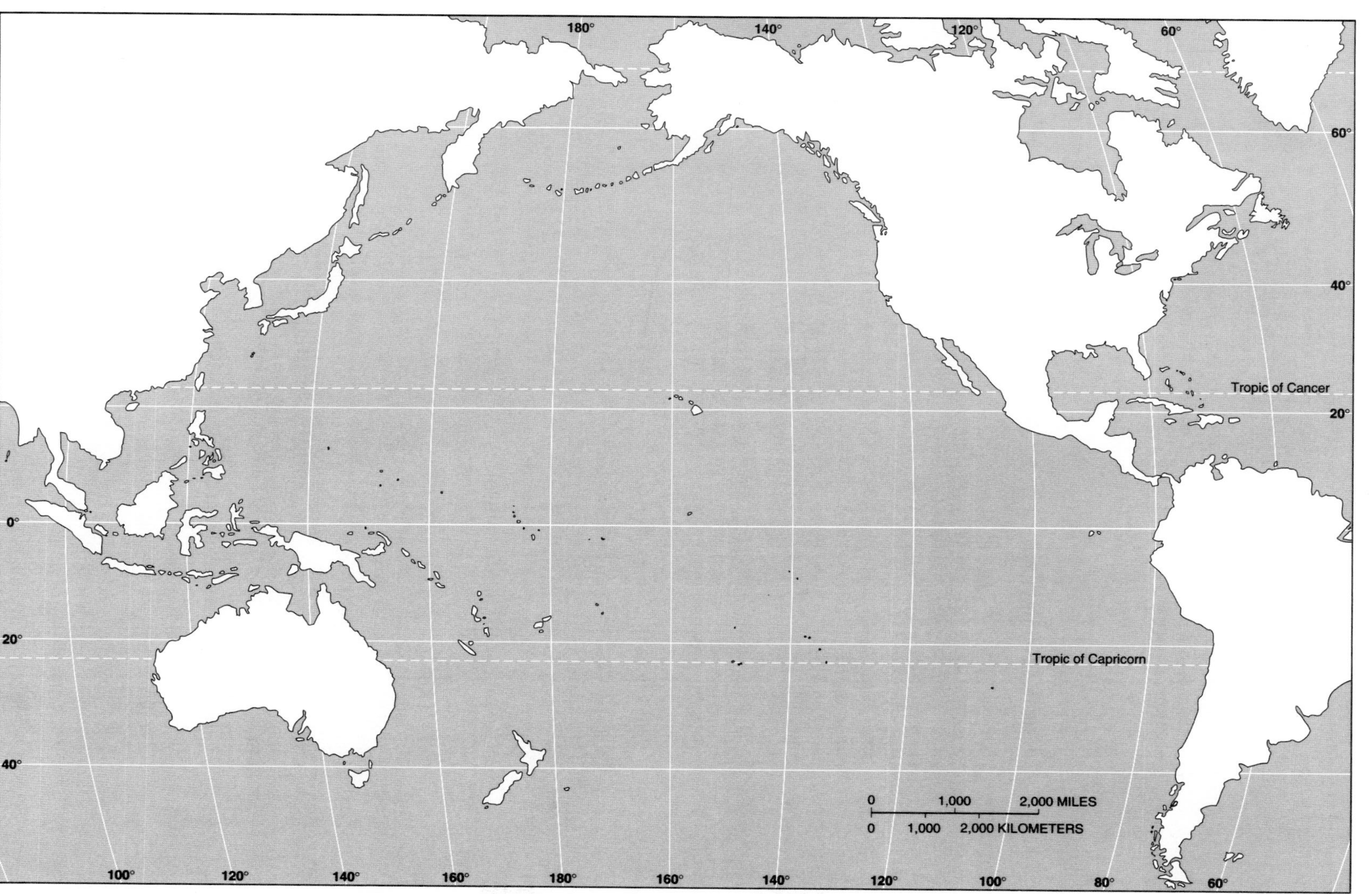


**FIGURE 3–9**
Pacific basin map.

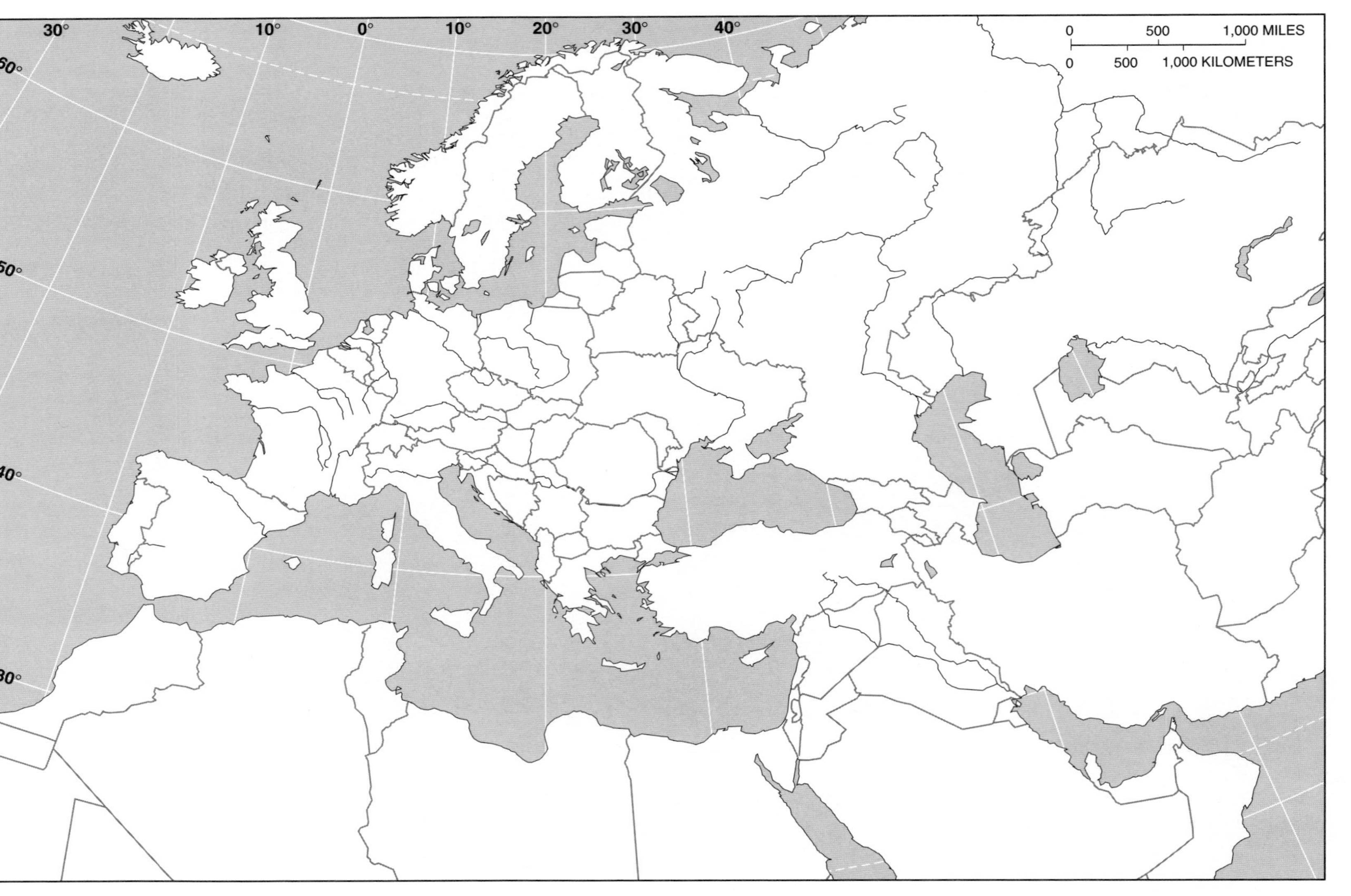


**FIGURE 3–10**
Europe map.

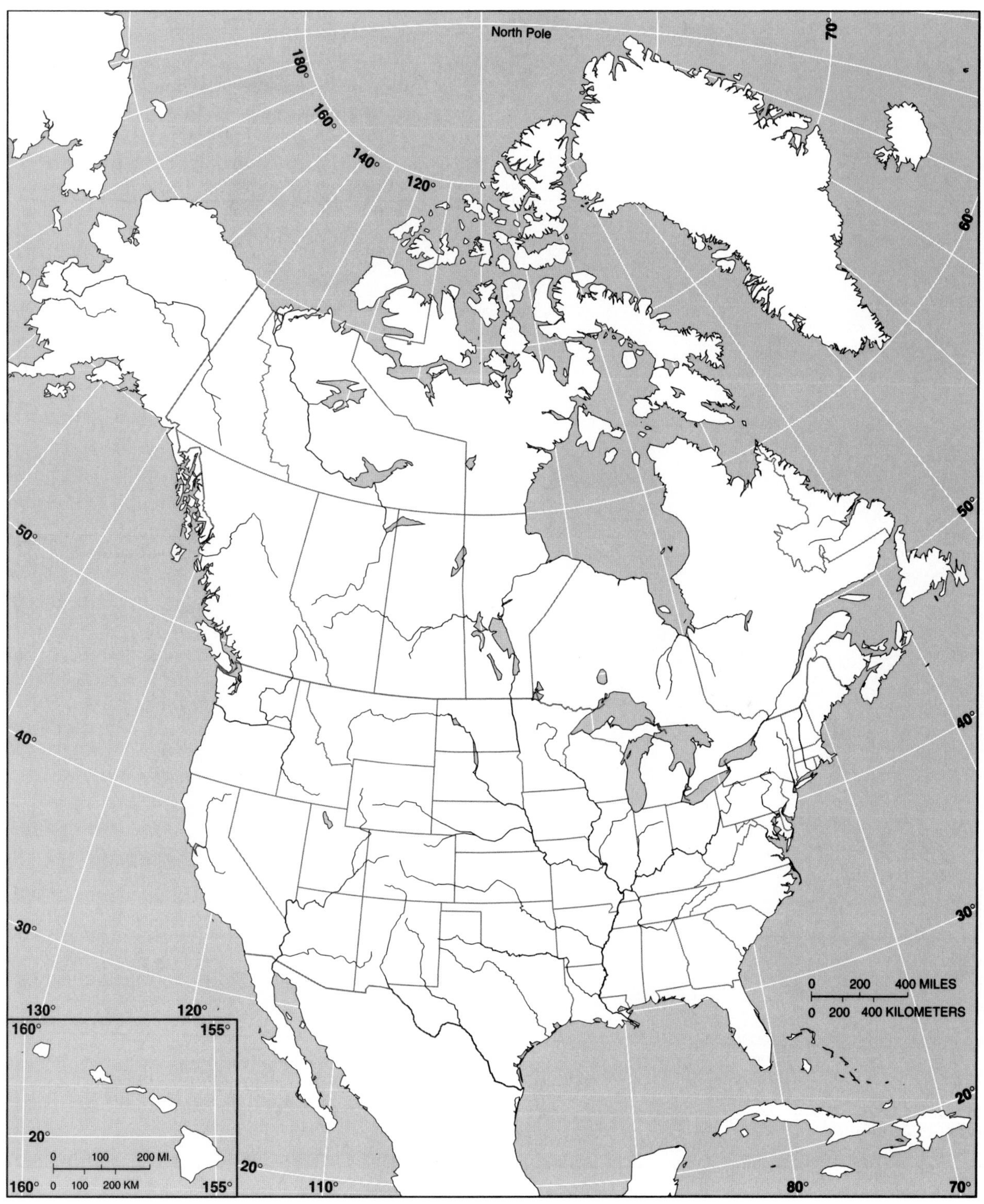


**FIGURE 3–11**
North America map.

## Africa

Locate and label these islands on the map of Africa in Figure 3–12:

Madagascar (Malagasy): 20° S, 47° W    Zanzibar: 6° S, 39° E

## South America

Locate and label these islands on the map of South America in Figure 3–13:

Tierra del Fuego: 55° S, 68° W    Falklands (Malvinas): 52° S, 60° W

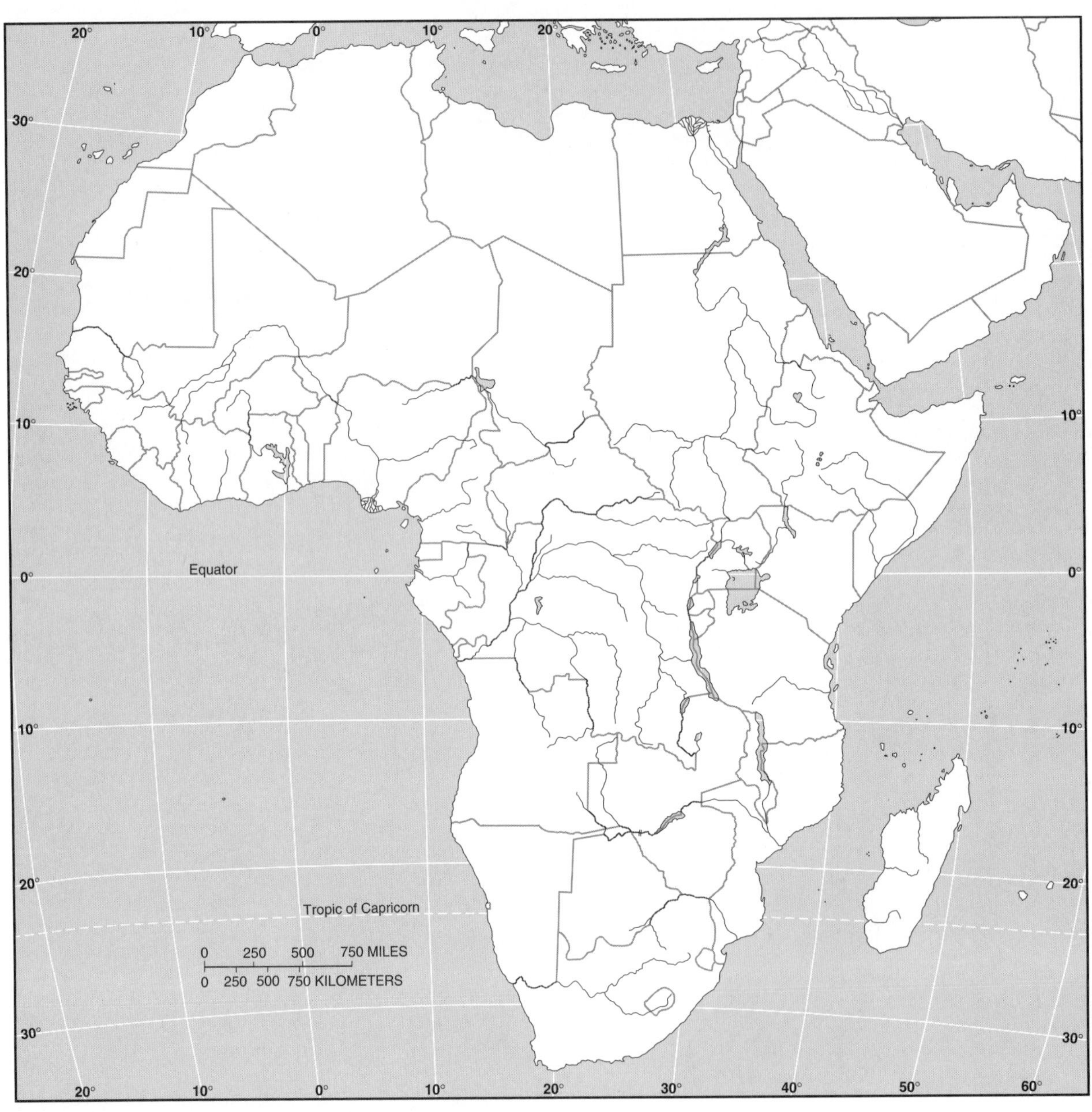


**FIGURE 3–12**
Africa map.

**FIGURE 3–13**
South America.

### Russia and the Newly Independent States

Locate and label these islands on the map of the newly independent states in Figure 3–14:

| | | | |
|---|---|---|---|
| Sakhalin: | 50° N, 142° E | Novaya Zemlya: | 75° N, 60° E |
| Kurils: | 45° N, 150° E | | |

### East Asia

Locate and label these islands on the map of East Asia in Figure 3–15:

| | | | |
|---|---|---|---|
| Honshu: | 35° N, 140° E | Taiwan: | 23° N, 122° E |
| Hokkaido: | 43° N, 143° E | Hainan: | 19° N, 110° E |
| Kyushu: | 32° N, 132° E | Cheju (Jeju): | 33° N, 127° E |
| Shikoku: | 33° N, 134° E | | |

### Southeast Asia

Locate and label these islands on the map of Southeast Asia in Figure 3–16:

| | | | |
|---|---|---|---|
| New Guinea: | 5° S, 143° E | Mindanao: | 7° N, 125° E |
| Borneo (Kalimantan): | 0° , 115° E | Paracel Islands: | 17° N, 112° E |
| Sumatra: | 0° , 102° E | Palawan: | 10° N, 118° E |
| Celebes (Sulawesi): | 20° S, 120° E | Mindoro: | 13° N, 121° E |
| Java (Jawa): | 70° S, 110° E | Samar: | 12° N, 125° E |
| Timor: | 9° S, 125° E | Panay: | 11° N, 122° E |
| Luzon: | 15° N, 122° E | | |

### Middle East and North Africa

Locate and label these islands on the map of the Middle East and North Africa in Figure 3–17:

| | | | |
|---|---|---|---|
| Cyprus: | 35° N, 33° E | Socotra: | 12° N, 54° E |

### South Asia

Locate and label this island on the map of South Asia in Figure 3–18:

| | |
|---|---|
| Ceylon (Sri Lanka): | 8° N, 81° E |

## MAJOR WATER BODIES

With water covering about 70% of the crust, the oceans are the planet's "final frontier." It is increasingly apparent that there must be international agreements and cooperation concerning management of the ocean's biological and mineral resources. Almost all the world's nations have agreed to ban or severely limit the harvest of endangered whale species, for example. International agreements limit ocean fishing for salmon, a species that begins life in freshwater rivers and returns to spawn in those same rivers after a year or more in the ocean. The advancing technology of undersea exploration opens new opportunities for seafloor mining, even in very deep water. Already, more than half the world's petroleum supply is pumped from under the ocean floor.

The territorial control of adjacent seas and seafloor by coastal countries is an aspect of international law that continually changes. Technology is extending exploitation further from shore, and ownership questions along with it. New laws of the seas have been proposed but not formally adopted by most major countries whose navies, merchant ships, fishing fleets, oil rigs, and robotic undersea

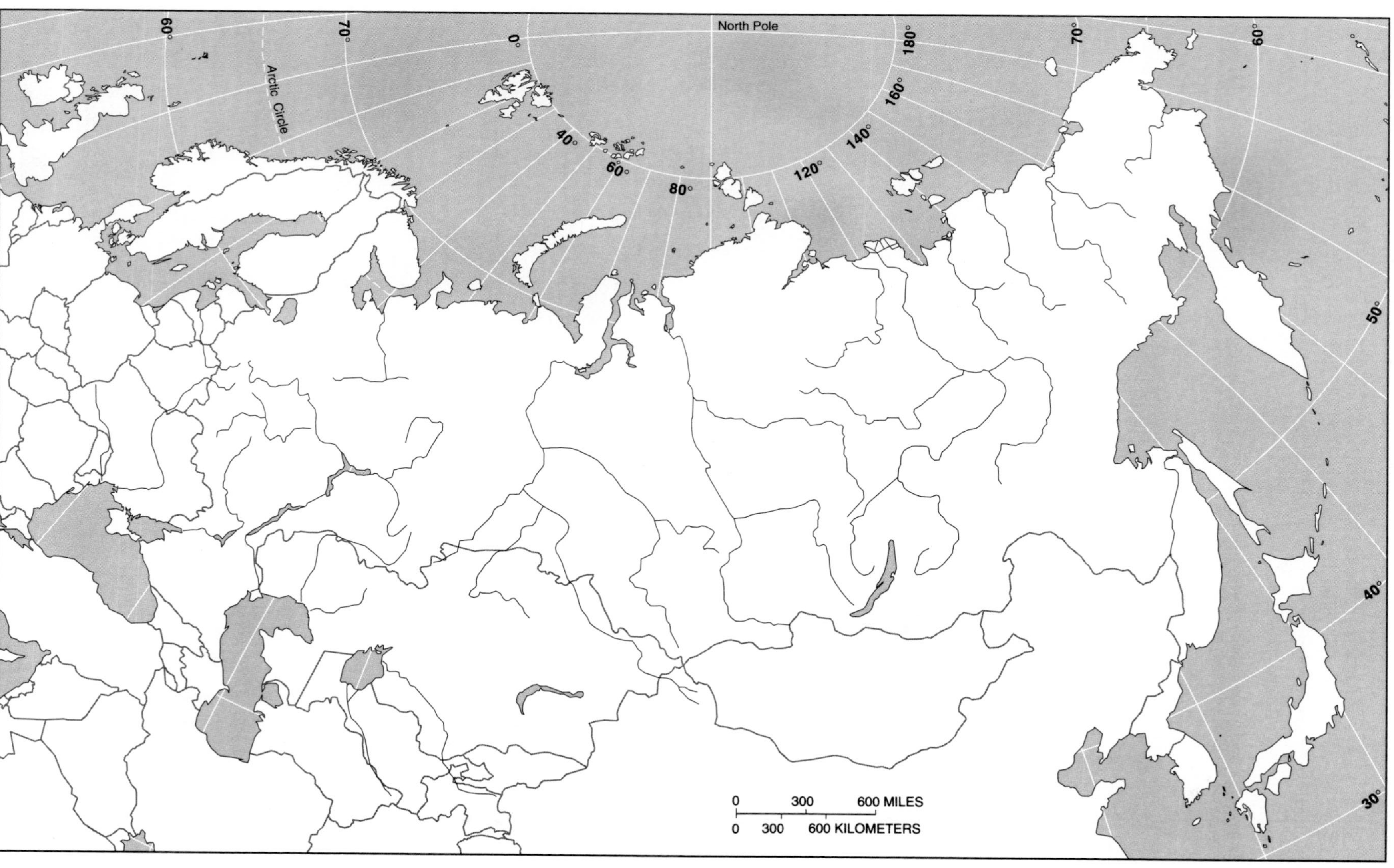


**FIGURE 3–14**
Russia, Transcaucasia, and Central Asia.

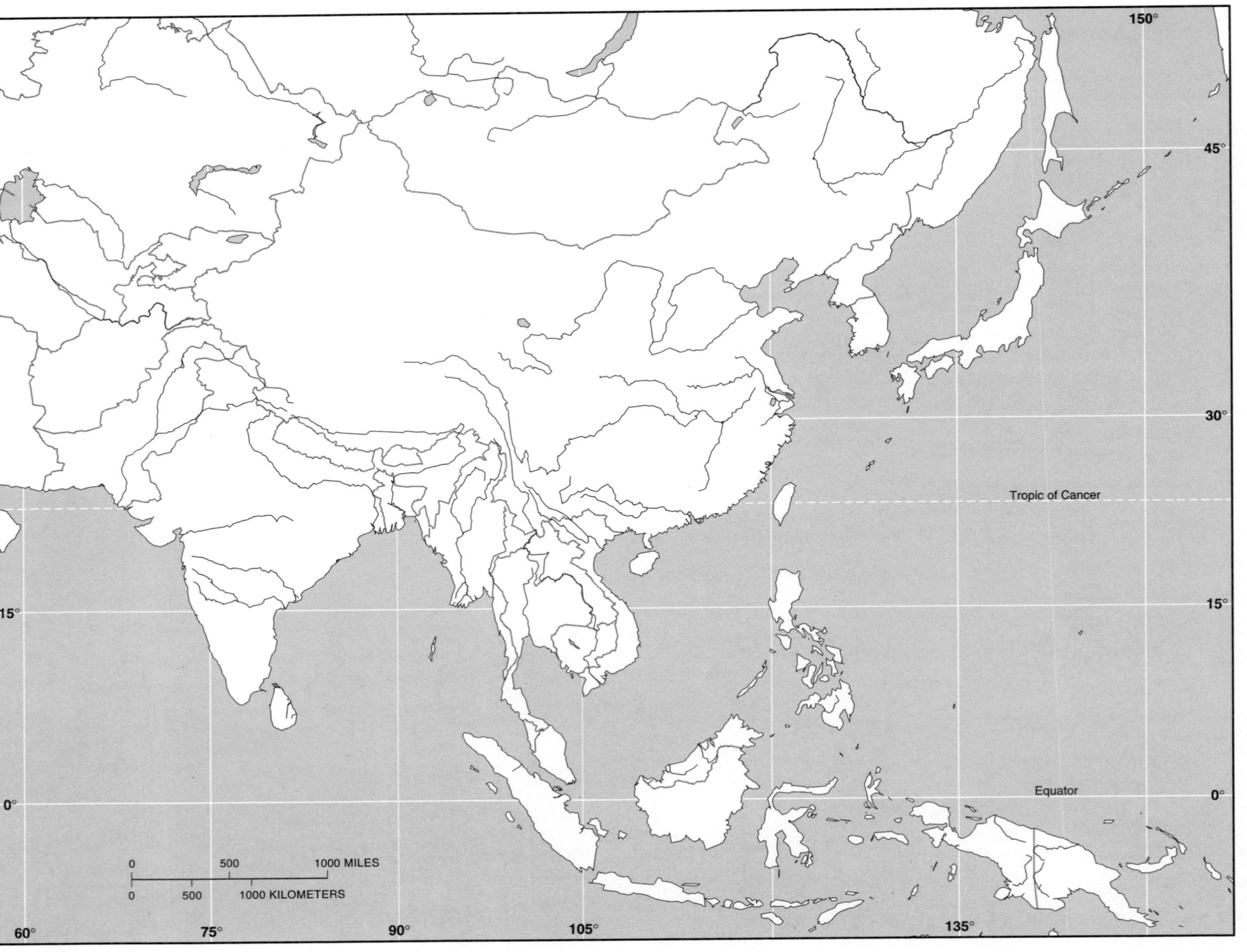


**FIGURE 3–15**
East Asia map.

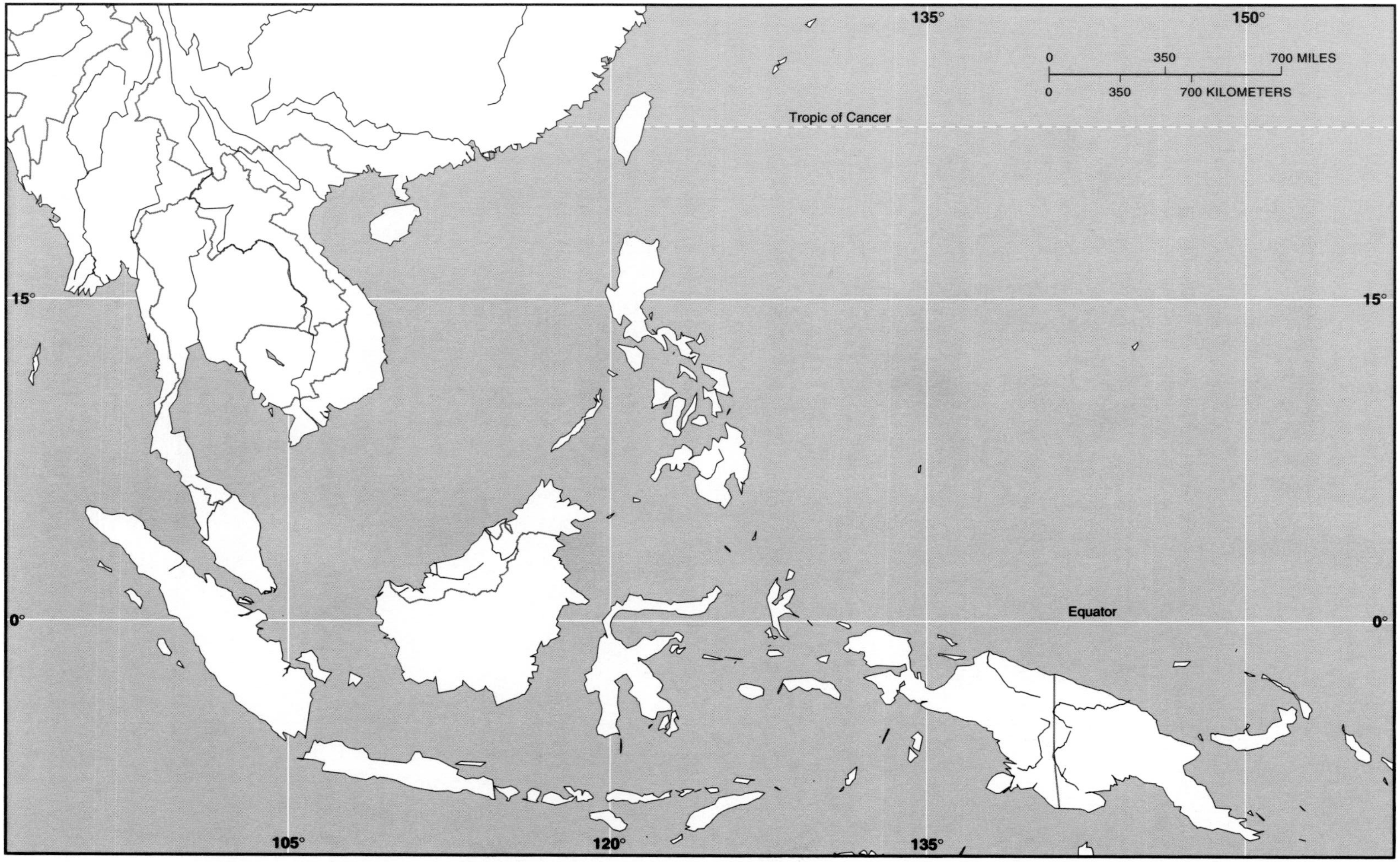

**FIGURE 3–16**
Southeast Asia map.

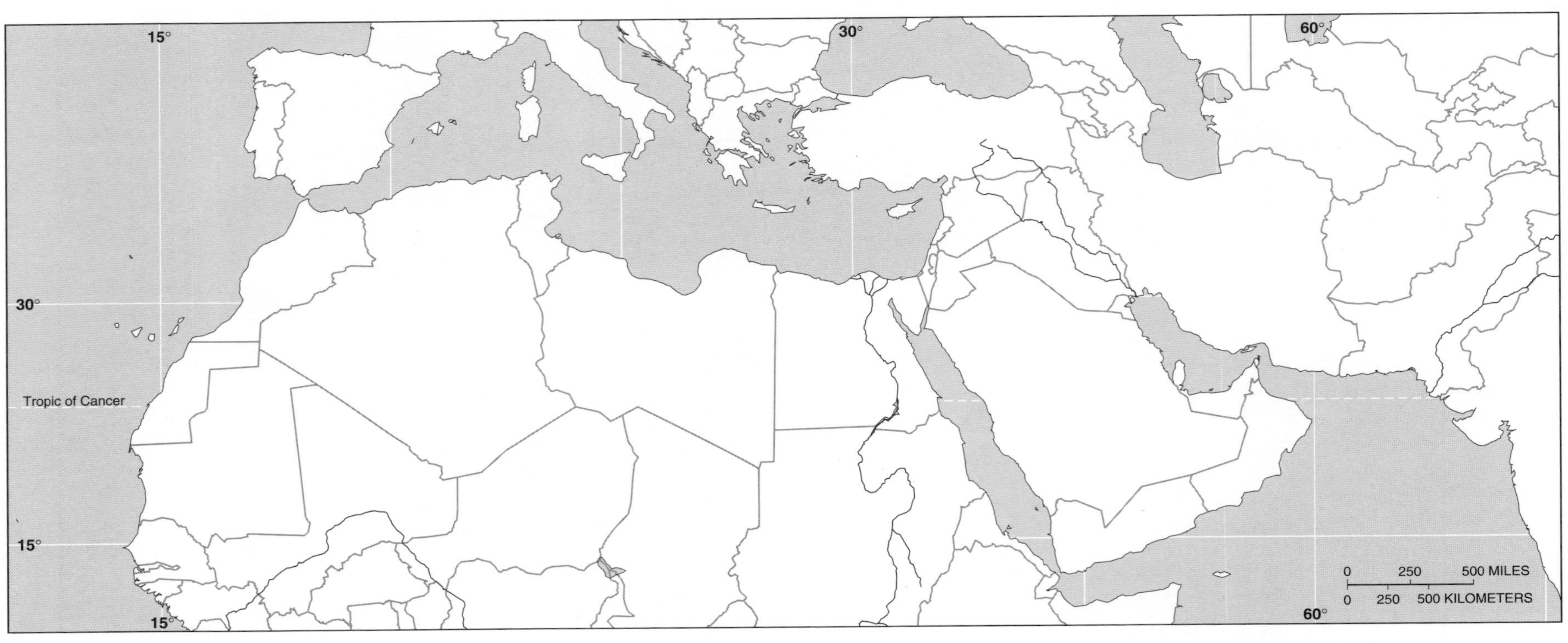


**FIGURE 3–17**
Middle East and North Africa map.

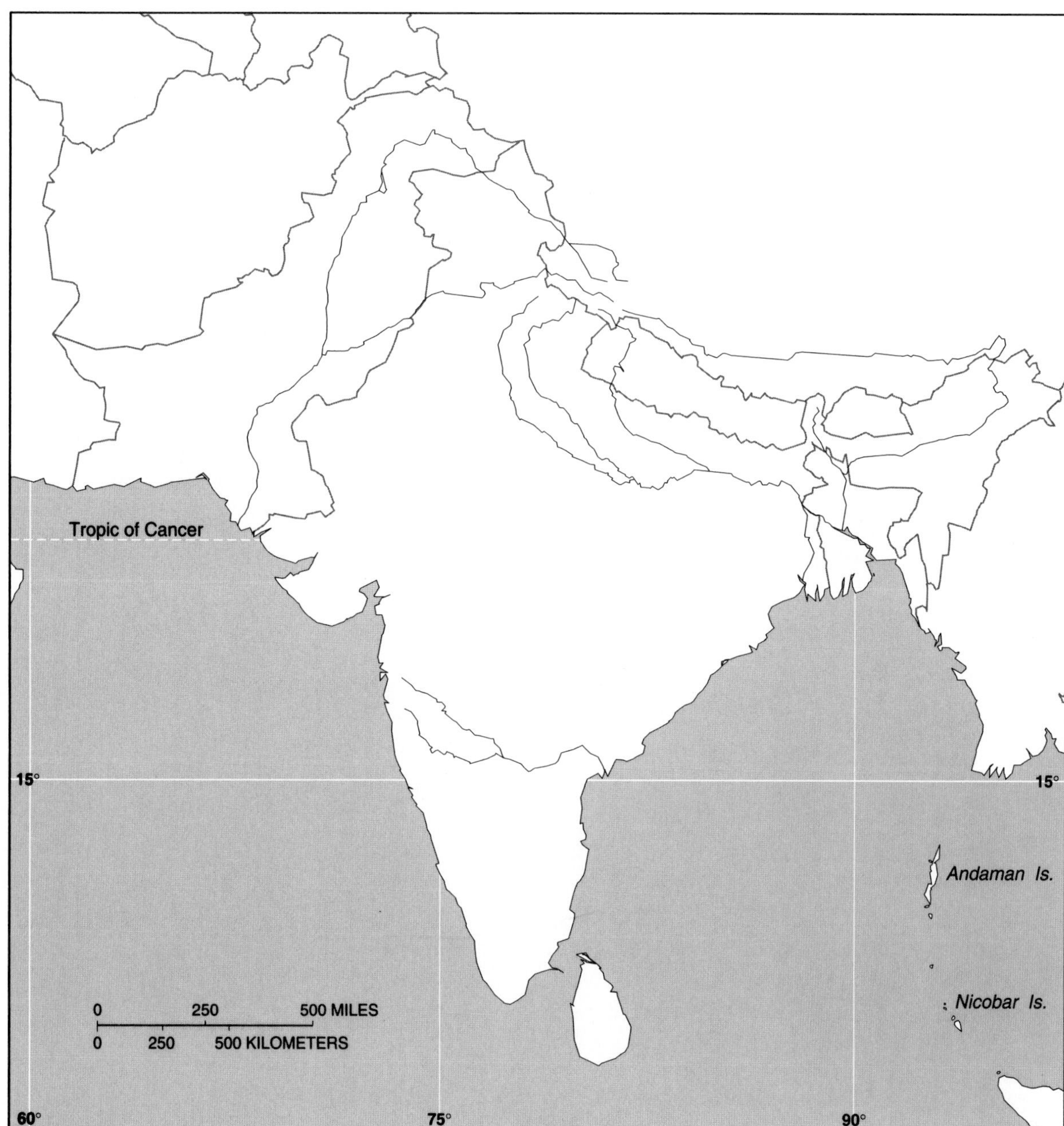


**FIGURE 3–18**
South Asia map.

mining operations would be restricted. Questions of national sovereignty over fishing rights, seafloor and subseafloor minerals, and even salvage rights to sunken ships will be of increasing importance to us all (see the following section, "Geoconcept: Territorial Seas and Ocean Sovereignty"). Treasures of the seas and seafloor make place knowledge of water bodies more important than ever before.

## The 15 Largest Seas, Gulfs, and Bays

On the world outline map in Figure 3–19, locate and label the seas listed in Table 3–1.

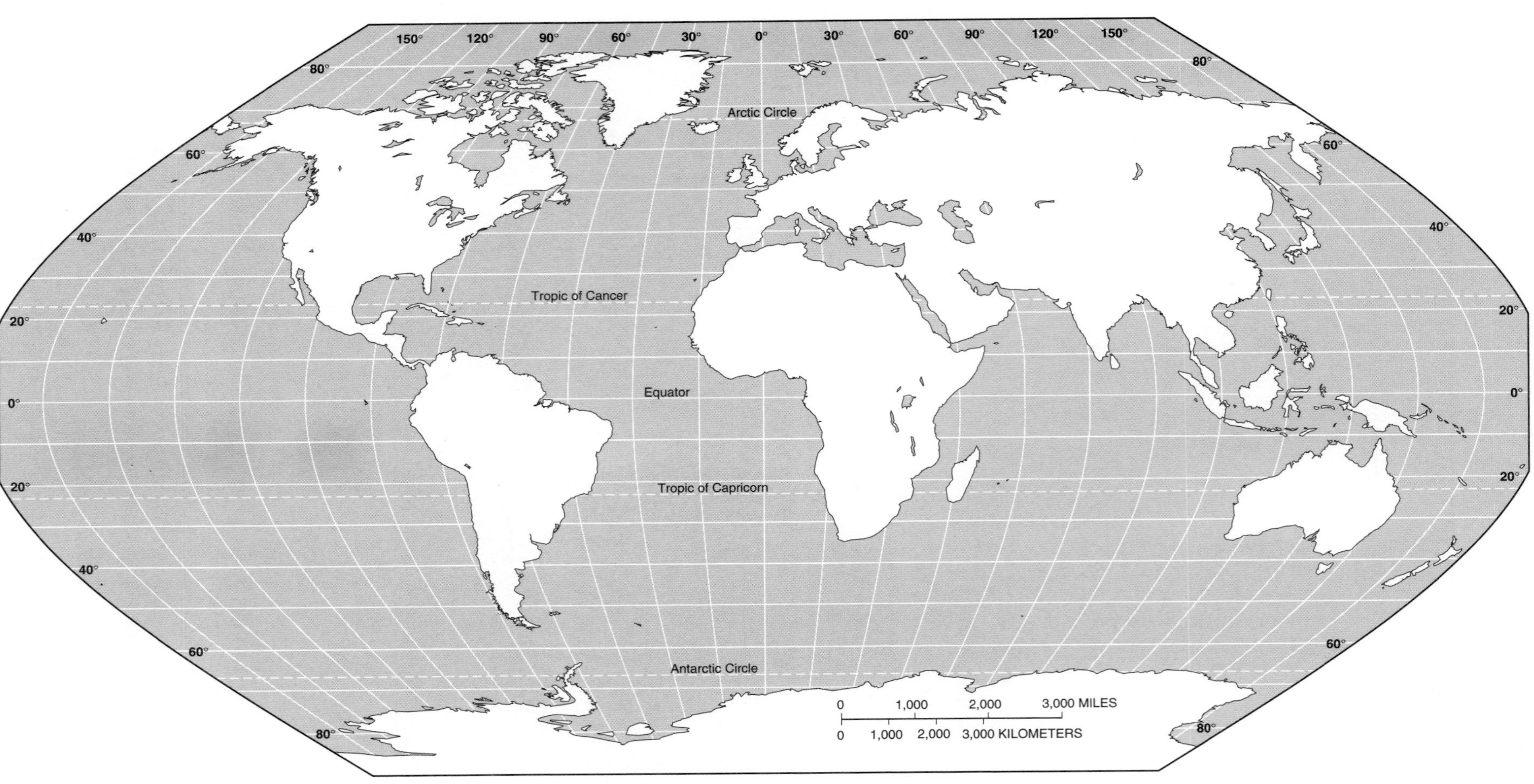


**FIGURE 3–19**
World map.

**TABLE 3–1** Earth's 15 largest seas, gulfs, and bays

| Water Body | | Area (square miles) | |
|---|---|---|---|
| Caribbean Sea: | 15° N, 75° W | 970,000 | (2,512,300 km$^2$) |
| Mediterranean Sea: | 33° N, 15° E | 969,000 | (2,509,710 km$^2$) |
| South China Sea: | 15° N, 112° E | 895,000 | (2,318,050 km$^2$) |
| Bering Sea: | 60° N, 175° W | 875,000 | (2,266,250 km$^2$) |
| Gulf of Mexico: | 23° N, 90° W | 600,000 | (1,740,000 km$^2$) |
| Sea of Okhotsk: | 55° N, 150° E | 590,000 | (1,528,100 km$^2$) |
| East China Sea: | 30° N, 125° E | 487,000 | (1,261,330 km$^2$) |
| Yellow Sea: | 37° N, 125° E | 480,000 | (1,243,200 km$^2$) |
| Hudson Bay: | 53° N, 85° W | 476,000 | (1,232,840 km$^2$) |
| Sea of Japan: | 40° N, 135° E | 389,000 | (1,007,510 km$^2$) |
| North Sea: | 55° N, 5° E | 222,000 | (574,980 km$^2$) |
| Black Sea: | 42° N, 35° E | 178,000 | (461,020 km$^2$) |
| Red Sea: | 22° N, 37° E | 169,000 | (437,710 km$^2$) |
| Baltic Sea: | 60° N, 20° E | 163,000 | (422,170 km$^2$) |
| *Caspian Sea: | 40° N, 52° E | 152,000 | (393,680 km$^2$) |

*Although called a "sea," the Caspian actually is the world's largest lake as no natural channel connects it with the world-ocean.

## United States and Canada

Locate and label the following lakes on the outline map of the United States and Canada in Figure 3–20:

| | | | |
|---|---|---|---|
| Lake Superior: | 47° N, 87° W | Great Slave Lake: | 62° N, 115° W |
| Lake Huron: | 45° N, 82° W | Lake Erie: | 42° N, 82° W |
| Lake Michigan: | 42° N, 87° W | Lake Winnipeg: | 53° N, 98° W |
| Great Bear Lake: | 66° N, 120° W | Lake Ontario: | 42° N, 77° W |

## Russia and the Newly Independent States

Locate and label these bodies of water on the outline map of the newly independent states in Figure 3–21:

| | | | |
|---|---|---|---|
| Caspian Sea: | 40° N, 52° E | Lake Ladoga: | 61° N, 32° E |
| Aral Sea: | 45° N, 60° E | Lake Balkash: | 46° N, 74° E |
| Lake Baykal (Baikal): | 54° N, 107° E | Lake Onega: | 62° N, 35° W |

## Africa

Locate and label these lakes on the outline map of Africa in Figure 3–22:

| | | | |
|---|---|---|---|
| Lake Victoria: | 2° S, 33° E | Lake Nyasa (Lake Malawi): | 14° S, 35° E |
| Lake Tanganyika: | 6° S, 29° E | Lake Chad: | 140° N, 14° E |

## Large Rivers

On the appropriate regional drainage outline map, locate and label the rivers listed in Table 3–2.

# CHECK UP

There is a common but incorrect belief that rivers flow southward, eastward, or westward, but never northward. To end that misconception, note that the longest river in the world, the Nile, flows northward! Rivers, of course, flow downhill, in whatever direction that may be.

**FIGURE 3–20**
U.S.—Canada map.

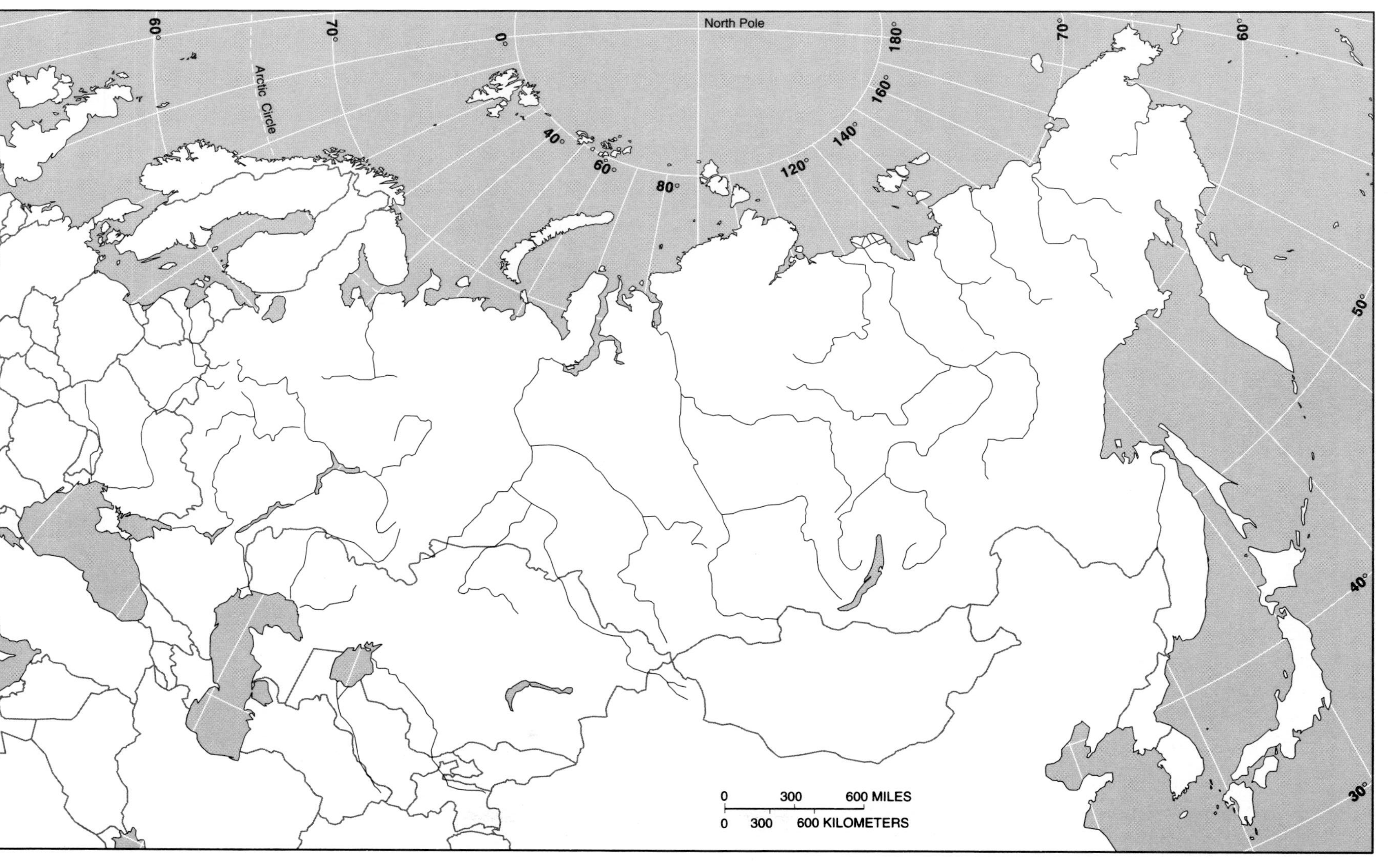


**FIGURE 3–21**
Russia, Transcaucasia, and Central Asia.

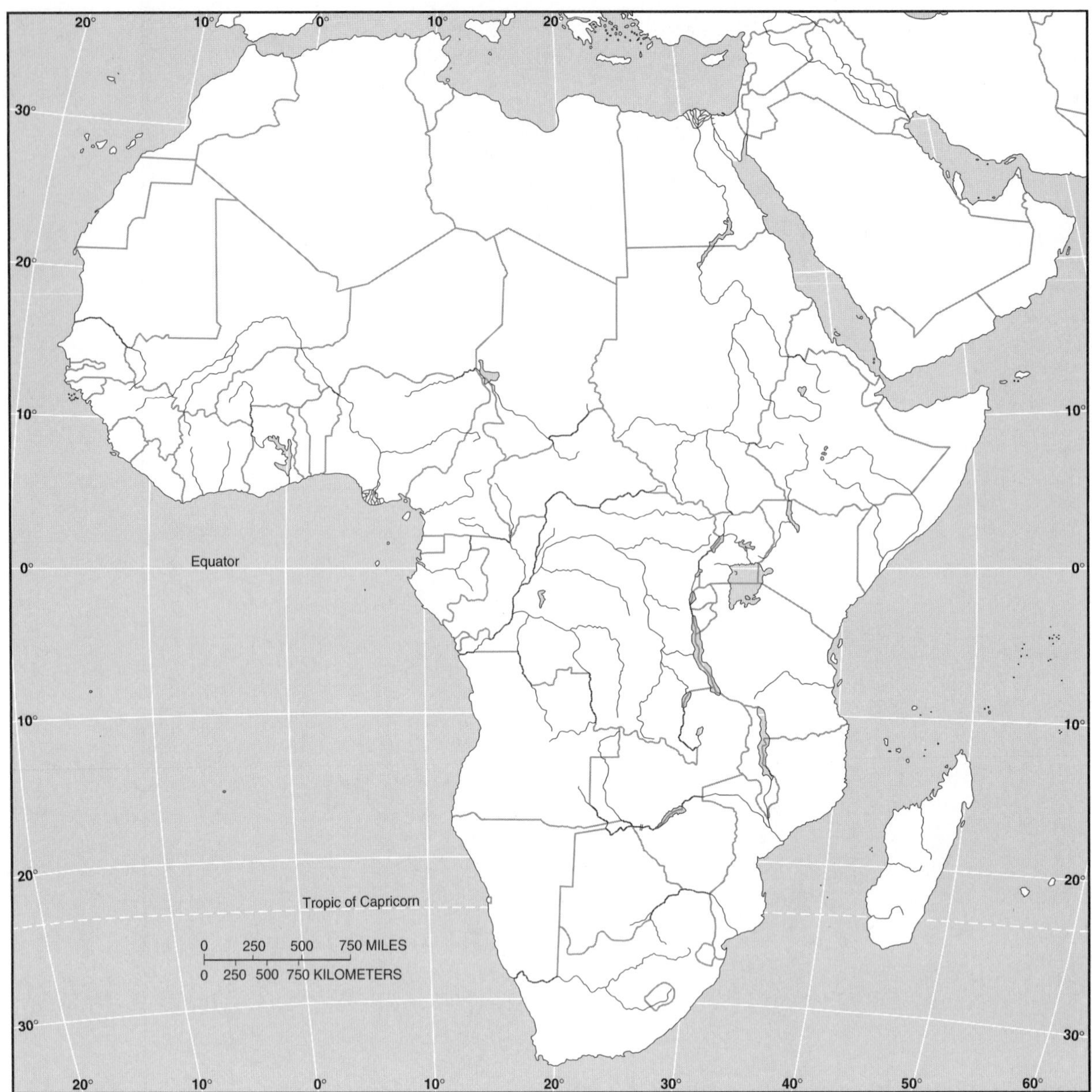


**FIGURE 3–22**
Africa map.

**3–10** Circle the major river that does *not* flow primarily northward: Rhine: 51° N, 7° E; Lena: 68° N, 124° E; Ob-Irtysh: 62° N, 67° E; Mackenzie-Peace: 63° N, 124° W; Congo (Zaire): 1° N, 18° E.

**3–11** Which is the smallest ocean? ____________________

**3–12** Which is the largest ocean? ____________________

**3–13** Circle which sea does *not* lie next to Russia: Black Sea; Red Sea; White Sea; Baltic Sea.

**TABLE 3–2** Large rivers.

| River | | Continent |
|---|---|---|
| Nile: | 19° N, 32° E | Africa |
| Amazon : | 2° S, 53° W | South America |
| Chang Jiang (Yangtze): | 30° N, 117° E | Asia |
| Mississippi-Missouri: | 31° N, 91° W | North America |
| Ob-Irtysh: | 62° N, 67° E | Asia |
| Yenesey-Angara: | 67° N, 87° E | Asia |
| Huang He (Yellow): | 35° N, 114° E | Asia |
| Amur-Shilka-Onon: | 50° N, 127° E | Asia |
| Lena: | 68° N, 124° E | Asia |
| Congo (Zaire): | 1° N, 18° E | Africa |
| Mackenzie-Peace: | 63° N, 124° W | North America |
| Mekong: | 17° N, 104° E | Asia |
| Niger; | 6° N, 6° E | Africa |
| Parana-Plata: | 32° S, 61° W | South America |
| Murray-Darling: | 34° S, 142° E | Australia |
| Volga: | 47° N, 46° E | Europe |
| Madeira: | 7° S, 63° W | South America |
| Purus: | 7° S, 64° W | South America |
| Yukon: | 62° N, 143° W | North America |
| St. Lawrence: | 48° N, 70° W | North America |
| Rio Grande: | 26° N, 99° W | North America |
| Syr Darya: | 44° N, 66° E | Asia |
| Sao Francisco: | 9° S, 40° W | South America |
| Indus: | 27° N, 68° E | Asia |
| Danube: | 44° N, 24° E | Europe |
| Salween: | 27° N, 98° E | Asia |
| Brahmaputra: | 27° N, 93° E | Asia |
| Shatt al Arab: | 30° N, 48° E | Asia |
| Tigris: | 34° N, 44° E | Asia |
| Euphrates: | 36° N, 40° E | Asia |
| Tocantins: | 3° S, 49° W | South America |
| Xi: | 23° N, 110° E | Asia |
| Amu Darya: | 37° N, 67° E | Asia |

**3–14** Circle the sea in which you could *not* go swimming from an Australian beach: Tasman Sea, Coral Sea, Philippine Sea, Timor Sea.

**3–15** If rich oil deposits are found under the floor of the South China Sea, which countries might claim at least part ownership of that oil on the basis of controlling part of the South China Sea's coastline? ____________________

______________________________________________

______________________________________________

**3–16** Which gulf is larger than many seas? ____________________

**3–17** Which bay is larger than many seas? ____________________

**3–18** Which sea would you cross in sailing from Italy to Libya? ____________________

**3–19** Which sea would you cross in sailing from Somalia to Pakistan?

______________________________________________

**3–20** Which sea would you cross in sailing from the Philippines to Vietnam?

______________________________

**3–21** If you wished to surf in the world's largest ocean, circle which beaches you would visit: California, Florida, Sri Lanka, France.

**3–22** If you were standing in the geographic center of the world's second-largest country (in area), circle the language you would most likely see on the sign identifying that location: Chinese, Russian, English, Spanish.

**3–23** If you were visiting the largest country (in area) in South America, the national flag on the flagpole would be that of (Circle the correct answer.): Mexico, Argentina, United Kingdom, Brazil.

**3–24** If you were visiting the southernmost *mainland* part of the Asian continent, circle the name of the nation you would be in: India, Indonesia, Malaysia, Vietnam.

**3–25** If you wanted to stand on the shores of the world's largest lake, circle the country you would *not* visit: Iran, Russia, Canada.

**3–26** Which *two* continents are the only ones located entirely in the Southern Hemisphere? ______________________________

**3–27** If you were to fly around the world following the Equator, you would fly over the national territories of which three South American countries?

______________________________

Which six African states? ______________________________

______________________________

Which Asian nation? ______________________________

**3–28** In which general direction (north, south, east, or west) would you fly to go from Argentina to Venezuela? ______________ From Japan to Australia? ______________ From Ireland to Poland? ______________ From Russia to Germany? ______________ From Zaire to Kenya? ______________

**3–29** Which ocean would you cross to go from Australia to South Africa? ______________ From South Africa to Brazil? ______________ From Canada to the United Kingdom? ______________

**3–30** If you were fishing in the world's longest river, which continent would you be on? ______________________________

## GEOCONCEPT

### Territorial Seas and Ocean Sovereignty

Water bodies that are completely surrounded by a country's national territory—Lake Michigan in the United States or Lake Baykal in Russia—are clearly controlled by those countries. Where countries share a water body as their mutual boundary, as is the case with the United States and Canada sharing Lake Erie, the actual international boundary usually runs through the middle of that water body. And where a river forms an international boundary, the actual boundary may run down the middle of the river or the middle of the deepest channel.

But how far out into an ocean or sea does a coastal state's control extend? By long tradition, a coastal state's "territorial sea"—that part of a neighboring sea considered to be completely within the jurisdiction of that country—extended three miles (4.8-km) outward from its shores. A complicated set of rules evolved over time to take into account the size and shapes of bays, islands, and internal water bodies in determining territorial seas (Figure 3–23). Supposedly, the three-mile (4.8-km) rule originated at a time when shore-based cannon could hit ships within three miles (4.8-km) of shore, but no farther.

The "high seas," or open seas beyond any one country's limits of territorial control, generally were open to the use by ships of any country, for peaceful navigation or fishing. Peaceful navigation, including that of warships not in conflict with the country concerned, usually involved the right to sail through narrow straits even when the three-mile (4.8-km) limit would otherwise block those

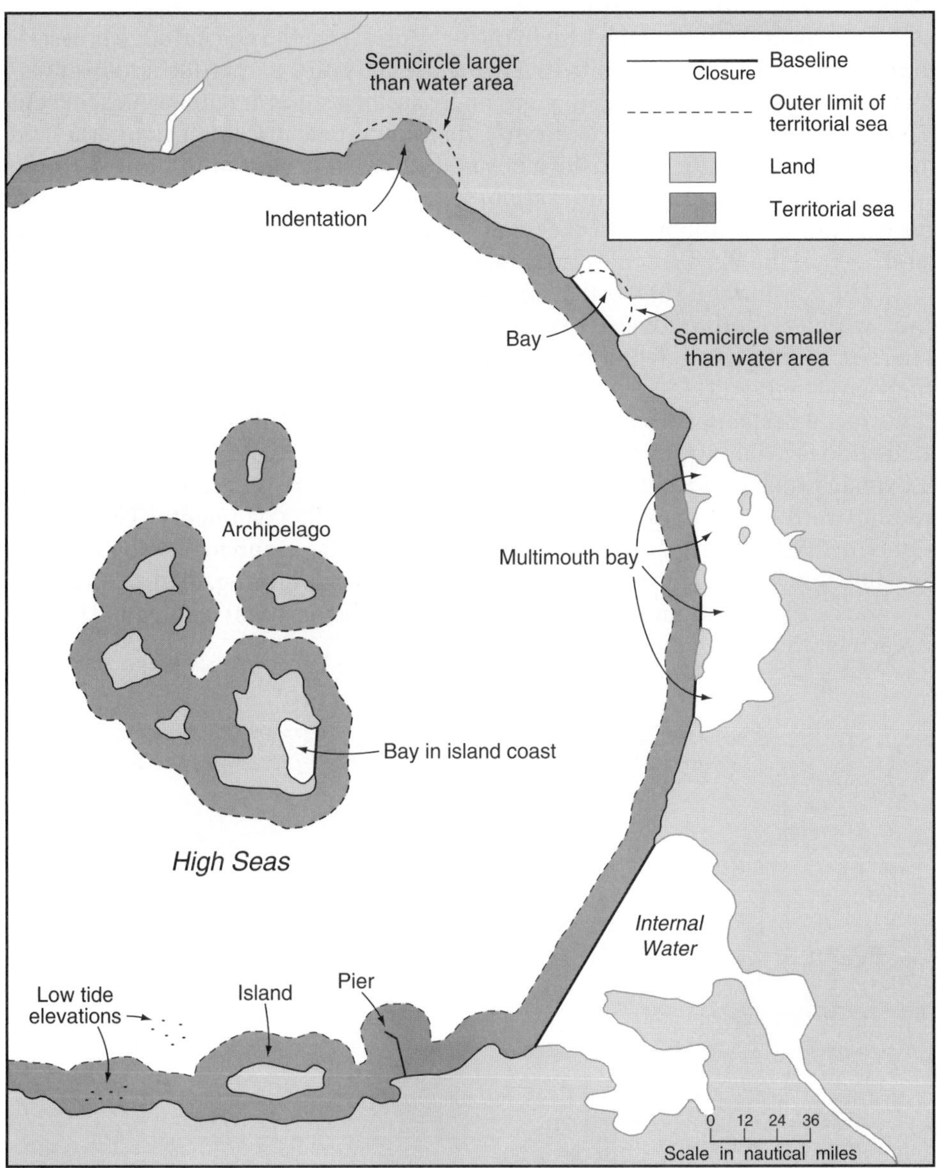


**FIGURE 3–23**
The baseline from which the territorial sea is measured. Most territorial sea limits have been extended now to 12 miles (19.3 km). (*Source*: U.S. Department of State.)

straits to foreign ships. International shipping between Russian Black Sea ports and the rest of the world can freely use the Turkish territorial waters of the Bosporus and Dardanelles straits, for example, as long as the Russians and Turks are not at war with each other.

Most coastal states now claim a 12-mile (nautical) (19.3-km) territorial sea and a 200-mile (322-km) "exclusive economic zone" in which the coastal state controls all exploitation of mineral and biological resources.

International law is a series of generally agreed-upon principles that, with luck, most countries honor most of the time. Some countries claim and try to enforce much more ambitious territorial limits at sea. To protect its fishing industry, Peru claims a 200-mile (322-km) limit, for example. The United Nations recommends a uniform 12-mile (19.3-km) limit, but many countries claim more, and some claim less. Coastal states have the right to establish a 12-mile (19.3-km) contiguous zone beyond the territorial sea in which they can enforce customs regulations and pollution control.

Whatever the claimed width of the territorial sea, the coastal state is asserting complete sovereignty within it. All of these activities may be forbidden or regulated by the state concerned: navigation by surface ships or submarines (except passage through certain narrow straits); fishing for finfish (which swim freely in the water); fishing for shellfish (whose adult existence often is limited to the seafloor); whaling; scraping the seafloor for mineral deposits; or drilling for oil or gas beneath the seafloor. However, some ownership rights and powers may extend beyond the limits of the territorial sea.

The *continental shelf* is that submerged land, geologically part of the continent, which extends outward from the coast as a gently sloping, shallow seafloor. This continental shelf, which hardly exists off some coasts whereas others have shelves 800 miles (1287 km) wide, ends where the gentle gradient of the shelf becomes a much steeper descent toward the ocean depths. This sharp break in angle of decline is called the *continental slope* (Figure 3–24). In 1945, President Harry S Truman proclaimed that the continental shelf adjacent to U.S. territorial waters, and outward to the edge of the shelf, was under U.S. sovereignty. This sovereignty, or "right to rule," did not, however, necessarily extend to the *waters* above the shelf and beyond the territorial sea limits. Seafloor mining, drilling for oil or gas, and catching bottom-dwelling shellfish like lobsters are all controlled by this claim over the continental shelf.

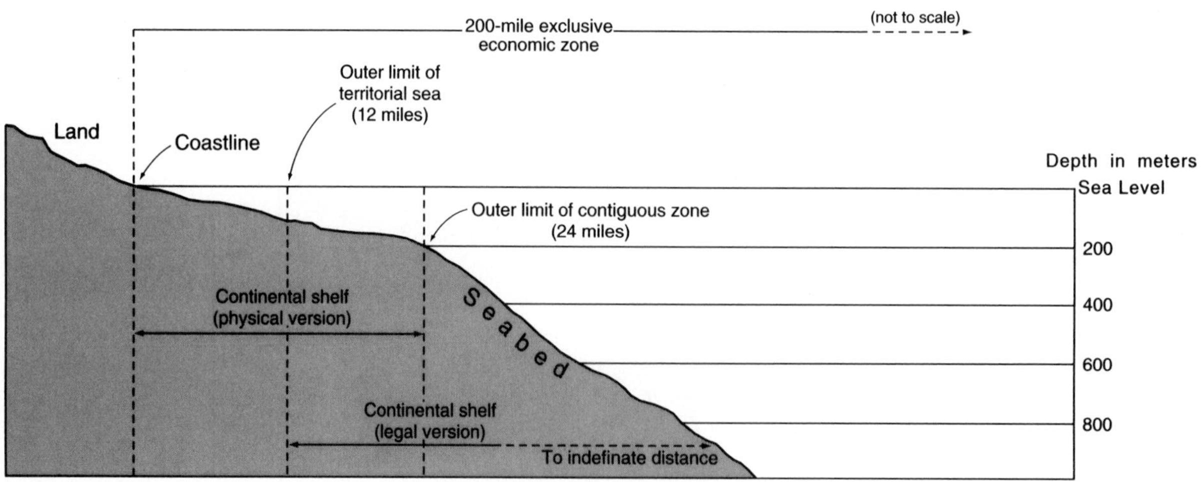


**FIGURE 3–24**
Continental shelf (vertical scale exaggerated). (*Source*: U.S. Department of State.)

# 4

# The United States and Canada

## INTRODUCTION

The United States and Canada are by far the largest two countries on the continent of North America. As a physical unit, the North American continent is the third largest, accounting for a little over 16% of the planet's land area. North America lies between the Atlantic Ocean on the east and the Pacific to the west of the landmass, and between the Arctic Ocean to the north and the Isthmus of Panama to the south, where North America meets the continent of South America. North America is a relatively lightly populated continent, holding only 8% of the world's population.

Culture, economy, and government often are more significant to people than which landmass they live on. Thus, it is customary to organize major world regions on the basis of these cultural, economic, and political factors rather than by continent. The United States and Canada commonly are recognized as a major region, based on strong similarities in culture, economy, and political philosophy. Sometimes referred to as Anglo-America (in contrast to Latin America), the United States and Canada share English as their predominant, though not only, language.

## PHYSICAL GEOGRAPHY

Canada is the world's second-largest country in territory (Russia is the largest), and the United States ranks fourth in size (following third largest People's Republic of China). Greenland, the world's largest island, usually is grouped with North America (the U.S.–Canada region) by default. Despite its ties to Denmark, Greenland is physically not a part of Europe.

The U.S.–Canada region is fairly compact (Figure 4–1). It can be simplified on a sketch map as a huge four-sided figure. This immense area is narrower on its southern side (the border with Mexico and the Gulf of Mexico) than on its northern edge (bounded by the Arctic Ocean).

Hudson Bay is the largest water body penetration into the heart of the landmass, and Canada's Arctic coast is fringed by an archipelago of islands. Newfoundland, off Canada's east coast, is the fifteenth largest island in the world.

The Mississippi–Missouri is the region's largest river system and is the fourth largest in the world. Canada's MacKenzie-Peace River system is Canada's largest, the region's second largest, and the world's tenth-ranked river in length.

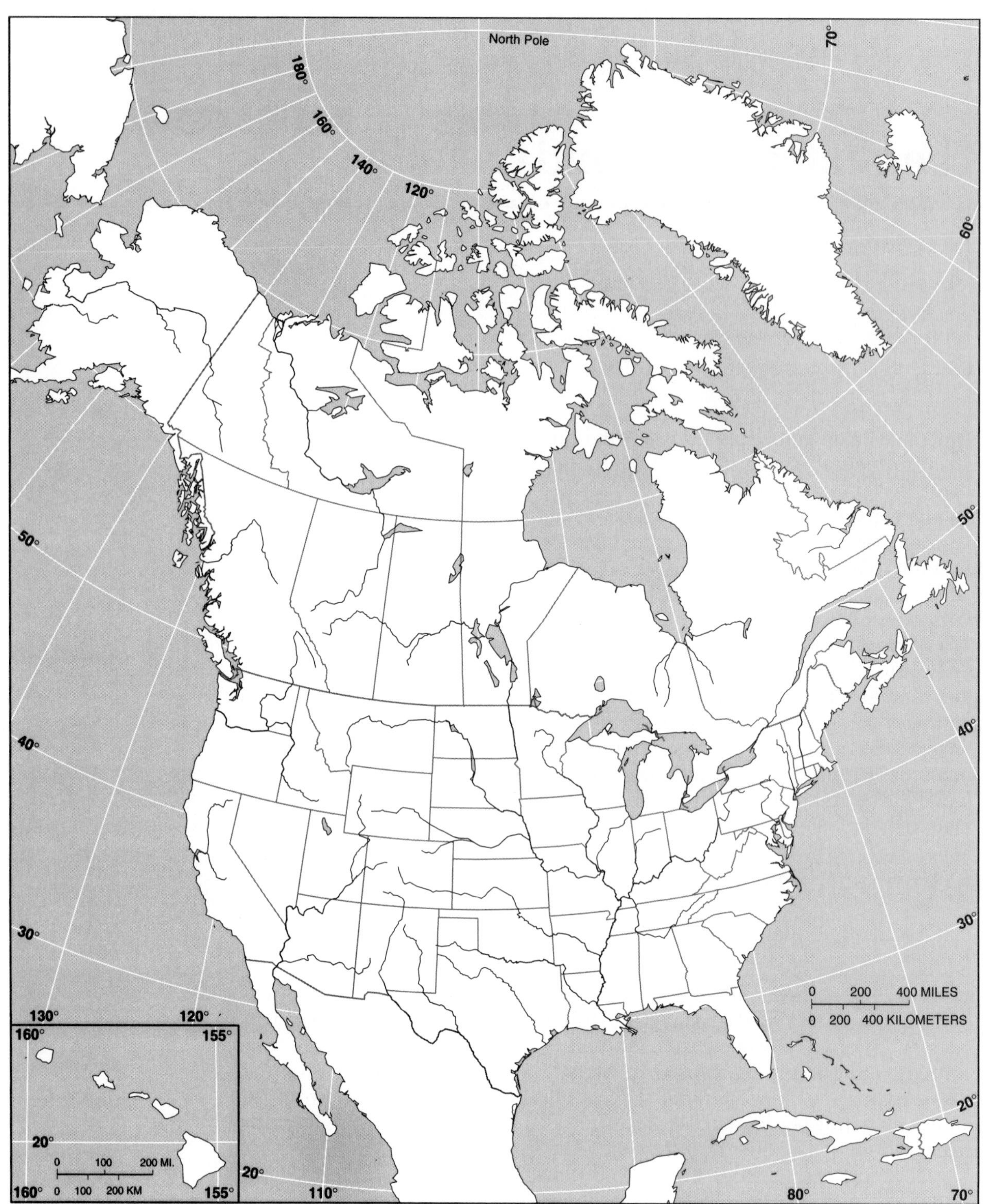


**FIGURE 4–1**
U.S.–Canada region–Physical Features 1.

Most mountain chains in the United States and Canada run essentially north–south, paralleling either the Atlantic or Pacific seacoasts. There are a few exceptions in the Midwest and Alaska that run east–west. Alaska's Mount McKinley is the continent's, and therefore the region's, highest peak.

On the outline map in Figure 4–1, locate and label the following physical features:

| | | | |
|---|---|---|---|
| Atlantic Ocean | | Beaufort Sea: | 73° N, 140° W |
| Pacific Ocean | | Labrador Sea: | 55° N, 55° W |
| Arctic Ocean | | Gulf of Alaska: | 58° N, 145° W |
| Gulf of Mexico: | 30° N, 90° W | Greenland: | 70° N, 40° W |
| Arctic Archipelago: | 75° N, 100° W | Hudson Bay: | 60° N, 85° W |
| Baffin Bay: | 70° N, 60° W | Hudson Strait: | 63° N, 72° W |
| Davis Strait: | 66° N, 60° W | Vancouver Island: | 50° N, 120° W |
| Aleutian Islands: | 52° N, 170° W | Newfoundland Island: | 49° N, 56° W |
| Baffin Island: | 65° N, 70° W | Lake Superior: | 47° N, 87° W |
| Victoria Island: | 70° N, 110° W | Lake Michigan: | 42° N, 87° W |
| Ellesmere Island: | 80° N, 80° W | Lake Huron: | 45° N, 82° W |
| Great Bear Lake: | 66° N, 120° W | Lake Erie: | 42° N, 82° W |
| Great Slave Lake: | 62° N, 115° W | Lake Ontario: | 42° N, 77° W |
| Lake Winnipeg: | 53° N, 98° W | Hawaiian Islands: | 20° N, 158° W |
| Lake Manitoba: | 51° N, 98° W | Prince Edward Island: | 46° N, 63° W |
| Gulf of St. Lawrence: | 48° N, 62° W | Cape Breton Island: | 46° N, 61° W |

To avoid producing an illegible map due to a large number of placenames, use a second outline map, Figure 4–2, to locate and label the following drainage features:

| | |
|---|---|
| Mississippi River: | 32° N, 91° W |
| Missouri River: | 40° N, 96° W |
| Ohio River: | 37° N, 88° W |
| St. Lawrence River: | 48° N, 69° W |
| Rio Grande: | 38° N, 107° W |
| Columbia River: | 46° N, 123° W |
| Snake River: | 46° N, 118° W |
| Gila River: | 32° N, 114° W |
| Yukon River: | 62° N, 163° W |
| Mackenzie River: | 63° N, 124° W |
| Peace River: | 56° N, 117° W |
| Red River (of the North): | 47° N, 97° W |
| Red River: | 35° N, 100° W |
| Pecos River: | 31° N, 103° W |
| Tennessee River: | 35° N, 88° W |
| Platt River: | 41° N, 100° W |
| Hudson River: | 42° N, 74° W |
| Connecticut River: | 44° N, 72° W |
| Ottawa River: | 46° N, 77° W |
| Arkansas River: | 35° N, 95° W |
| Sacramento River: | 40° N, 122° W |
| Saskatchewan River: | 53° N, 103° W |
| Chesapeake Bay: | 38° N, 76° W |
| Bay of Fundy: | 45° N, 66° W |

Lightly sketch in and label:

| | |
|---|---|
| Rocky Mountains: | 40° N, 107° W |
| Appalachian Mountains: | 38° N, 80° W |
| Sierra Nevada (California): | 38° N, 118° W |
| Cascades (Oregon and Washington): | 45° N, 123° W |
| Ozark-Ouachita Mountains: | 38° N, 93° W |

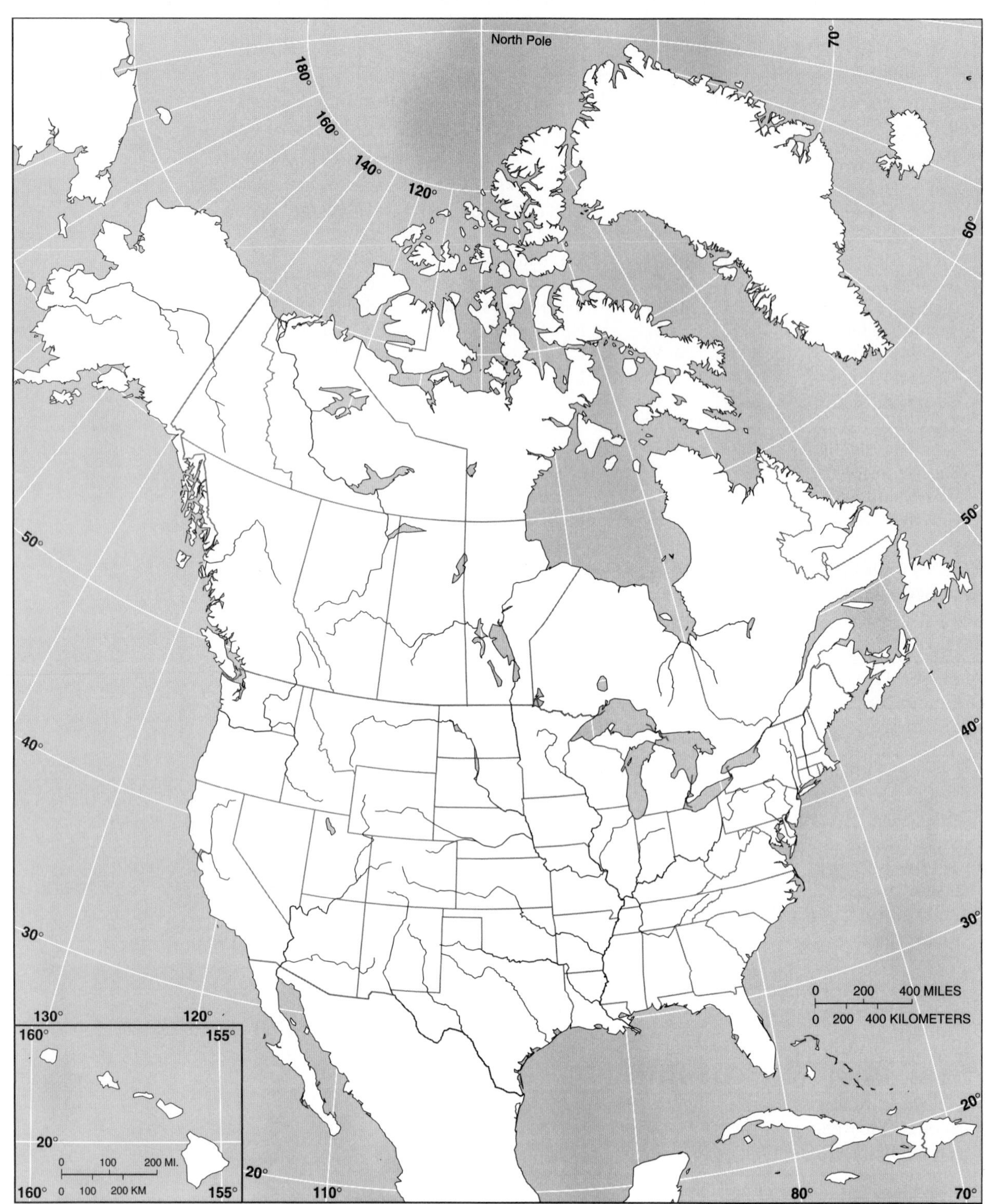


**FIGURE 4–2**
U.S.–Canada region–Physical Features 2.

## OBJECTIVES AND STUDY HINTS

The national units of the United States and Canada are among the largest countries in the world. Their citizens are, on average, among the richest of the world's people. The United States enjoys even greater variety in its range of physical environments than does Canada. Both possess expanding, energetic populations busily developing large stocks of natural resources, including productive farmlands, great forests, and varied mineral and fuel deposits. Both nations are heavily urbanized. This does not mean that most Americans and Canadians live in the older, densely occupied cores of metropolitan areas—far from it. Many Americans and Canadians live in the scattered suburbs and small-town satellites of the suburban fringe. Suburbia, where countryside and the edge of the city meet, is the new frontier of settlement.

The economy of both countries shows a major and continuing shift in employment. In the past, employment emphasis has shifted from farming to manufacturing, and now economic activity is shifting emphasis from manufacturing to the service sector. Furthermore, the diminishing importance of heavy industries (steel, railroad equipment, heavy machinery, ships, autos) means that new jobs are being created with little regard for the geography of industrial raw materials. Many service-economy jobs are just as logically located in suburbs or small towns as in the hearts of cities.

The geographic concepts stressed in this chapter are *megalopolis*—the growth of sprawling urban-suburban regions linking big cities, and *regional capitals*—the rise of business, administrative, and cultural "capitals" of subregions in these two huge national units. Watch for shifts and changes in geographic patterns and geographic trends. How are the decline of central-city populations and the continuing growth of suburbs linked with the megalopolis phenomenon? How are shifts from heavy industry to services and "high-tech" industry linked to the Sun Belt expansion versus Snow Belt or Rust Belt decline?*

Remember to look beyond this region to place it in a global context. Can the marked shift in U.S.–Canadian trade relations and the marked shift in sources of immigrants from the trans-Atlantic focus of the past toward a newer emphasis on trans-Pacific interaction constitute a "West Coast Challenge" to the formerly dominant East Coast cities? Why are western U.S. and Canadian cities such as Los Angeles, San Diego, Seattle, and Vancouver gaining populations in their metropolitan areas faster than most East Coast centers? Look for the links among geographic phenomena.

## POLITICAL GEOGRAPHY

Both the United States and Canada are the political offspring of Britain (the United Kingdom). The United States was the headstrong, rebellious child, fighting for its independence. Canada, the loyal child, simply signed for its independence when its constitution formally was "returned" from the United Kingdom by Queen Elizabeth II. Both are federal states, that is, some powers of government

*The Sun Belt, that group of states characterized by both subtropical climates and rapid population growth, commonly is understood to include California, Arizona, New Mexico, Texas, and Florida. Sometimes Nevada, Utah, Colorado, Louisiana, Mississippi, Alabama, South Carolina, and Georgia are added. In contrast, the Snow Belt consists of the old manufacturing core of the Northeastern United States—from Minnesota, Iowa, and Missouri on the west through Wisconsin, Illinois, Indiana, Michigan, Ohio, West Virginia, Pennsylvania, Maryland, Delaware, New Jersey, and New York to the six New England states. Because some of the older industrial plants in the Snow Belt are obsolete and even abandoned, the Snow Belt's alternative name is the Rust Belt.

are exercised by states or provinces rather than all the power being concentrated at the national level. Generally, Canada is considered the weaker federation, with more power granted on the provincial level than is granted at the state level in the United States. In the U.S. Constitution, powers not specifically assigned to the federal government are the privileges and responsibilities of the states. In Canada, it is the opposite; any powers not specifically assigned to the provinces become federal government concerns. Canadian provinces exercise greater power relative to Ottawa than do U.S. states relative to Washington, D.C. Thus, the province of Quebec, threatening to secede from Canada, acts almost like an independent country in negotiating with the Canadian government.

The political map of Canada changed in 1999 with the creation of a third territory, Nunavut. Nunavut was created from the eastern and northern parts of the old Northwest Territory. Nunavut now constitutes 61% of the former Northwest Territory, while the new Northwest Territory is 39% of the former. Nunavut makes up 21% of Canada's land but contains only about 28,000 people, mostly Inuit. Nunavut, "our land" in Inuit, was created to satisfy Inuit (formerly Eskimo) land claims by offering local autonomy but not sovereignty.

**4–1** Which Canadian provinces would be isolated from the bulk of Canadian territory if Quebec were to secede as an independent country (Figure 4–3)? ____________________

____________________

**4–2** Why are Canada's far northern Yukon, Nunavut, and Northwest territories not yet provinces? How is their status related to population density?

____________________

____________________

**4–3** Capitals of federal states are frequently located on a regional or cultural boundary. Canada's capital, Ottawa, is on what kind of cultural divide?

____________________

On the outline map of the United States and Canada in Figure 4–3, locate and label the provinces, capitals, and major cities listed in Table 4–1.

**TABLE 4–1** Canadian provinces, territories, capitals, and major cities.

| Province | Capital | Other Major Cities |
|---|---|---|
| Alberta | Edmonton | Calgary |
| British Columbia | Victoria | Vancouver |
| Manitoba | Winnipeg | |
| New Brunswick | Frederickton | |
| Newfoundland | St. John's | |
| Northwest Territories | Yellowknife | |
| Nova Scotia | Halifax | |
| Nunavut | Iqalut | |
| Ontario | Toronto | |
| Prince Edward Island | Charlottetown | |
| Quebec | Quebec City | Montreal |
| Saskatchewan | Regina | |
| Yukon Territory | Whitehorse | |
| | | Ottawa (federal capital) |

**FIGURE 4–3**
Canadian provinces and capitals.

On the outline map of the United States and Canada in Figure 4–4, locate and label the states and capitals listed in Table 4–2.

## ECONOMIC GEOGRAPHY

On the outline map of the United States and Canada in Figure 4–5, draw a line from Portland, Maine, to Toronto, and continue west to Minneapolis; then south to St. Louis; then eastward to Norfolk, Virginia, and finally up the East Coast back to Portland. This crude trapezoid generally corresponds with the traditional manufacturing belt of the northeastern and midwestern United States and adjacent portions of Canada.

When first described in the 1920s, this region enclosed most of the manufacturing activity of the United States and Canada. Not only were major manufacturing cities like Detroit, Cleveland, and Pittsburgh imbedded within this belt, but hundreds of small manufacturing centers were located here as well. Few significant centers of manufacturing lay beyond the boundaries of this belt at that time. By contrast, today many important manufacturing centers have risen in the Southeast and on the West Coast. The region outlined on the map is likely to be scorned as the Rust Belt, implying that it has become a region of declining, obsolescent heavy industries while newer, high-tech activities blossom in the Sun Belt.

**TABLE 4–2** The United States and its state capitals.

| State | Capital | State | Capital |
|---|---|---|---|
| Alabama | Montgomery | Missouri | Jefferson City |
| Alaska | Juneau | Montana | Helena |
| Arizona | Phoenix | Nebraska | Lincoln |
| Arkansas | Little Rock | Nevada | Carson City |
| California | Sacramento | New Hampshire | Concord |
| Colorado | Denver | New Jersey | Trenton |
| Connecticut | Hartford | New Mexico | Santa Fe |
| Delaware | Dover | New York | Albany |
| (District of Columbia; |  | North Carolina | Raleigh |
| Federal Capital) | Washington | North Dakota | Bismarck |
| Florida | Tallahassee | Ohio | Columbus |
| Georgia | Atlanta | Oklahoma | Oklahoma City |
| Hawaii | Honolulu | Oregon | Salem |
| Idaho | Boise | Pennsylvania | Harrisburg |
| Illinois | Springfield | Rhode Island | Providence |
| Indiana | Indianapolis | South Carolina | Columbia |
| Iowa | Des Moines | South Dakota | Pierre |
| Kansas | Topeka | Tennessee | Nashville |
| Kentucky | Frankfort | Texas | Austin |
| Louisiana | Baton Rouge | Utah | Salt Lake City |
| Maine | Augusta | Vermont | Montpelier |
| Maryland | Annapolis | Virginia | Richmond |
| Massachusetts | Boston | Washington | Olympia |
| Michigan | Lansing | West Virginia | Charleston |
| Minnesota | St. Paul | Wisconsin | Madison |
| Mississippi | Jackson | Wyoming | Cheyenne |

**FIGURE 4–4**
The United States and state capitals.

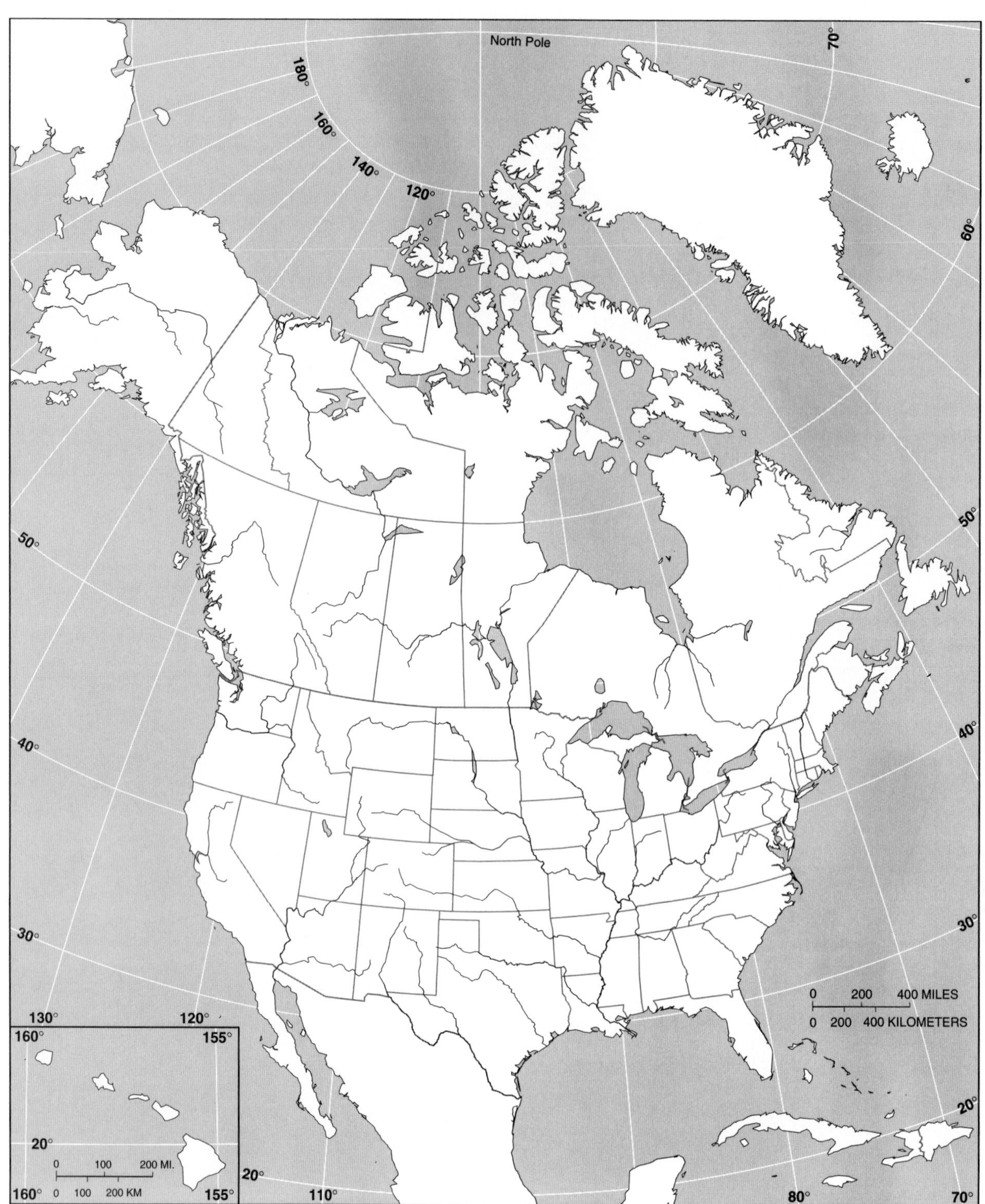


**FIGURE 4–5**
Traditional manufacturing belt.

## CULTURAL GEOGRAPHY AND DEMOGRAPHICS

Conduct an informal survey by asking at least 20 respondents which states they consider Sun Belt states—the fast-growing states with a reputation for pleasant climates. From the responses, devise two categories of states: often-mentioned (11 to 20 mentions) and less-mentioned (1 to 10). On the outline map in Figure 4–6, color in the often-mentioned states. Then use a line pattern to highlight the less-mentioned states. Leave blank all states not mentioned.

**4–4** In your opinion, does this map show states that are attractive to migrants from other states?

________________________________________

________________________________________

As a whole, the U.S. population is growing at a little over 1% a year. Some states far surpass this average, while others lag behind or even lose population. Are the "Sun Belt" states uniformly fast-growing?

Using the data in Table 4–3, construct a map of U.S. states' population growth rates, Figure 4–7. First, identify the 10 states with the highest growth rates (those above 15%, 1990–1998). On an outline map, Figure 4–7, locate these high-growth states and color them a dark, solid color. Now identify the 10 next-fastest growing states (7.99 to 15%) with a cross-hatch pattern (two sets of parallel lines intersecting at right angles). Next, locate the third-ranked group of 10 (6.4 to 7.98%) and label these states with a diagonal line pattern. The fourth-ranked set of 10 states (3.3 to 6.3%) should be labeled on the map with small dots. The fifth ranked group of 10 should be left blank, except those three states showing negative growth: label these with a large minus sign. Remember to add a legend on this map showing each map pattern next to its definition.

**TABLE 4–3** U.S. States—Percentage Change in Population, 1990–1998 (estimates)

U.S. average 8.7

| | | | | | | | |
|---|---|---|---|---|---|---|---|
| AL | 7.7 | IN | 8.7 | NE | 5.3 | SC | 10.0 |
| AK | 11.6 | IA | 3.1 | NV | 45.4 | SD | 5.1 |
| AZ | 27.4 | KS | 6.1 | NH | 6.8 | TN | 11.3 |
| AR | 7.98 | KY | 6.8 | NJ | 4.7 | TX | 16.3 |
| CA | 9.7 | LA | 3.5 | NM | 14.6 | UT | 21.9 |
| CO | 20.5 | ME | 1.3 | NY | 1.0 | VT | 5.0 |
| CT | –0.4 | MD | 7.4 | NC | 13.8 | VA | 9.7 |
| DE | 11.6 | MA | 2.2 | ND | –0.1 | WA | 16.9 |
| FL | 15.3 | MI | 5.6 | OH | 3.3 | WV | 1.0 |
| GA | 18.0 | MN | 7.99 | OK | 6.4 | WI | 8.0 |
| HI | 7.6 | MS | 6.9 | OR | 15.5 | WY | 6.0 |
| ID | 22.0 | MO | 6.3 | PA | 1.0 | | |
| IL | 5.4 | MT | 10.2 | RI | –1.5 | | |

**4–5** Of those states entirely or mostly within the traditional manufacturing belt (Figure 4–5), which had the highest growth rates? (Table 4–3 and Figure 4–6) ________________________________

________________________________________

**4–6** Which two manufacturing belt states showed *negative* growth rates?

________________________________________

**FIGURE 4–6**
The perceived Sun Belt states.

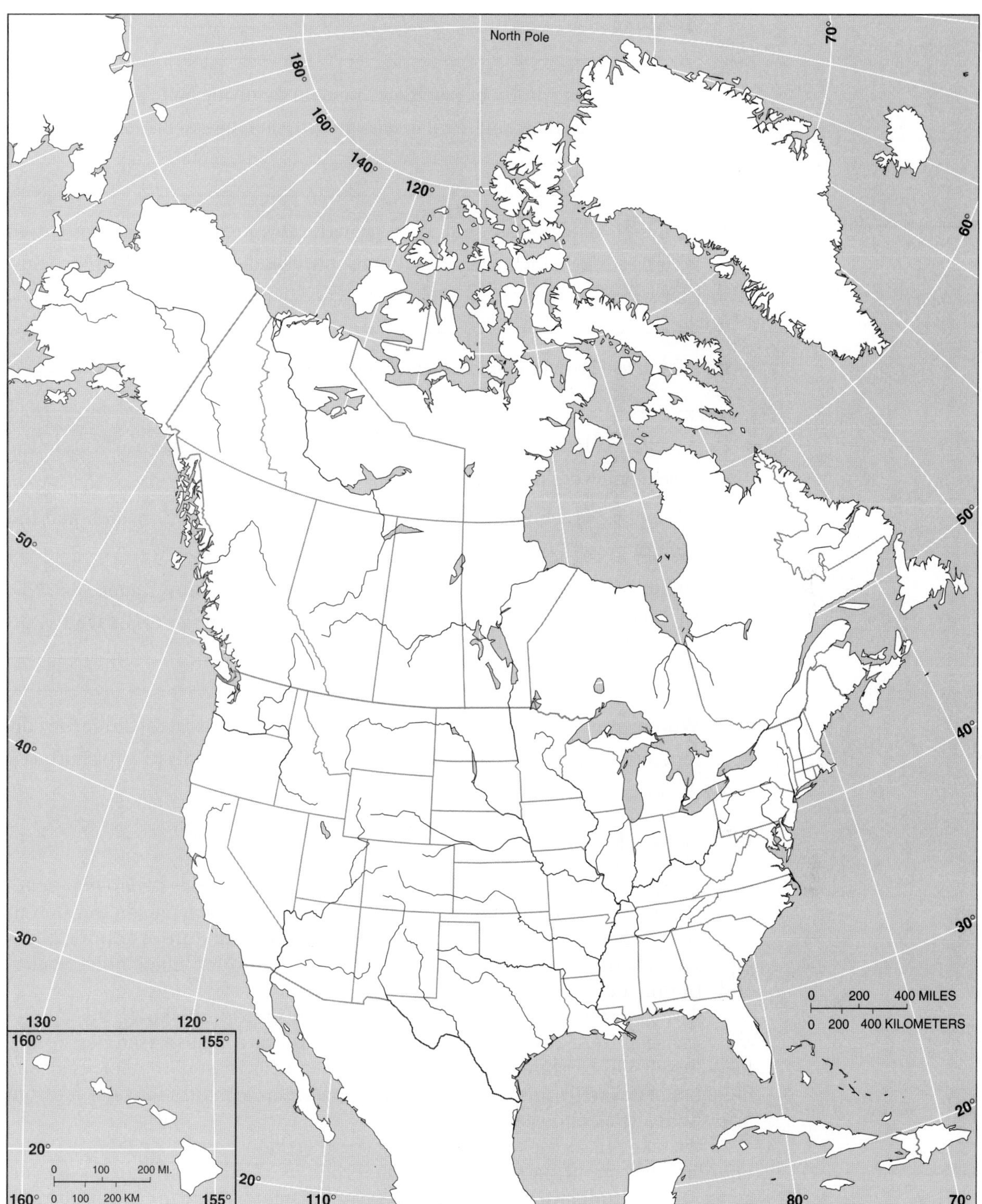


**FIGURE 4–7**
Population growth rates for U.S. states.

**4–7** Which state *outside* the traditional manufacturing belt had a negative growth rate? ____________

____________

**4–8** Of those states wholly or partly in the manufacturing belt, how many (list them) had less than half the national growth rate (half of 8.7, or 4.35)?

____________

____________

**4–9** Of the 10 states with the highest growth rates (over 15%) in Figure 4–6, which scored below (over 50% of those surveyed mentioned them) as "Sun Belt states"? ____________

____________

**4–10** Which fast growing Sun Belt state is *not* on a seacoast? ____________

____________

**4–11** As a generalization, which U.S. region has the fastest growth? (Circle the correct answer.) northeast; southwest; north central

**4–12** Of the six New England states (CT, ME, MA, NH, RI, VT), which had the highest growth rate? ____________ The lowest? ____________

**4–13** Of the five states wholly or mostly within the great plains region (KS, ND, NE, SD, OK), are any growing at or above the national rate? (List if any.)

____________

____________

**4–14** Which group of coastal states is growing at the slowest rate? (Circle the correct answer.) Pacific coast; northeast Atlantic coast (ME to NJ); southeast Atlantic coast (VA to FL)

## Perception Maps

Any map that reflects people's opinions, attitudes, or biases is a *map of perception*. Perception is what people believe or visualize, and each person is different from every other, because each individual has a unique mixture of cultural backgrounds, educations, beliefs, and values. Thus, perception maps reflect belief and opinion, not necessarily fact.

Perhaps you have seen playful perception maps that show "the United States as seen from Texas" (or any other state). If it is a Texas-based perception map, Texas will be shown outlandishly huge in comparison to other states. Its neighbor states will appear smaller than on an accurate map, and states far from Texas will shrink almost to invisibility.

We take this as a joke because we recognize that we really do tend to prefer familiar places and perhaps a few glamorous, well-advertised places farther away. Distant, unfamiliar places that we regard as unappealing in climate, terrain, or economy or that are very limited in recreational opportunities recede into the background in our perception when compared with the large and prominent place our imagination assigns to the familiar and desirable.

Your survey produced a perception map of the Sun Belt states. Here is another interesting perception map you can make, using the outline map of the United States and Canada in Figure 4–8. Ask friends and acquaintances to name the states and/or provinces that they would like to live in or at least visit. (They

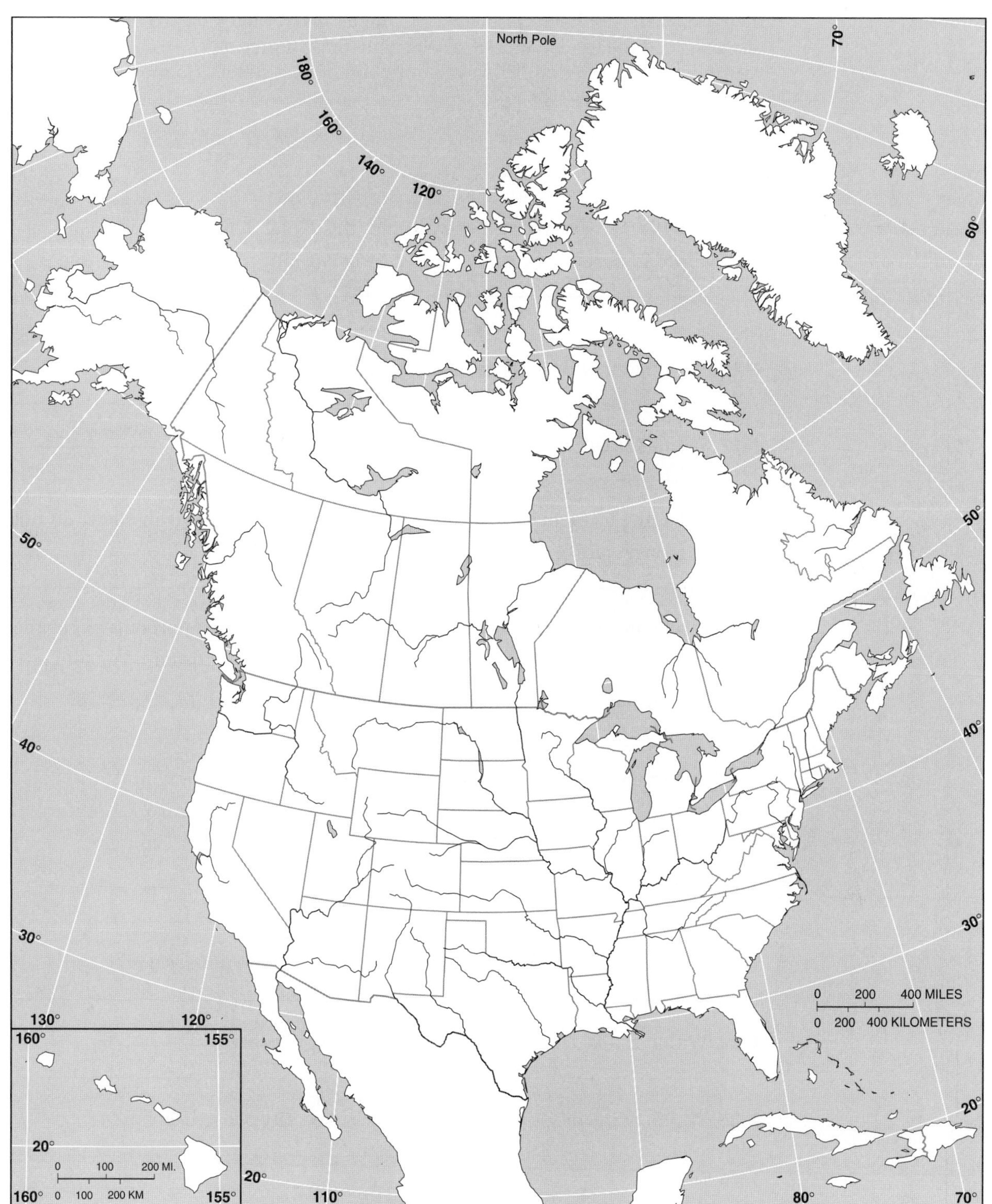


**FIGURE 4–8**
Perception map of preferred states and provinces.

should assume that they can get a good job there, so money concerns won't distort their preferences.) Each person can name as many states or provinces as they wish. Record the total "votes" for each state or province on the outline map. Did your home state rank high? How about neighboring states? Do most people like the "Sun Belt" states?

## CHECK UP

**4–15** Via the St. Lawrence Seaway and the Great Lakes, certain states are afforded ports that can be reached by ocean shipping. Name these states: __________

**4–16** Which four U.S. states meet at a single common point? __________

**4–17** Which Canadian provinces have a seacoast? __________

**4–18** Referring to a detailed map of landforms, which American states do not have any significant area of mountain terrain? __________

**4–19** Circle the city that is *not* a seaport, Great Lakes port, or port on a navigable river: Boston, Baltimore, St. Louis, Denver, Chicago, Toronto, Vancouver.

**4–20** Which state has an Atlantic seacoast *and* a Gulf of Mexico coastline? __________

**4–21** Referring to an atlas map showing the population of cities, circle the states in which the largest city *also* is the state capital: New York, California, Massachusetts, Georgia, Alaska, Texas, Florida, Arizona, Colorado.

**4–22** Which is the smallest Canadian province (in territory)? __________

**4–23** Which Canadian provinces and territories border Hudson Bay? __________

**4–24** Which states use the Mississippi River as a border? __________

**4–25** Which of the original 13 U.S. states* lacked any ocean beaches? __________

**4–26** If you wished to visit a Pacific Ocean seacoast for some surf fishing, which five states could you visit? __________

*Connecticut, Delaware, Georgia, Maryland, Massachusetts, New Hampshire, New Jersey, New York, North Carolina, Pennsylvania, Rhode Island, South Carolina, and Virginia.

**4–27** Lobsters thrive in cold ocean water. Which Canadian province would you visit for the freshest lobster dinner? Alberta, Nova Scotia, Saskatchewan?

______________________________

**4–28** Which state would be involved in regulating oil rigs operating just offshore in the Beaufort Sea? ______________________________

______________________________

**4–29** An airplane leaving Washington, D.C. and flying directly west to San Francisco would fly over which states? ______________________________

______________________________

**4–30** Which U.S. states share a border with Canada? ______________________________

**4–31** Which two eastern Canadian provinces consist of a relatively large island and a portion of the mainland? ______________________________

______________________________

# GEOCONCEPTS

## Megalopolis

*Megalopolis* literally means "giant city" or "super city." In a book first published in 1961, geographer Jean Gottmann described an urban-dominated region stretching from Boston's New Hampshire suburbs to the Virginia suburbs of Washington, D.C. and called "Boswash." Within this sprawling region, which still contains a large acreage of open farmland and forest, already-large cities were growing toward one another. The interconnecting highways, railroads, turnpikes, and communications lines that link these cities form the infrastructure supporting new suburban growth. Former open spaces between the cities gradually are filling loosely with housing, shopping malls, and industrial parks as people discover the geographic advantages of locating between big cities and astride busy highways (Figure 4–9).

The original northeastern megalopolis turned out to be the forerunner of other, similar urban-suburban regions. The phenomenon is an international one—similar city-chains of interconnected cities and suburbs, frequently strung along seacoasts, river valleys, or lakeshores, can be seen on detailed maps of Japan (southwest from Tokyo toward Nagoya, Kyoto, and Osaka), Germany (Rhine Valley), the Netherlands (the famed "ring city," a circle connecting Amsterdam, Utrecht, Rotterdam, and the Hague), and the United Kingdom (the London–Birmingham–Liverpool axis).

**4–32** Where are other megalopoli evolving? ______________________________

______________________________

**4–33** Which of these states is *not* wholly or partly within the original megalopolis? (Circle the correct answer.)

a. CT

b. TN

c. MA

d. RI

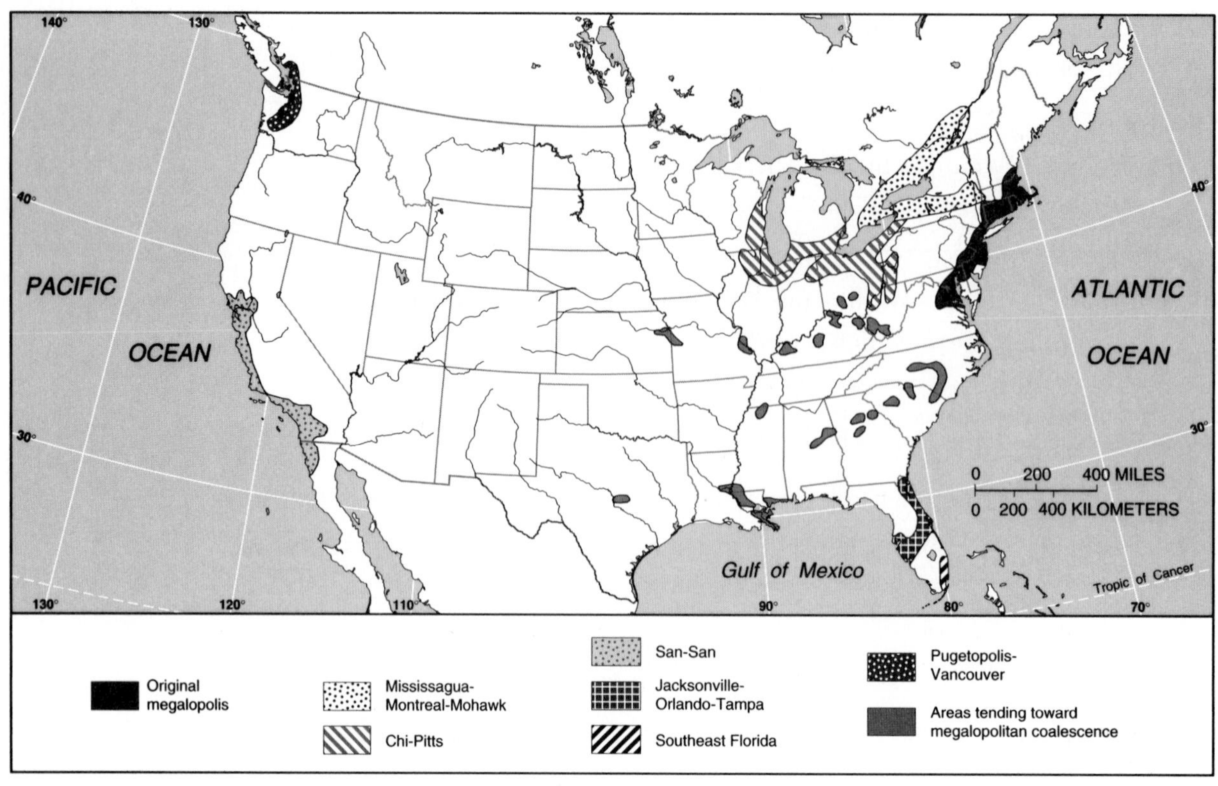


**FIGURE 4–9**
North American megalopoli. A series of similar urban giants is emerging in other parts of the United States as growth expands along major transportation routes.

**4–34** Which two states, part of the original megalopolis, have *negative* growth rates? ________________________________

## Regional Capitals

Capital, as used here, is any administrative headquarters, not necessarily limited to government. For example, Sears, Roebuck and Company has its "capital" near Chicago; the Ford Motor Company's is in Dearborn, Michigan. Corporations operating on national and global scales often find it inefficient to have a single administrative headquarters through which all routine decisions must be channeled.

Most huge countries are federated units, with some government functions dispersed among provinces or states (e.g., the United States, Canada, Russia, Brazil, Mexico, Australia). Similarly, many large companies set up regional or district administrative headquarters. Smaller corporations, whose activities and interests are concentrated in a particular region, will also choose a headquarters or capital in the leading metropolis of the region. These subnational headquarters or administrative capitals center on a "regional capital" (Figure 4–10). The requirements for a regional capital are (1) a major metropolitan center, the largest in its part of the country; (2) excellent transportation and communications links with its surrounding region, the nation, and the world; and (3) a location reasonably central to its region. The regional capital designation refers to corporate, trade, communications, and cultural functions; it also may or may not be a state capital. Atlanta and Denver are excellent examples of regional capitals. The most obvious

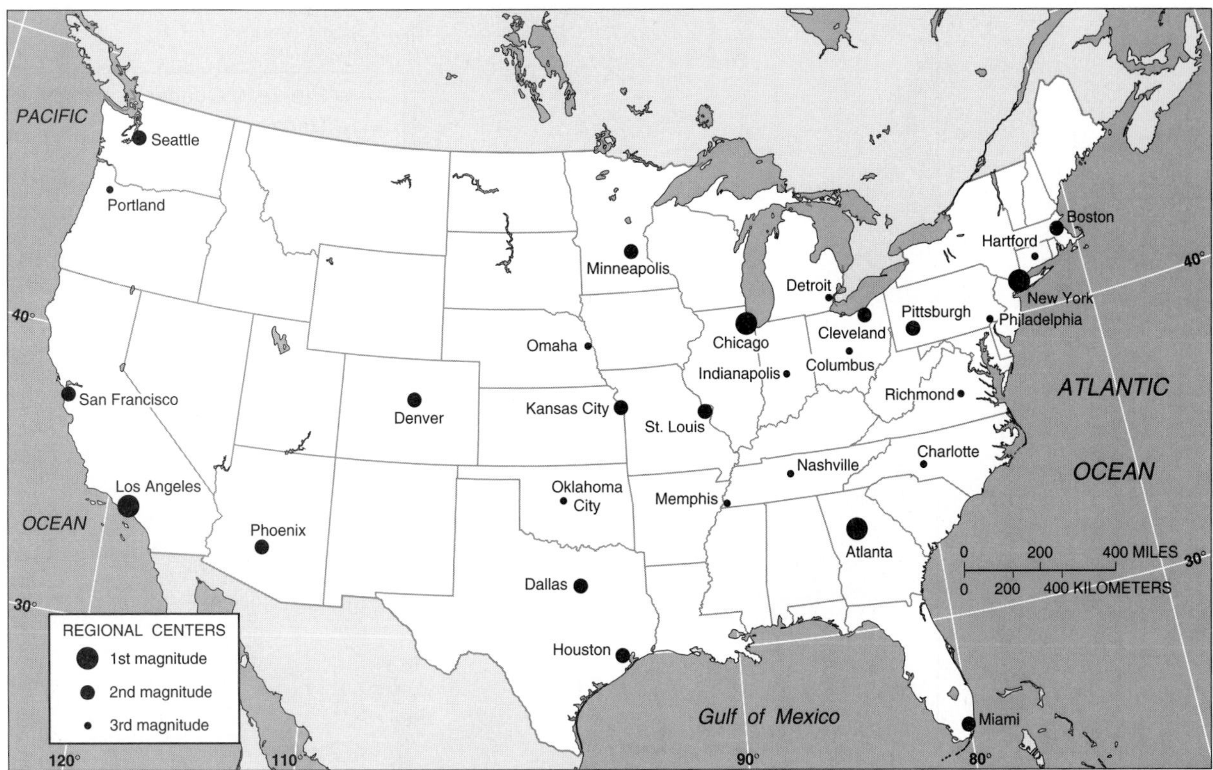


FIGURE 4–10
Regional centers of the United States. In large political units, the sheer scale of the economy merits a network of regional office and service centers. Such a network of sales, service, headquarter, and distribution centers is particularly evident in the United States and Canada.

clues to regional capital status (in addition to population size) are a busy international "hub" airport and a "nodal" position in the web of interstate highways.

## An Independent Quebec

If, as many Quebec citizens advocate, Quebec Province of Canada secedes as an independent country, new boundaries may appear on the map. The northern two-thirds of the present province, "Rupert's Land," was not transferred from the British Empire to Canada until 1870. Rupert's Land (most of which now is known as Ungava to Native Canadians) was never historically part of French Canadian settlement and was assigned by the UK to Canada, not Quebec. See Figure 4–11. Native Canadians understandably are not impressed by the French Canadian "we were here first" (compared with the British) argument for taking the entire province out of Canada; Native Canadians threaten to secede Ungava from an independent Quebec. Canada's highest court has ruled recently that it would be unconstitutional for Quebec to secede unilaterally.

**4–35** Why have the provincial governments of Newfoundland, New Brunswick, Nova Scotia, and Prince Edward Island demanded an "All-Canadian Corridor," not controlled by Quebec, connecting Ontario and New Brunswick should Quebec secede? ______________________________

______________________________

**4–36** Which Canadian province or territory includes the "Canadian Archipelago" of islands in the Arctic Ocean? ______________________________

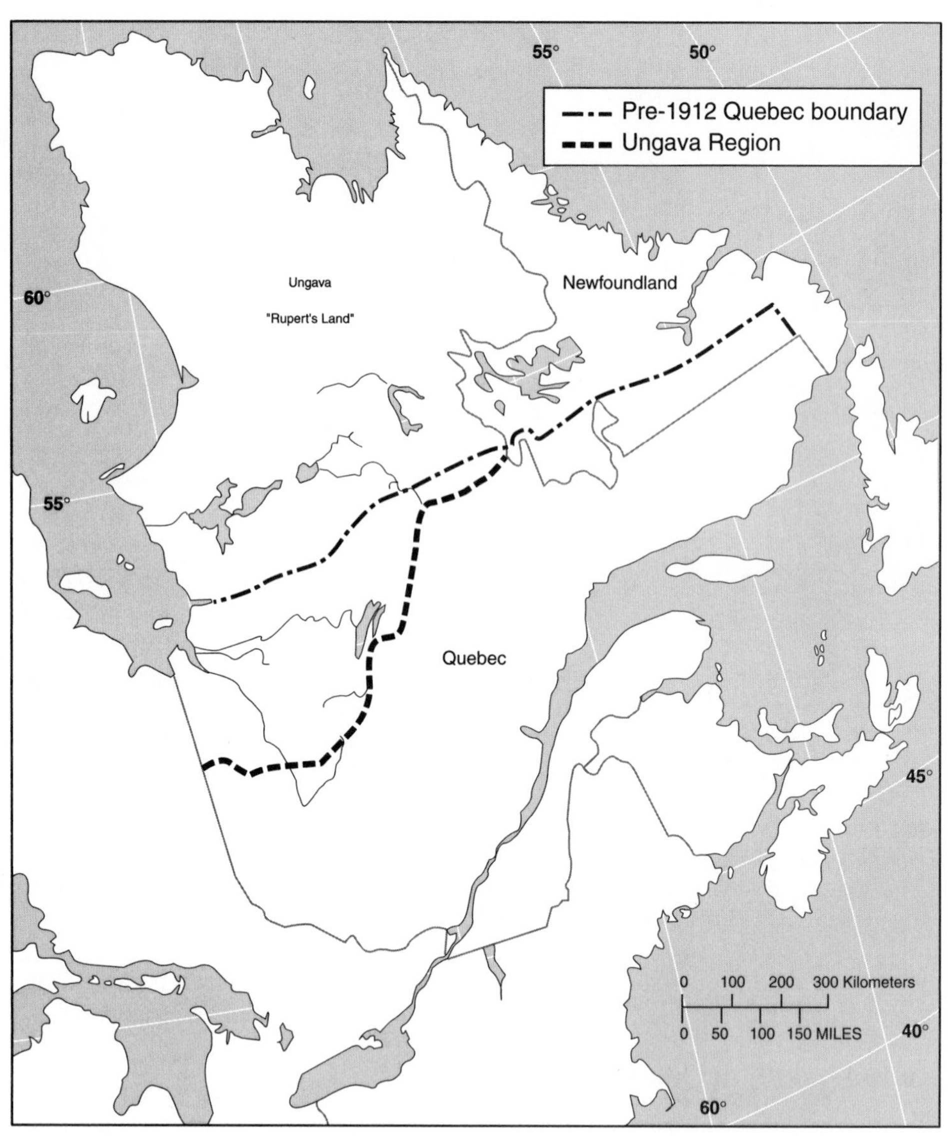


**FIGURE 4–11**
Quebec: Historical boundary changes.

## REGIONAL WATCHLIST

Progress through the ten regions will reveal repeated examples of two apparently opposite trends. In the political world, there has been an outburst of ethnic nationalism. Ethnic, linguistic, and cultural minorities seek to create new states whose boundaries would more nearly coincide with the geographic distribution of the group aspiring to independence. The result can be political *fragmentation*. In the economic world, however, the trend seems to be the *integration* of national economies into larger free trade zones. The North American Free Trade Agreement, for example, links Mexico with the already merging U.S. and Canadian Free Trade Zone and creates a new economic entity in North America that suggests a future U.S., Canada, and Mexico region. On the political map, Quebec's separatist movement could create a new country based on its unique cultural, linguistic, and ethnic identity.

# 5

# Europe

## INTRODUCTION

Europe is a good example of a region based on culture rather than physical geography. Europe is a continent by virtue of its historical and cultural identity, not any dramatic physical boundary, a fact that has confused students for years. Many geographers identify Earth's continents as Eurasia (combining Europe and Asia), Africa, North America, South America, Antarctica, and Australia, in descending order of size. Viewed on a physical geography map, which emphasizes landforms rather than political units, Europe is only a peninsula of the huge landmass of Eurasia, which accounts for nearly 37% of the world's land surface. This European peninsula, or westward projection of the world's greatest landmass, is itself fragmented into a series of smaller peninsulas (Scandinavian, Iberian, Balkan, Italian) and large islands (Britain, Ireland, Sicily, Iceland, Sardinia).

Europe is accorded continental status, or at least recognized as a continental-scale region, by courtesy and custom. The ancient Greeks recognized that the nations and empires east of the straits connecting the Aegean Sea and the Black Sea were notably different from Greece and its neighbors to the west of these waters, and thus labeled these as two different continents. A larger view, less influenced by the European viewpoint, sees Europe and Asia as one uninterrupted landmass.

As a continent, Europe's eastern boundary extends to Russia's Ural Mountains. As a cultural and economic region, Europe's eastern boundary is commonly defined as Russia's western frontier. The breakup of the Soviet Union is cited as the justification for including six former members of the Union (Estonia, Latvia, Lithuania, Belarus, Ukraine, and Moldova) with their European-tradition neighbors, advancing the regional boundary to the western borders of Russia.

When the Soviet Union disintegrated in 1991, each of its 15 former "member republics" declared itself a sovereign state. The six newly independent states to the west and southwest of Russia fall into two groups. The Baltic Republics—Estonia, Latvia, and Lithuania—once part of the tsarist empire, emerged in 1919 from the chaotic post-revolutionary period as independent republics. In 1940, they were forcibly reincorporated into the Soviet Union. The United States never officially recognized the Soviet annexation of the Baltic Republics; all three always were determined to regain their independence. Their strong sense of national identity, recent history of independence, and rapidly evolving economic and cultural links with the Europe region all argue for including them in the Europe region.

Ukraine is a historic state with its own language, culture, and history of resistance to Russian domination. Belarus and Moldova (earlier known as Moldavia) also have distinctive ethnic identities and languages. All three of these states share a "western" orientation and outlook, and so also have been added to the Europe region.

## PHYSICAL GEOGRAPHY

In terms of human settlement and use of the land for food production, Europe is among the must usable, and used, of world regions. Long occupancy by a dense population of hardworking and ingenious people has meant that people have made productive use of nearly every last acre of potential utility. Marshes have been drained, barren lands fertilized, and meadows high in the mountains used to pasture animals in summer. A great European plain, sometimes of gently rolling rather than flat land, stretches northward and eastward from the flanks of the Pyrenees Mountains along France's southern border, deep into Russia, widening as it sweeps eastward. About one-third of Europe is a fairly level plain, with the balance in uplands. Perhaps the most important aspect of Europe's physical geography is its complex coastline. Here, land and sea don't just meet; they intermingle. Relatively few parts of Europe are more than 125 miles (201 km) from the sea, and great navigable river systems, interconnected by canals, provide even land locked countries like Switzerland and Austria with access to the sea.

Europe's climate is strongly affected by the surrounding seas that penetrate into and flank the region. Westerly winds (winds are named for the direction from which they come) help transfer the humidity and temperature characteristics of relatively warm offshore ocean currents onto northwestern Europe. Thus, Britain, at approximately the same latitude as Labrador (part of Newfoundland), successfully grows wheat.

On the outline map of Europe in Figure 5–1, locate and label the following:

### Saltwater Bodies

| | | | |
|---|---|---|---|
| Atlantic Ocean | | Adriatic Sea: | 42° N, 15° E |
| Mediterranean Sea: | 33° N, 15° E | Norwegian Sea: | 67° N, 5° E |
| North Sea: | 55° N, 5° E | Aegean Sea: | 38° N, 25° E |
| Baltic Sea: | 60° N, 20° E | Tyrrhenian Sea: | 40° N, 13° E |
| Black Sea: | 42° N, 35° E | Bosporus (strait): | 42° N, 28° E |
| Dardanelles (strait): | 41° N, 27° E | Kattegat: | 57° N, 12° E |
| English Channel: | 50° N, 2° W | Irish Sea: | 53° N, 5° W |
| Strait of Gibraltar: | 36° N, 6° W | Bay of Biscay: | 45° N, 3° W |
| Gulf of Finland: | 59° N, 25° E | Strait of Messina: | 37° N, 16° E |
| Gulf of Bothnia: | 63° N, 20° E | Strait of Otranto: | 40° N, 19° E |
| Skagerrak: | 57° N, 7° E | Gulf of Riga: | 57° N, 23° E |
| Strait of Dover: | 52° N, 30° E | | |

### Islands

| | | | |
|---|---|---|---|
| Isle of Britain: | 55° N, 30° W | Cyprus: | 35° N, 33° E |
| Iceland: | 65° N, 19° W | Balearic Islands: | 39° N, 3° E |
| Ireland: | 53° N, 7° W | Shetland Islands: | 60° N, 2° W |
| Sicily: | 38° N, 14° E | Faeroe Islands: | 63° N, 7° W |
| Sardinia: | 40° N, 9° E | Malta: | 36° N, 14° E |
| Corsica: | 43° N, 8° E | Crete: | 35° N, 25° E |

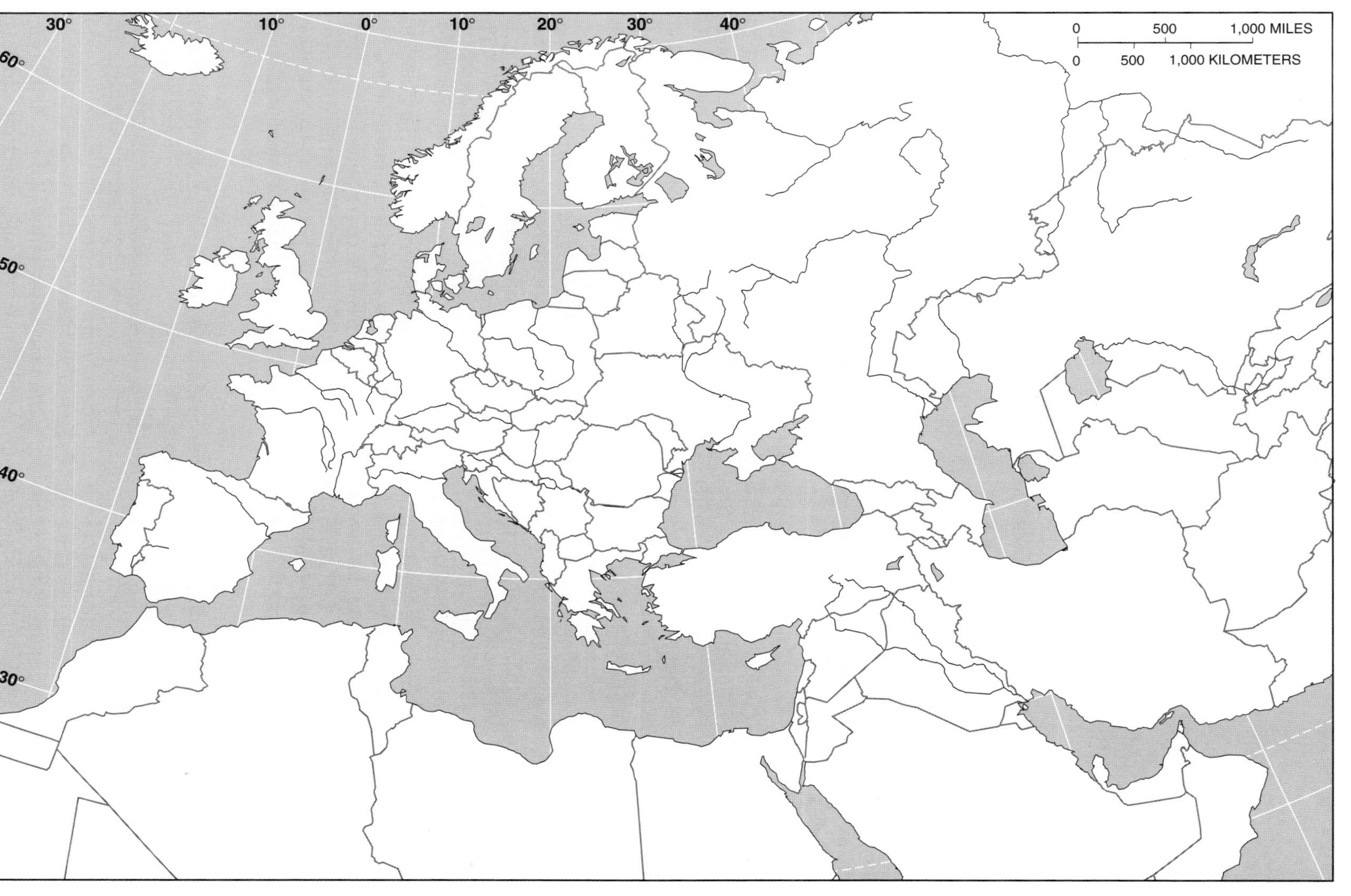


**FIGURE 5–1**
Outline map of Europe.

### Rivers and Lakes

| | | | |
|---|---|---|---|
| Rhine River: | 49° N, 8° E | Po River: | 45° N, 12° E |
| Danube River: | 44° N, 24° E | Vistula River: | 53° N, 19° E |
| Thames River: | 52° N, 0° E | Seine River: | 48° N, 3° E |
| Rhone River: | 45° N, 5° E | Ebro River: | 41° N, 1° W |
| Lake Geneva: | 47° N, 7° E | | |

### Peninsulas

| | | | |
|---|---|---|---|
| Peninsula of Jutland: | 57° N, 9° E | Iberian Peninsula: | 40° N, 5° E |
| Italian Peninsula: | 43° N, 12° E | Scandinavian Peninsula: | 62° N, 12° E |

### Mountains

| | | | |
|---|---|---|---|
| Pyrenees Mountains: | 43° N, 0° | Dinaric Alps: | 44° N, 16° E |
| Apeninnes: | 43° N, 12° E | Balkan Mountains: | 43° N, 25° E |
| Pennines: | 54° N, 2° W | Carpathians: | 46° N, 25° E |
| Massif Central (France): | 45° N, 3° E | Alps: | 47° N, 10° E |

## OBJECTIVES AND STUDY HINTS

To Americans and Canadians, travel in Europe dramatizes the smaller scale of European countries compared with the relative immensity of Canada and the United States. If Europe is counted as a separate continent, in area it ranks sixth out of seven. It is less than half the size of North America and almost 30% smaller than Antarctica. Yet, Europe contains 43 independent nations, one colony (British-controlled Gibraltar), a small portion of Turkey, and most of the population (but not the territory) of Russia. Included are six microstates—Andorra, San Marino, Monaco, Malta, Liechtenstein, and Vatican City.

Though the political map of Europe is not likely to lose many independent countries and thus not likely to see many present borders disappear, the *significance* of these borders is not the same as a half-century ago. Another geographic concept developed in this chapter is that of *regional economic communities.* The members of the European Community (EU) effectively have eliminated the previous economic impact of borders.

It is useful in studying the placename geography of Europe to envision this complex region not just as 40-plus national units, but as a somewhat smaller number of cultural, language, and religious groups. Broader patterns of culture can be seen to transcend national borders, keeping in mind that there are many local exceptions in always-complicated Europe.

Most Europeans, for example, speak a language that is a member of one of three great language families—Germanic, Slavic, and Latin-based Romance. Whereas in some cases the common use of a particular language ends abruptly at a national border, in other instances a language or dialect (a regional variation of the language) will continue across the border. Similarly, Europe can be generalized into broadly defined regions where the major religious tradition is Roman Catholicism, Protestantism, or Eastern Orthodox Christianity (although important concentrations of Jews and Muslims are also present to complicate this generalization).

Learning to think of broader, culture-based regions in Europe helps bring order to this highly varied region. Ask this question in connection with each country you study: Which of its neighbors is most different, and which the least different, in all aspects of culture, economy, and system of government? This may help you understand past and present tensions and rivalries and make sense of a veritable mosaic of cultures and nations.

## POLITICAL GEOGRAPHY

Some European states, like France, have been unified within approximately the same boundaries for many centuries. Others, such as Germany and Italy, were unified from many smaller states late in the nineteenth century. Still others, like Yugoslavia, appeared only in this century, heirs to fragments of old empires and defeated powers.

A few, Poland among them, disappeared from the political map temporarily as a *state* (a self-governing area with a territory, a population, and national goals and policies), but never disappeared as a *nation* (a culturally distinctive group, conscious of its heritage, occupying a traditional homeland but not necessarily constituting a sovereign state). The happiest combination is a *nation-state*—one in which the territory occupied by the nation coincides exactly with the boundaries of an independent state. Ethnic minorities within a state that actively seek an independent state of their own, or a nation divided among two or more sovereign states, can mean trouble. These situations were common in Europe in the past, and today's news reports bring daily reminders that the problems continue to flare as in Yugoslavia's civil war.

Yugoslavia has split into at least five independent states—Slovenia, Croatia, Bosnia-Herzegovina, Macedonia, and Yugoslavia (which presently includes Serbia, Vojvodina, Kosovo, and Montenegro). The boundaries among these states, the internal subdivisions, and even the continued existence of these states all are likely to change in this area of tragic chaos.

Locate and label on the outline map in Figure 5–2 the nations, capitals, and important cities listed in Table 5–1.

Throughout this book, the standard English spelling of the short form of the official name of each country is used. Thus, it is the *Netherlands,* not Holland. And it is the *United Kingdom,* not the United Kingdom of Great Britain and Northern Ireland (the full title), nor England (a historical entity, not a sovereign state), nor Britain (the name of the island containing England, Wales, and Scotland).

**TABLE 5–1** Nations, capitals, and important cities of Europe.

| Country | Capital | Other Important Cities |
|---|---|---|
| Albania | Tirana | |
| Austria | Vienna | |
| Belarus | Minsk | |
| Belgium | Brussels | Antwerp |
| Bosnia-Herzegovina | Sarajevo | |
| Bulgaria | Sofia | |
| Croatia | Zagreb | |
| Czech Republic | Prague | |
| Denmark | Copenhagen | |
| Estonia | Tallinn | |
| Finland | Helsinki | |
| France | Paris | Lyon, Marseille |
| Germany | Berlin | Munich, Hamburg, Bonn |
| Greece | Athens | |
| Hungary | Budapest | |
| Iceland | Reykjavik | |
| Ireland | Dublin | |
| Italy | Rome | Milan, Venice |
| Latvia | Riga | |
| Lithuania | Vilnius | |
| Luxembourg | Luxembourg City | |

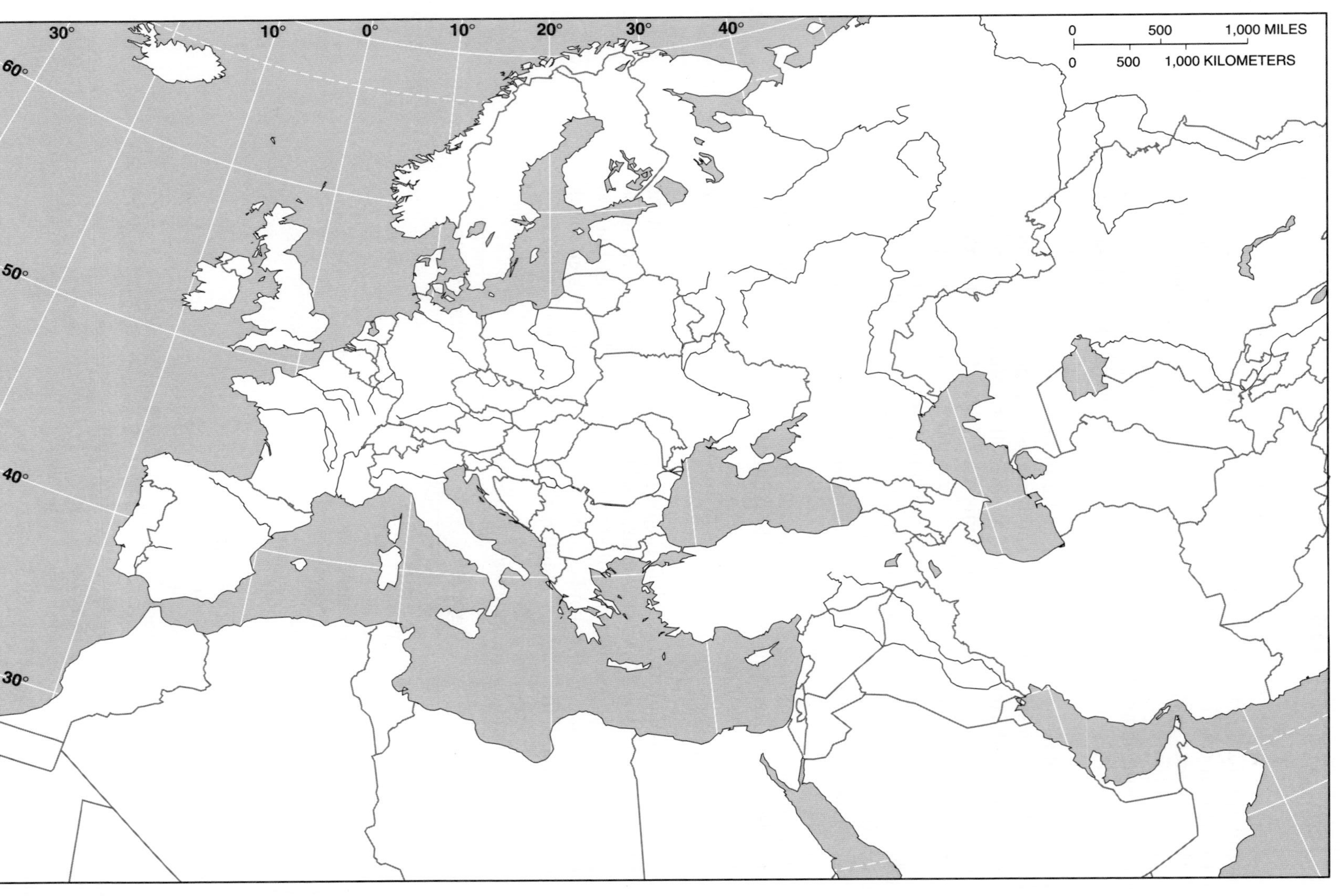


**FIGURE 5–2**
European states, capitals, and important cities.

**TABLE 5–1** Nations, capitals, and important cities of Europe. *(continued)*

| Country | Capital | Other Important Cities |
|---|---|---|
| Macedonia | Skopje | |
| Malta | Valletta | |
| Moldova | Chisinau | |
| Netherlands | The Hague | Amsterdam, Rotterdam |
| Norway | Oslo | |
| Poland | Warsaw | Gdansk |
| Portugal | Lisbon | |
| Romania | Bucharest | |
| Russia (part)* | Moscow | St. Petersburg |
| Slovakia | Bratislava | |
| Slovenia | Ljubljana | |
| Spain | Madrid | Barcelona |
| Sweden | Stockholm | |
| Switzerland | Bern | Zurich |
| Ukraine | Kiev | Odesa, Kharkov |
| United Kingdom | London | Birmingham, Edinburgh |
| Yugoslavia** | Belgrade | |

*Part of the continent of Europe, but considered a part of another region.
**Yugoslavia includes Serbia, Vojvodina, Montenegro, and Kosovo.

## ECONOMIC GEOGRAPHY

The complex of technological, organizational, and sociological changes called the *Industrial Revolution* first made its dramatic impact upon northwestern Europe. The United Kingdom, northern France, Belgium, the Netherlands, and western Germany were among the first modern industrial areas in the world, followed closely by the northeastern United States. The use of *inanimate power* (direct water power, followed by steam generated by burning coal) replaced human and animal muscles. Workers were gathered into factories to perform specialized tasks, and the *assembly line* was introduced to produce large volumes of goods cheaply. Machines were applied to increase efficiency in all sectors of the economy—transport, agriculture, mining, construction, and communications. The happiest product of the Industrial Revolution was a general rise in living standards as nations, or parts of nations, progressed through the Industrial Revolution.

Although Europe as a region certainly is classed as industrialized, having moved through the Industrial Revolution process to a modern society, there are definite variations in degree of modernization and living standards. Using the outline map of Europe in Figure 5–3, create a living-standards map using *gross national product per capita* data (GNP per capita from the Appendix). First, determine the *range* of this income data for Europe's countries (the highest and lowest figures) and then divide that range into equal thirds (the lowest-income third, the highest-income third, and the middle-income third). Use a solid pattern or darkest color to show those nations in the highest-income third, a cross-hatch pattern or medium color for the middle-income third, and a diagonal-line pattern or light color to identify countries in the lowest income group.

**5–1** Which are the relatively wealthy nations? ______________________________

______________________________________________

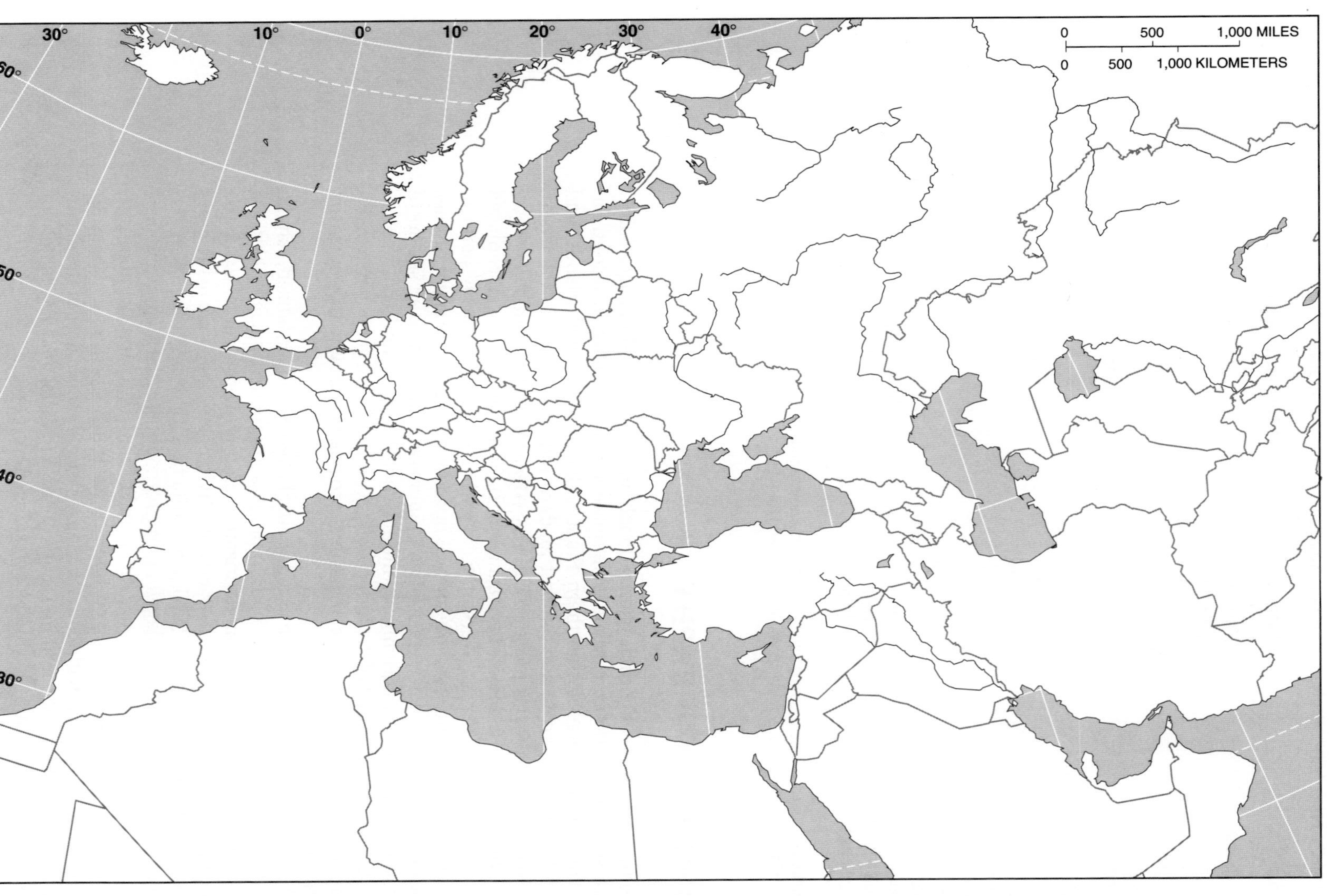


**FIGURE 5–3**
European living standards.

**5–2** Which are the poorest? ______________________________

______________________________

**5–3** Is any particular subregion uniformly rich or poor? ______________

**5–4** Identify the sharpest differences in standard of living among neighboring states: ______________________________

______________________________

**5–5** Are the original centers of industrialization, which have had longer to modernize and longer to accumulate wealth, necessarily still the richest? Give some examples of older and newer industrialized countries.

Older: ______________________________

______________________________

Newer: ______________________________

______________________________

## CULTURAL GEOGRAPHY AND DEMOGRAPHICS

Europe is highly attractive to U.S. and Canadian tourists for historical, cultural, and personal reasons. The three most widely spoken languages in the United States and Canada—English, French, and Spanish—are all European languages. Those Americans who are not direct descendants of Europeans, along with those who are, live in societies with a strong and obvious cultural inheritance from Europe. Language, religious custom and belief, political systems and economic philosophy, and basic technology all show important, though not exclusive, links between Europe and the United States and Canada. U.S. and Canadian education emphasize European history more than, say, the history of Asia or Africa. Then, too, many Americans go to Europe to visit relatives or to seek their roots—to walk the streets and climb the hills their ancestors knew.

Use the world outline map in Figure 5–4 to show your personal links with other world regions. Identify the approximate origin area or country of your ancestors. For example, if an ancestor immigrated from northern Ireland to the United States, draw a line from northern Ireland to the new home in North America. If, like the author of Roots, you can trace ancestry to the vicinity of the Gambia River in West Africa, draw the appropriate line from Gambia to whatever state or province your ancestors came to. Draw as many lines on this map as you can to identify ancestral homes.

An outstanding characteristic of Europe is the rich mixture of many different peoples and cultures in a relatively compact area. To appreciate this great variety and to identify some of the broad, transnational cultural groups involved, create a highly generalized language map of Europe. On the outline map of Europe in Figure 5–5, use a different color to indicate each of Europe's three major language groups:

1. *Germanic language group.* Include Germany, Britain, and Ireland (except the extreme western coasts of Ireland, Wales, and Scotland, where Celtic languages survive); the Netherlands; the northern half of Belgium; Luxembourg; the eastern two-thirds of Switzerland (except on the Italian border); Austria; Denmark; Sweden; Norway; and Iceland.

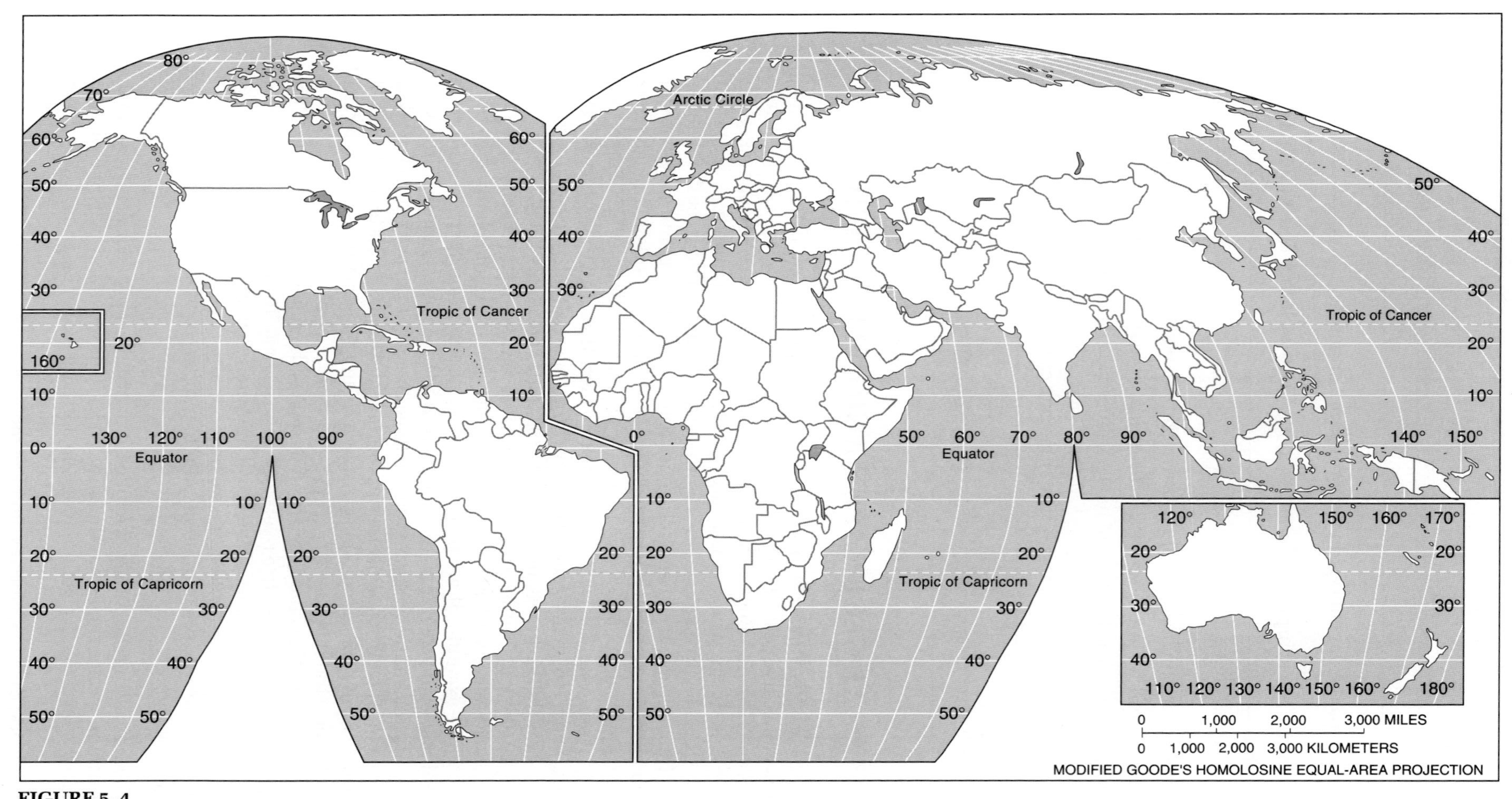


**FIGURE 5–4**
World outline map–(Ancestral Migrations).

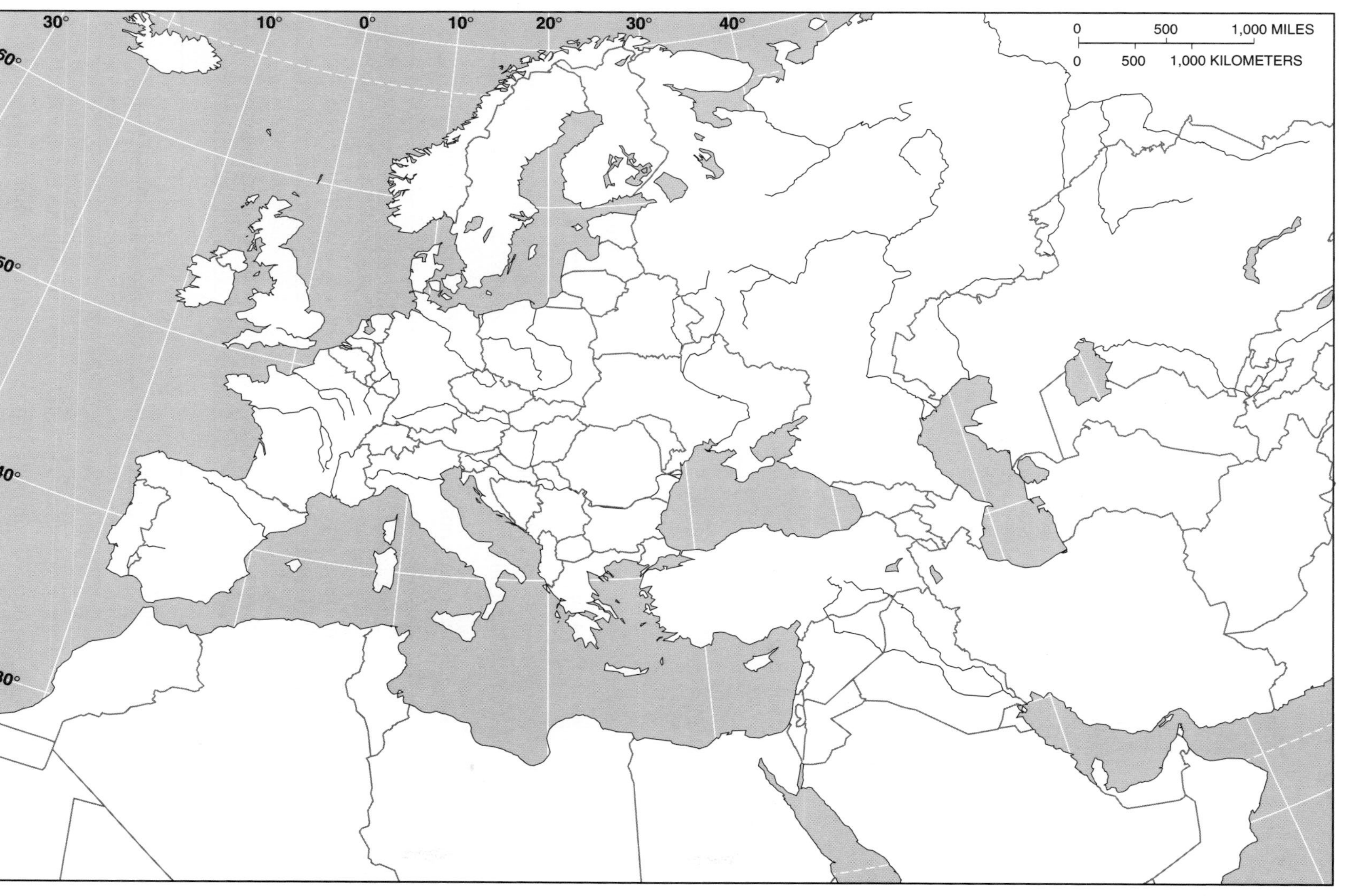


**FIGURE 5–5**
Major language groups of Europe.

2. *Romance language group.* Include France (except the Breton-speaking Celtic descendants in Brittany); Spain (except the Basque-language area straddling the border with France along the Bay of Biscay); the western third of Switzerland, and the Swiss border zone with Italy; Portugal, Italy, and Romania.
3. *Slavic language group.* Include Poland, Russia, Slovakia, Belarus, Ukraine, Czech Republic, Bulgaria, Macedonia, Slovenia, Bosnia-Herzagovina, and Croatia.

The balance of the map, the areas having languages that are not members of the three broad groups, includes Greek, Finnish, Albanian, Basque, and Celtic languages.

## CHECK UP

**5–6** On what sea would Polish bathing beaches front? ______________

**5–7** List the countries that share the Scandinavian peninsula:

______________________________

**5–8** Which European countries share a common land border with Russia?

______________________________

**5–9** Of the larger islands in the Mediterranean, which belongs to France? ______________ Which two are part of Italy? ______________

**5–10** If you were driving from Lisbon to Rotterdam by the shortest route, list the countries you would drive through: ______________

**5–11** Europeans looking for great ski slopes (high mountains, cold winters) would most likely visit which of their neighbors? ______________

**5–12** List the European countries that front on the Mediterranean Sea:

______________________________

______________________________

**5–13** Which European state's national territory (not a colony) extends farthest south? ______________

**5–14** Omitting the microstates, which European states are landlocked?

______________________________

______________________________

**5–15** In which country are French, German, and Italian all official languages?

______________________________

**5–16** An *enclave* is a piece of political territory, usually an independent state, that is surrounded completely by the territory of another country. Italy has two such enclaves: ______________ and ______________.

**5–17** If you traveled due west from the southern tip of Sweden, what country would you land in? ______________

**5–18** Which European capitals are also seaports, reachable by large ocean-going ships? __________

**5–19** Which European country, excluding Russia, extends farthest north?

__________

**5–20** Which is the westernmost European country? __________

# GEOCONCEPTS

## Ethnicity, States, and Chaos: The Collapse of Yugoslavia

Yugoslavia was a federal state and, like India and the former USSR, the member republics and their boundaries were intended to reflect ethnic homelands. Figure 5–6 shows these boundaries. As evident in Figure 5–7, ethnic distributions are far more complex than the political boundaries, so that most of the five now-independent, states of former Yugoslavia have a complicated pattern of intermixed, different ethnic groups. Civil war within Bosnia-Herzagovina (usually known as Bosnia) is an outgrowth of long-standing ethnic rivalries and tensions.

**FIGURE 5–6**
Political boundaries of former Yugoslavia. The state presently known as Yugoslavia includes Serbia and has two autonomous provinces, Kosovo, and Vojvobina. (*Source*: Charles Stanfield and Chester Zimolzak, *Global Perspectives: A World Regional Geography* Columbus OH: Merrill Publishing Co. 1990)

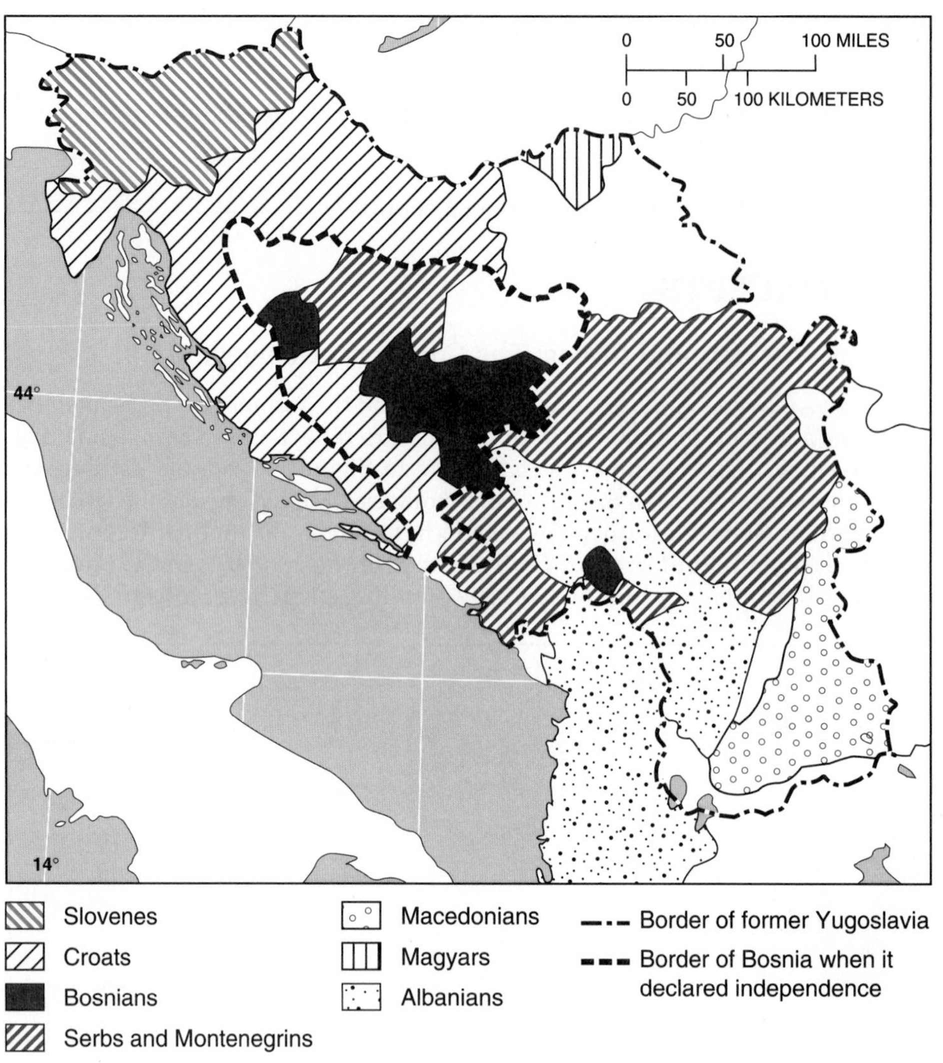


**FIGURE 5–7**
Ethnic distributions in the former Yugoslavia.

**5–21** How many different ethnic groups live in Bosnia, whose original boundary is shown in Figure 5–7? ______________________

**5–22** Is it any surprise to you that the most stable of former Yugoslavia's states is Slovenia? Why? ______________________

## Regional Economic Communities

The relatively small size of most European countries has been noted. Although there are some local advantages of crossing boundaries, international borders also pose many negatives for trade and economic development. International boundaries typically separate countries having very different laws, policies, regulations, and tax structures. Furthermore, these taxes and regulations frequently are designed specifically to reduce or eliminate competition from another country's agricultural products, manufactured goods, industrial raw materials, or even labor. At best, borders are a hassle with red tape; at worst, they form a barrier to trade and movement. Borders can have a dark side.

Just after World War II, it occurred to some European leaders that one reason for the spectacular economic strength of the United States was that its state borders very rarely functioned as obstacles to the free movement of people, raw materials, manufactured products, energy, capital, agricultural products, or anything else. Indeed, many Americans commute across state borders, hardly noticing them. The only indication of many state borders is a roadside welcome sign. Perhaps only very large territories and populations with no border taxes or restrictions are able to take advantage of the most efficient, large-scale production of goods by providing unrestricted access to the largest possible market.

The European solution was to reduce the negative economic impact of borders without affecting their political significance. France and Germany, who have historically been enemies, launched the beginnings of the economic-community idea. Northeastern France has a lot of iron ore but lacks good-quality coal, the other major ingredient required for steel industries. Nearby, across the German border, lies plenty of coal, but the Germans lack iron ore. Why not remove all tax and regulatory barriers so that French iron ore can move easily to German steel mills and German coal can flow to French steel mills? The price of steel would be lower in both countries; everybody would win. This Franco-German "coal and steel community" was so successful it helped convince six nations to work together to reduce the negative economic impacts of their mutual borders.

The original six were France, (West) Germany, Italy, Belgium, Luxembourg, and the Netherlands. On the outline map in Figure 5–8, apply a dark color to the original six nations. Membership has since expanded to include the United Kingdom, Ireland, Spain, Portugal, Denmark, Sweden, Finland, Austria, Former East Germany, and Greece. Apply a lighter color to these more recent members of the European Community. The EU collectively out-produces every nation on Earth except the United States. Its success will lead to requests from other nations to join. Switzerland, Turkey, Cyprus, and Malta are in the process of applying.

The European Free Trade Association (EFTA), presently including Iceland, Norway, and Switzerland, has, for practical purposes, merged with the EU. Use a diagonal, parallel line pattern to label the EFTA states.

The former Soviet satellites of eastern Europe once were grouped into a Communist counterpart to the EU. Known as the Warsaw Pact nations, they once included (East) Germany, Poland, Czechoslovakia, Hungary, Romania, and Bulgaria. Traditionally, eastern Europe has been less developed than the western portion. Economic progress may be complicated by the rise of ethnic nationalism and its impact on the unity of Romania, Hungary, Bulgaria, and Albania. Poland, Slovakia, the Czech Republic, Hungary, Romania, and Bulgaria expect to join the EU (European Union) eventually.

Does membership in a regional economic community such as the EU or NAFTA (North American Free Trade Association) automatically guarantee escalated economic progress? Table 5–2 lists GNP per capita data for 1992 and 1998, with percentage change over those six years. Obviously, there are many, and complex, economic, political, and demographic factors which have an impact on a national economy. These may or may not be related to membership in the EU or NAFTA. Also, a six-year timespan may not show a true picture of long-term trends. Nevertheless, data in Table 5–2 suggest that recent growth rates in GNP per capita among the 15 full members of the EU vary enormously. Although nine EU members' growth rates exceeded the average for the developed world, six did not.

**5–23** Which three EU members achieved the *highest* growth rates? ______________

______________________________________________________________

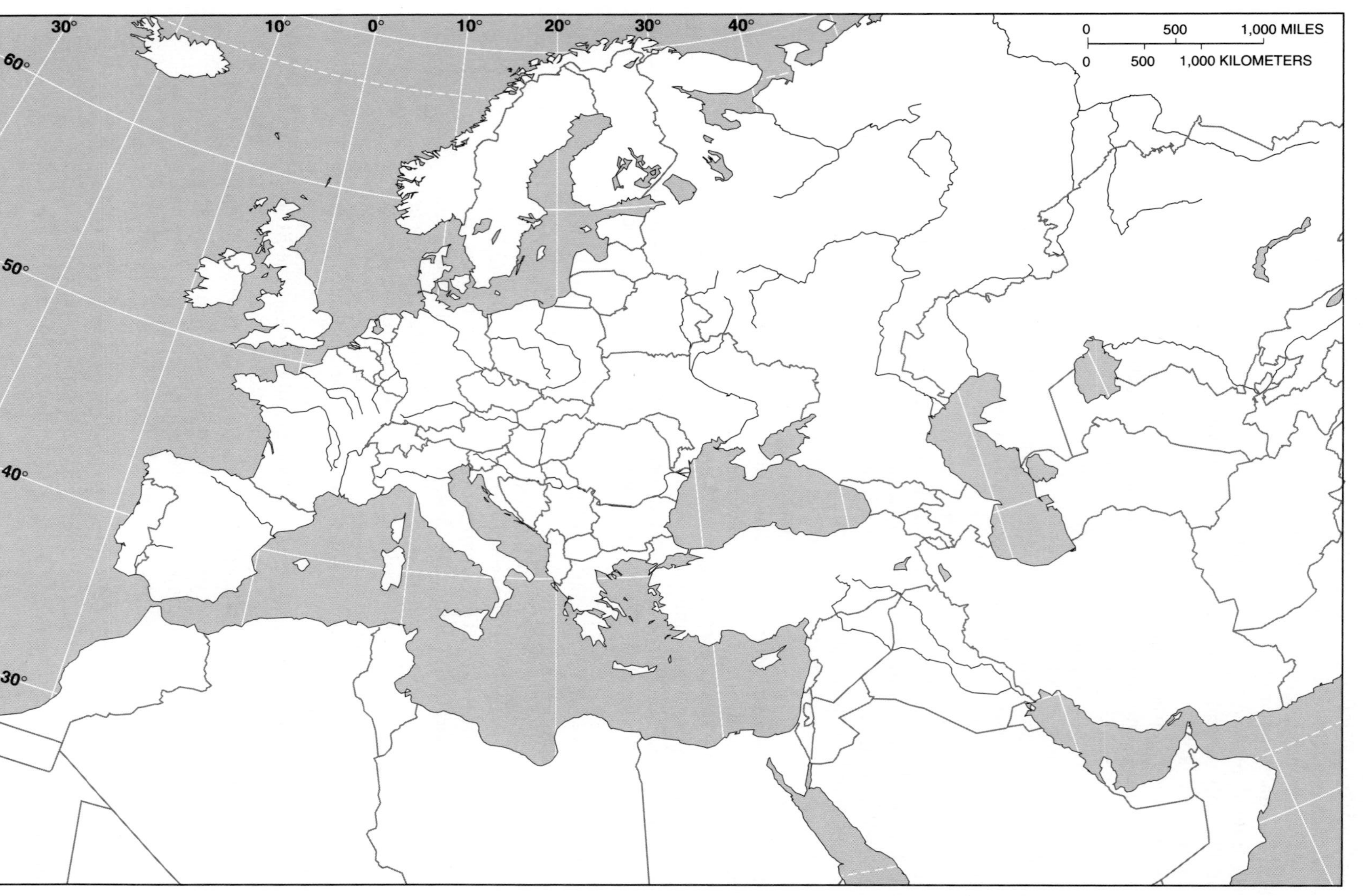


**FIGURE 5–8**
The European Community (EU).

**5–24** Which two EU members had *negative* growth, 1992–1998? ____________

**5–25** Which EU member led the group in GNP/capita both in 1992 and 1998?

____________

**5–26** Which EU member had the *lowest* GNP/capita in 1992? ____________

In 1998? ____________

**5–27** By comparison, Japan, the only large wealthy industrial nation *not* part of any regional economic community, had a 1998 GNP/capita above those of how many EU countries? ____________

**5–28** By comparison, has membership in NAFTA produced a higher-than-world average gain in GNP/capita for Mexico? ____________

**TABLE 5–2** GNP/Capita 1992 and 1998; Percentage Change (U.S. dollars)

| | 1992 | 1998 | % Change |
|---|---|---|---|
| **World** | 4340 | 4890 | 13% |
| **More Developed World** | 16610 | 19480 | 17% |
| **EU Members:** | | | |
| Austria* | 22110 | 26830 | 21% |
| Belgium | 20880 | 25380 | 21% |
| Denmark | 25930 | 33040 | 27% |
| Finland* | 22980 | 24280 | 6% |
| France | 22300 | 24210 | 8% |
| Germany | 23030 | 26570 | 15% |
| Greece | 7180 | 11740 | 63% |
| Ireland | 12100 | 18710 | 55% |
| Italy | 20510 | 20090 | –2% |
| Luxembourg | 35260 | 45100 | 28% |
| Netherlands | 20590 | 24780 | 20% |
| Portugal | 7450 | 10670 | 43% |
| Spain | 14020 | 14100 | .5% |
| Sweden* | 26780 | 25580 | –4% |
| UK | 17760 | 21410 | 20% |
| **EFTA Members:** | | | |
| Iceland | 23670 | 27830 | 17% |
| Norway | 25800 | 34310 | 33% |
| Switzerland | 36230 | 39980 | 10% |
| **NAFTA** Members:** | | | |
| Canada | 20320 | 19170 | –6% |
| Mexico | 3470 | 3840 | 11% |
| US | 23120 | 29240 | 26% |
| **For Comparison Only:** | | | |
| Russia | 2680 | 2260 | –16% |
| Japan | 28220 | 32350 | 15% |

*Entered EU on Jan. 1, 1995
**NAFTA went into effect Jan. 1, 1994

## REGIONAL WATCHLIST

The political and economic maps of Europe have undergone fundamental and crucial changes in the early 1990s. It is likely that the full impact of these changes has yet to be realized, and it is possible that further boundary changes will occur. The situation in former Yugoslavia is likely to remain fluid before mutually acceptable national boundaries there can be finally determined. Other southeastern European boundaries could be affected by the ambitions of national minorities in Hungary, Romania, Albania, and Bulgaria. Secessionist-minded Scots could even disunite the United Kingdom in this era of renewed ethnic consciousness.

# 6

# Russia, Transcaucasia, and Central Asia: A Region in Transformation

## INTRODUCTION

The profound and dramatic changes that have transformed the region once known as the USSR comprise an excellent argument for the value of studying cultural and political geography. The degree of political and economic independence achieved by the new republics is somewhat variable, and their role in international politics and in the global economy has not yet crystallized.

As noted in the definition of the Europe region in Chapter 5, the 1991 collapse of the Soviet Union presents regional geographers with a problem. Which, if any of the 14 ethnically based republics once dominated by Russia should remain grouped with Russia? The six to the west and southwest of Russia have been added to the Europe region.

The Transcaucasian ("across the Caucasus Mountains"—from Russia) republics of Armenia, Georgia, and Azerbaijan had independent status, historically and then briefly again when tsarist Russia convulsed in revolution. They briefly formed a confederation of their own before being absorbed by the Soviet state in 1922. Armenia and Georgia were ancient Christian kingdoms; their strong cultural identities forged under centuries of attack and repression. Azerbaijan, the eastern-most Transcaucasian Republic, is predominantly Muslim. This gives it more in common with the five Central Asian republics (Kazakhstan, Kyrgyzstan, Tajikistan, Turkmenistan, and Uzbekistan) than with its historic enemies, Armenia and Russia. The resurgence of Islam, particularly Islamic fundamentalism, has led to many Muslims reasserting their Muslim identity.

Transcaucasia and Central Asia have been retained in the same region with Russia, their onetime colonial master, on the basis of both their recent common history in the tsarist empire and Soviet eras, and the still-important transport and communications infrastructure that continue to link them with Russian markets and resources. While these primarily economic factors argue for a regional association with Russia, the cultural factors of Islam could one day become a sufficiently strong bond with the Middle East–North Africa region to lead to an eventual realignment of boundaries.

The USSR's successor states make up a unique region in several respects. Once the world's largest nation in territory, the USSR was an uneasy union of many different ethnic, linguistic, and religious groups. Ethnic strife divided and eventually dismembered it. Although the region possesses one of the most varied and abundant natural resource bases in the world, it lags behind the United States, the European Community, and Japan in most items of high-technology and in industrial capacity, except for steel production.

Several of the problems of the onetime Soviet Union have geographic origins. The former USSR faced a titanic struggle to effectively integrate its economy because its raw materials, energy resources, labor force, and markets were scattered over what was then the world's largest national territory. Much of the region is too cold, too dry, or too rugged for crop production. And the greatest challenge to former Soviet governments was the political and cultural integration of its many varied—and geographically dispersed—ethnic, racial, cultural, and religious groups. It was a challenge that the USSR ultimately failed, resulting in the declared independence of its former member republics.

## PHYSICAL GEOGRAPHY

This region stretches west to east from Europe's Baltic Sea to the Bering Strait that separates Siberia from Alaska. It stretches north to south from the Arctic Ocean coast to the subtropical deserts of central Asia. This immense region sprawls over 11 different time zones. When it is midnight in easternmost Siberia, it is only 2 P.M. in St. Petersburg, a major western city.

The topography of this sprawling region is as varied as its climate. The Great North European Plain sweeps eastward from Poland all the way to the Urals, relieved (but not interrupted) by such low, rolling terrain as the Moscow Hills. Many portions of the region's borders with Asian nations are mountainous and included some of the world's highest peaks near the central Asian borders with Afghanistan and western China.

On the outline map in Figure 6–1, locate and label the following:

| | | | |
|---|---|---|---|
| Arctic Ocean | | East Siberian Sea: | 75° N, 160° E |
| Pacific Ocean | | Sea of Okhotsk: | 55° N, 150° E |
| Barents Sea: | 70° N, 45° E | Bering Sea: | 60° N, 175° W |
| White Sea: | 66° N, 38° E | Laptev Sea: | 75° N, 125° E |
| Black Sea: | 42° N, 35° E | Kara Sea: | 75° N, 70° E |
| Caspian Sea: | 40° N, 52° E | Bering Strait: | 65° N, 170° W |
| Baltic Sea: | 60° N, 20° E | Aral Sea: | 45° N, 65° E |
| Atlantic Ocean | | Ural Mountains: | 55° N, 58° E |
| Lake Balkhash: | 46° N, 74° E | Caucasus Mountains: | 44° N, 45° E |
| Kamchatka Peninsula: | 55° N, 160° E | Lake Baikal (Baykal): | 54° N, 107° E |
| Novaya Zemlya (island): | 75° N, 58° E | Lake Ladoga: | 61° N, 32° E |
| Gulf of Finland: | 59° N, 25° E | Kuril Islands: | 46° N, 150° E |
| Gulf of Riga: | 57° N, 23° E | Tatar Strait: | 48° N, 141° E |
| Sakhalin Island: | 50° N, 143° E | Sea of Azov: | 46° N, 36° E |
| Volga River: | 47° N, 46° E | Sea of Japan: | 40° N, 135° E |
| Yenisey River: | 67° N, 87° E | Lake Onega: | 62° N, 35° E |
| Ob-Irtysh River system: | 62° N, 67° E | Amur River: | 50° N, 127° E |
| Lena River: | 68° N, 124° E | | |

## OBJECTIVES AND STUDY HINTS

The old Soviet political structure was incapable of containing nationalist movements among dissident ethnic groups. Geographic keys to understanding ethnic strife and nationalist movements among the different peoples of this region are the great regional and ethnic differences in population growth rates, interrepublic migration rates, and levels of economic development. Look for links among the geographic factors of relative location, distance, physical environments, natural resources, ethnic homelands, and economic and political problems.

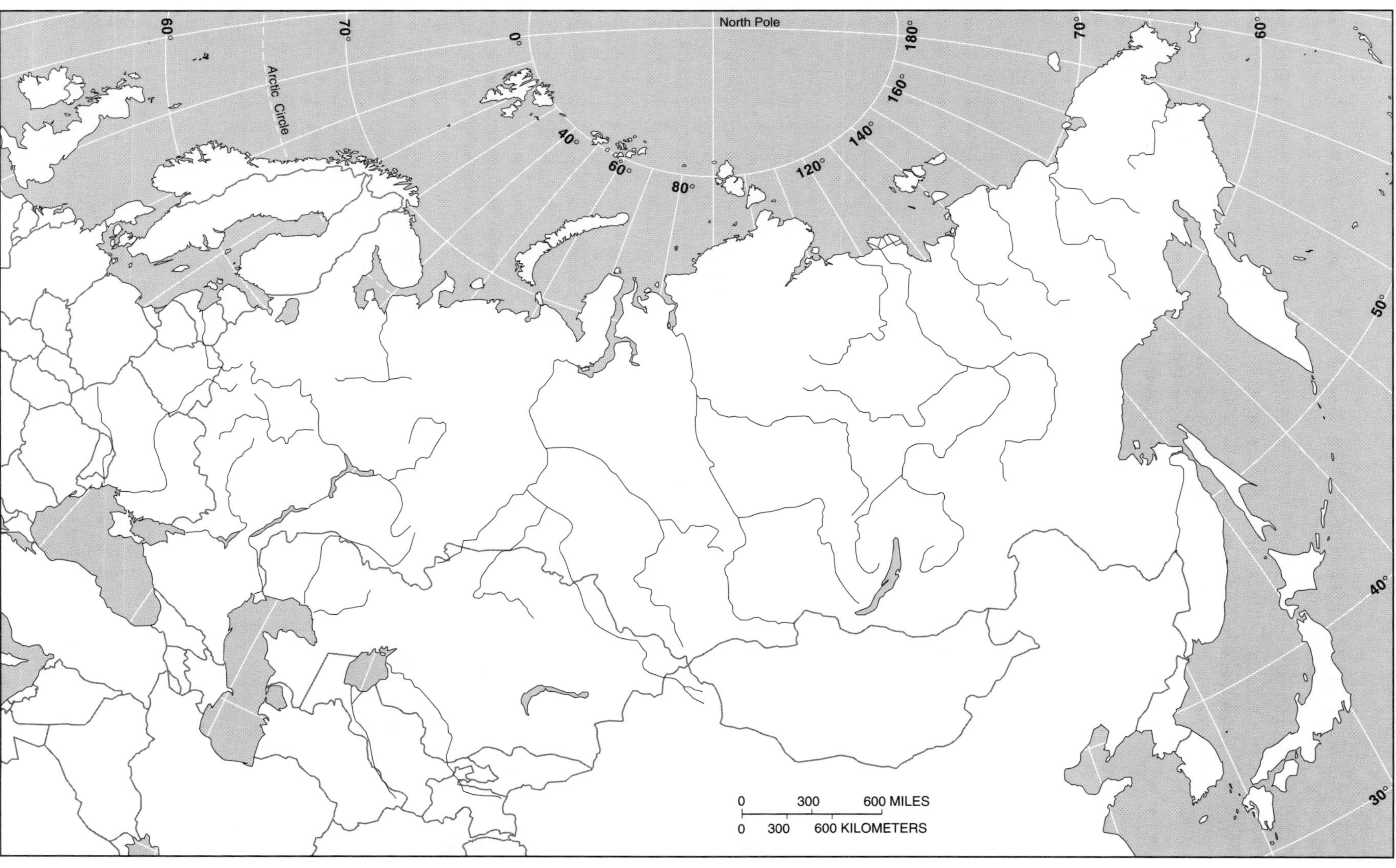


**FIGURE 6–1**
Outline map of Russia, Transcaucasia, and Central Asia.

## POLITICAL GEOGRAPHY

The former Soviet Union had been described as the world's last empire. There is some truth in this label, for an empire is a group of nations united only by military force and rigid internal control. The government of the USSR was the successor, or political heir, to the Tsarist Russian Empire. The pre-1991 boundaries of the Soviet Union included virtually all the territories conquered by the Tsarist armies, except Finland and part of eastern Poland. Just as in the days of the Tsars, the Soviet government commanded loyalty, or at least tried to prevent rebellion and secession, by the use of secret police and military power.

As pointed out in Chapter 5, a *nation* properly refers to a people, whereas a *state* is a political organization in control of a territory. Most of the ethnic groups for whom the republics are named—the Lithuanians, Estonians, Armenians, Ukrainians, and others—have had independent countries in the past.

The map of the expansion of the Russian state, Figure 6–2, shows a longtime pattern of territorial growth under both Tsars and Soviets. From its original core in the Russian heartland around Moscow, the Russian state had expanded outward in all directions. As Figure 6–2 indicates, however, annexation of territory evidently was easier, or more successful, in some directions than in others. Expansion eastward into Asia (Siberia and the Muslim-culture region known as Central Asia) carried Russian domination farther from Moscow than expansion westward into Europe. Compare this map of territorial expansion with the map of this region shown in Figure 6–3.

**6–1** Which Transcaucasian and Central Asian states' present territories were entirely or mostly acquired by Russia during the period from 1809 to 1904?________________________________________

________________________________________________________

At this time, get acquainted with the region. Using the outline map of the region in Figure 6–3, locate and label the places listed in Table 6–1.

**TABLE 6–1** Republics, capitals, and important cities of Russia, Transcaucasia, and Central Asia. Other than the giant Russian Federation, the new states are grouped by geographic location and historic cultural similarities.

| Republic | Capital | Other Important Cities |
|---|---|---|
| Russian Federation (Russia) | Moscow | St. Petersburg, Novosibirsk, Vladivostok, Omsk, Irkutsk |
| | *The Transcaucasian Republics* | |
| Armenia | Yerevan | |
| Azerbaijan | Baku | |
| Georgia | Tbilisi | |
| | *The Central Asian Republics* | |
| Kazakhstan | Astana | Karaganda |
| Kyrgyzstan | Bishkek | |
| Tajikistan | Dushanbe | |
| Turkmenistan | Ashkhabad | |
| Uzbekistan | Tashkent | Samarkand |

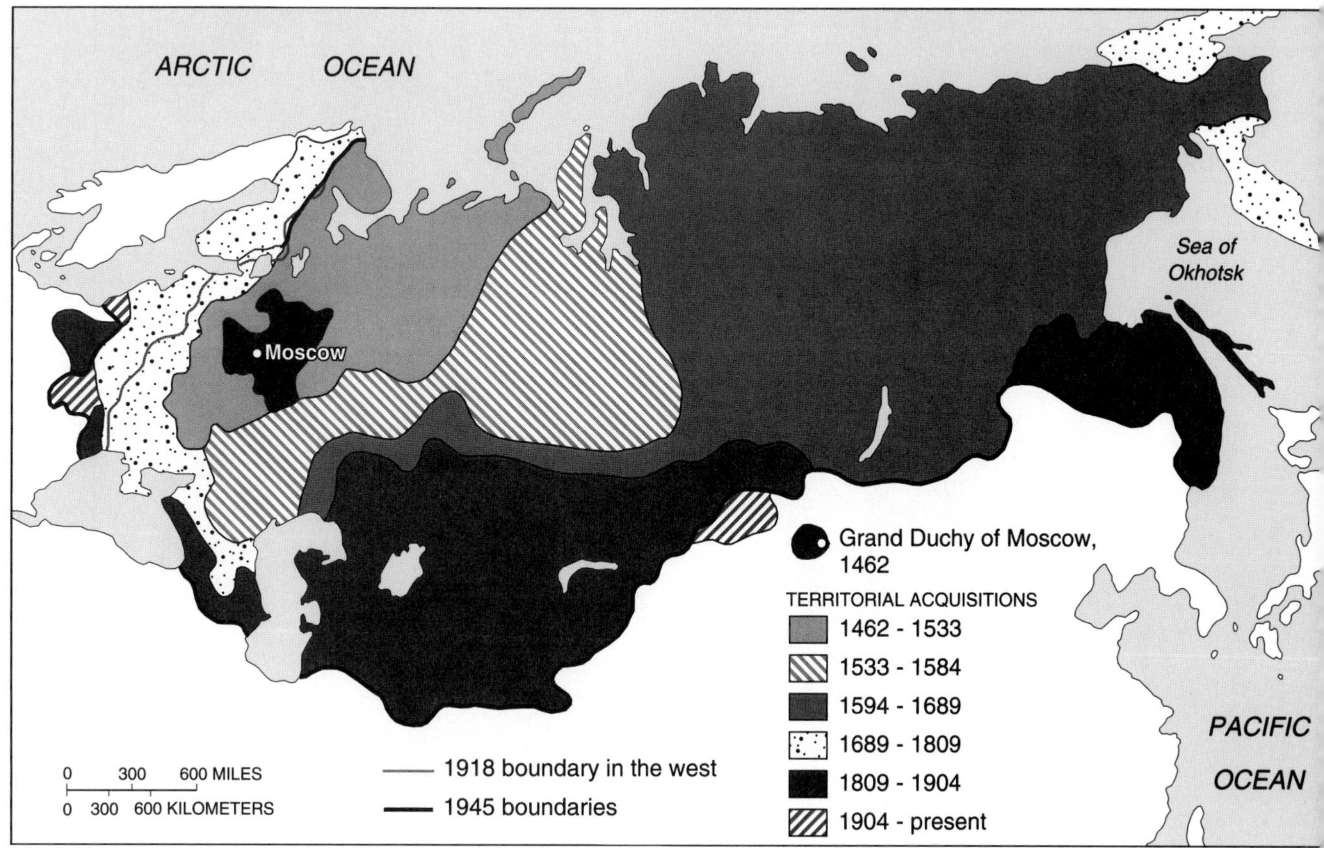


**FIGURE 6–2**
Expansion of the Russian Empire/USSR to 1991. Moscow was the focus for the subsequent expansion of the Russian states as the power of surrounding states was successfully challenged.

## ECONOMIC GEOGRAPHY

The United States cultivates 20% of its territory; Russia cultivates less than 10%. Although total farmlands are still huge in this vast region, the comparison says much. On the outline map of the region in Figure 6–4, draw a line along Russia's western border, southward from St. Petersburg along Russia's western borders to the Sea of Azov near Rostov-on-Don. From this point draw a line to Novosibirsk ("New Siberia"), and then a line from Novosibirsk back to St. Petersburg. You have enclosed the Russian heartland—the most productive agricultural lands, much of the population, most of the big cities, and much of the industry of this giant country. The shortest leg of the triangle (St. Petersburg–Sea of Azov) is bounded by international borders, borders not likely to change except through the reincorporation of the western, former member republics. To the southeast of the Sea of Azov–Novosibirsk line, agriculture is handicapped by increasing aridity and rugged mountains. Crops to the southeast of the triangle often are limited to fertile valleys and to areas where irrigation water is available. To the northeast of the Novosibirsk–St. Petersburg line, the intense cold of long winters becomes the limiting factor. Of these two environmental limits on agriculture, the shortage of water is more easily solved than the shortage of heat energy.

### Fund and Flow Resources

The Soviet administration was justifiably proud of having constructed many of the world's largest dams and hydroelectric projects. Hydropower is a "clean"

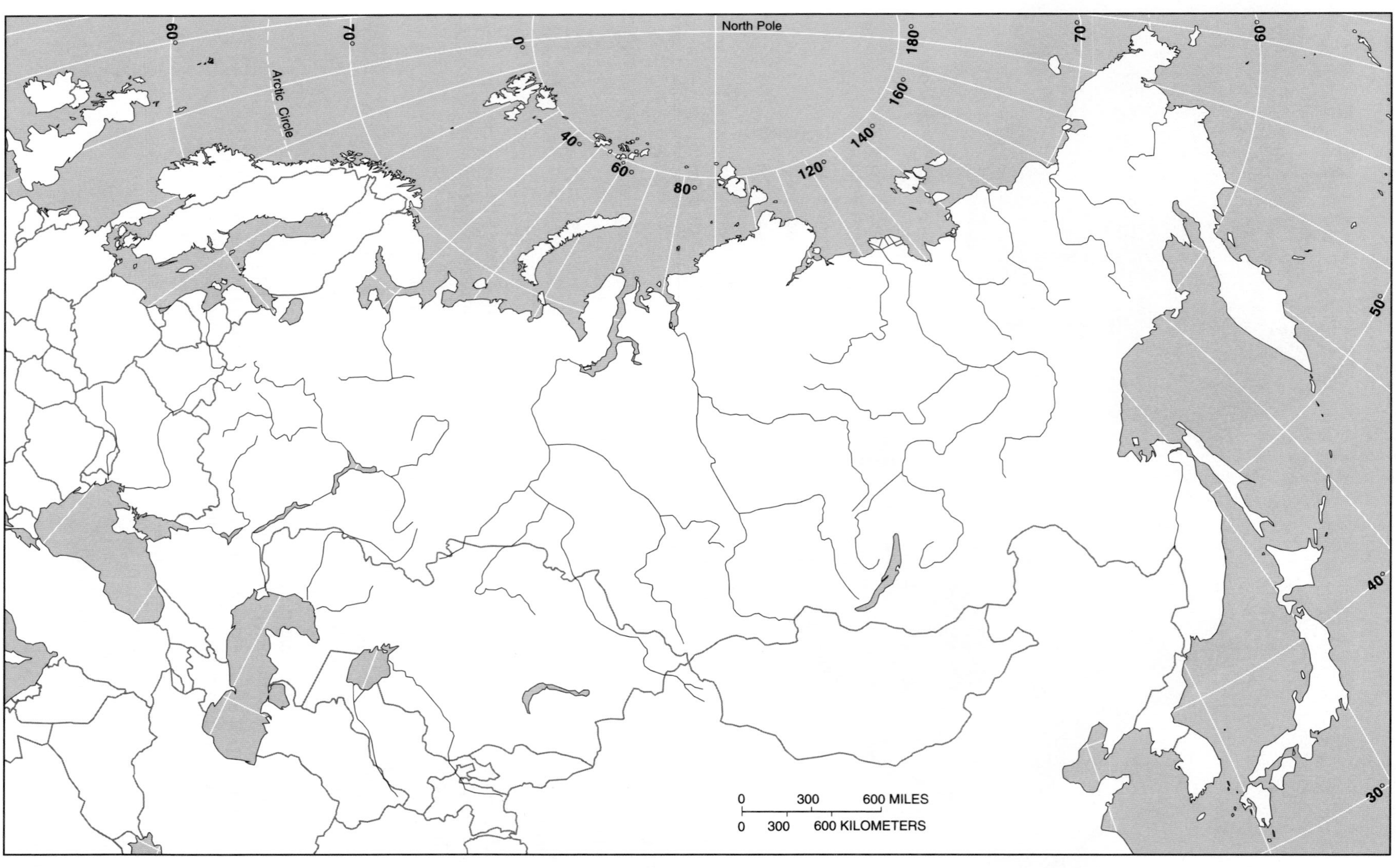


**FIGURE 6–3**
Map of republics, capitals, and important cities of Russia, Transcaucasia, and Central Asia.

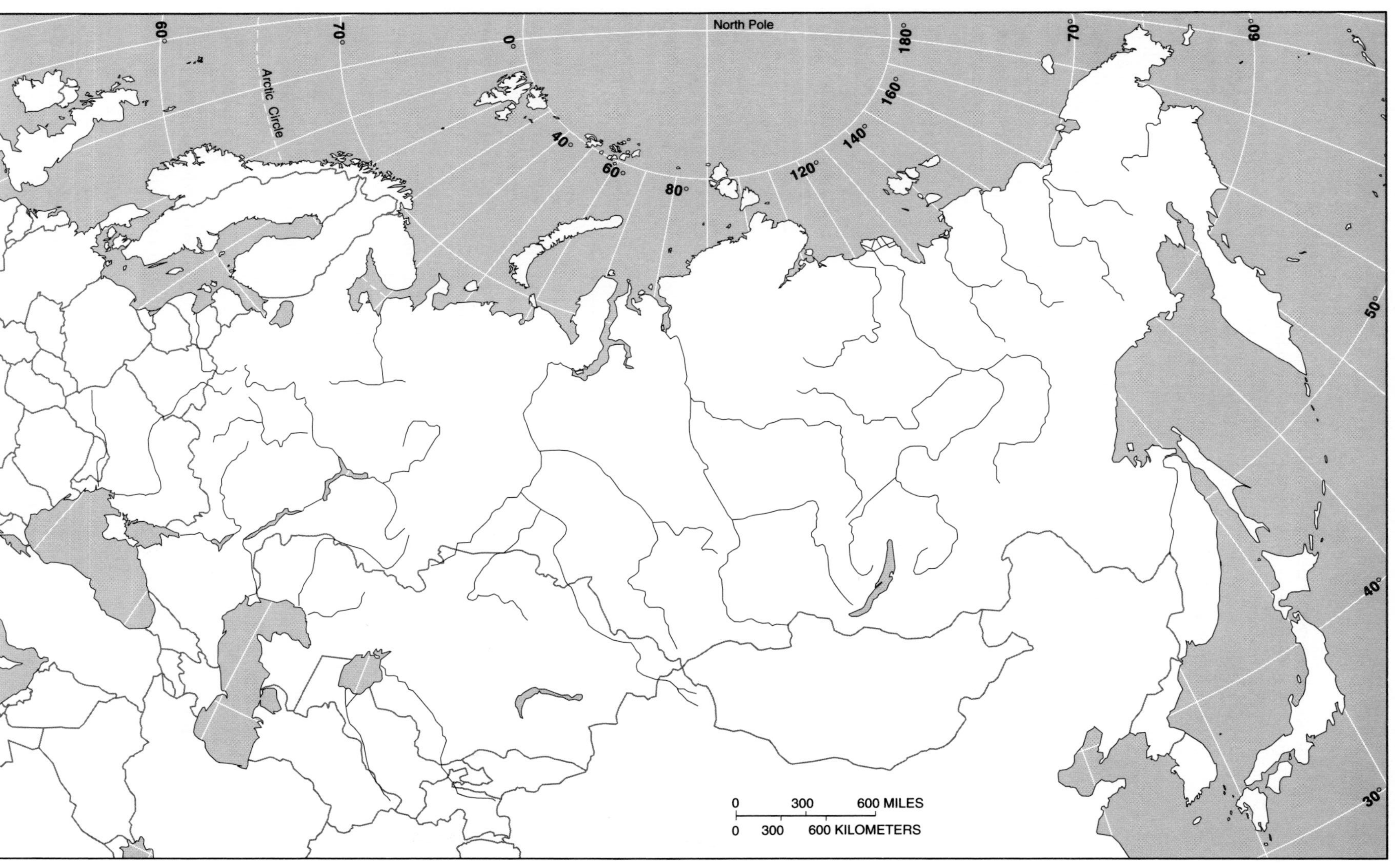


**FIGURE 6–4**
Russian heartland

source of energy. It does not pollute the air as does the burning of fossil fuels. Nor does it produce the radioactive waste of nuclear power generation.

Hydroelectricity, generated by the energy of falling water, is a good example of a *flow,* or *renewable,* resource. As long as a dam and its equipment are maintained properly, people can obtain energy from this source as long as water runs downhill. By contrast, other sources of energy, including mineral resources, are *fund* resources, or *nonrenewable* resources. Although it is arguable that geologic processes continue to create petroleum in Earth's crust, these processes are so slow in contrast to the rate of extraction that petroleum must be considered a fund resource. No doubt people will discover new deposits, but eventually we will run out of oil if we continue to use it. Fund resources are comparable to a bank account in which only an opening deposit has been made. No more deposits will occur, so the account will not last indefinitely if it is being drawn upon. Someday, depending on the size of the initial deposit and the rate of withdrawals, the account will be used up.

**6–2** Name some other examples of *flow* resources: ____________________

**6–3** Name some other examples of *fund* resources: ____________________

**6–4** Why does it make sense to use more flow-energy resources than fund-energy resources? ____________________

____________________

## Friction of Space

Geographers often talk about the *friction of space.* Distance (linear space) can only be crossed, or overcome, through the expenditure of time and energy and usually the use of capital equipment. Because transport technology reduces the time and labor costs of crossing space, advancing transport technology reduces the friction of space.

Three factors create Russia's fundamental transportation problem: the gigantic size of the nation; the fact that most of its big cities, people, farmland, and industrial centers lie in the "developed triangle"; and huge stocks of raw materials and energy resources are found in remote parts of Siberia. The massive effort to complete the Baikal-Amur Mainline (BAM) rail line produced an alternative to the very heavily used Trans-Siberian Railroad. The BAM and the Trans-Siberian railroads serve two purposes. Siberian coal, metal ores, other minerals, and forest products are transported west to European cities and industries (Figure 6–5). Siberian resources also flow eastward.

**6–5** Which highly industrialized but resource-poor country is the major customer for these eastbound resources? (The answer is obvious on the map.)

____________________

**6–6** Which two other nations with large territories strove to reduce the friction of space by constructing their own transcontinental railroads before the Russians completed the Trans-Siberian? ____________________

____________________

Reducing the friction of space also involves constructing canals to link navigable rivers and seas and to create navigable waterways where none currently exists. The Soviets developed an extensive canal system in the western USSR, making it possible to move by boat from Moscow to the Baltic, the White, the Black, or the Caspian seas. The mammoth Siberian River reversal scheme also will create canal links between European and Asian portions of the region.

**6–7** Which other two major world regions have developed canal links between navigable rivers, lakes, and seas? ______________________

## Economic Integration Versus Political Disintegration

When different geographers devise somewhat different regionalization schemes, it is not a case of one being "right" and the other(s) being "wrong." Undoubtedly, the major area of disagreement about the geographic character and boundaries of any of the traditional major world regions is what to do with the non-Soviet dis-Union.

Under both Tsarist and Soviet governments, there were continuing, determined efforts to unite and integrate the economies of all parts of the Russian Empire and the Soviet Union. Raw materials, like coal, oil, natural gas, and metal ores, flowed throughout the region. Communications and transportation systems were designed to interconnect all parts of the Russian-Soviet state, intensifying their economic integration. The map of railroads in the former Union, Figure 6–5, deliberately does *not* show the political boundaries *within* the region. As can be seen on this map, the rail net attempted to connect important cities and industrial regions in *all* 15 republics of the onetime Union.

**6–8** Which is the *longest* railroad constructed before the Revolution?

______________________

**6–9** Which mainline railroad was constructed under the Soviet administration to supplement the old Trans-Siberian? ______________________

**FIGURE 6–5**
Railroads of the former Soviet Union.

In addition to economic integration of this huge region via its railroad system, electricity transmission grids, oil and gas pipelines, barge and ship canals, irrigation canals and aqueducts, telephone and telegraph lines, and satellite relays, they all foster economic integration at the same time the area is experiencing political disintegration. Thus, this region reflects the two contrasting trends that typify many parts of the world: Ethnic nationalism encourages and supports political fragmentation, whereas the evident success of economic unions like the European Community suggest the increasing consolidation of larger and larger common markets.

**6–10** Which world region *other than* Russia and the newly independent states has had the most recent changes in national boundaries? ______________

______________________________

**6–11** Which three large countries in the Americas have moved toward economic integration? ______________________________

## CULTURAL GEOGRAPHY AND DEMOGRAPHICS

The population growth rate seems to be related to the level of industrialization in each country and region. Third World countries commonly have explosively rapid rates of increase, but the more technologically advanced, highly industrialized, and urbanized societies typically have relatively low rates of population expansion. (The geographic concept of the population explosion is discussed in Chapter 10.) Western Europe, Japan, and the United States are all in the slow-growth category.

World regions or even individual countries seldom have uniform rates of population growth in all geographic areas or among different ethnic or income groups. However, a few generalizations can be made. For example, middle-income families tend to have fewer children than poor families, and rural families tend to have more children than urban-suburban families. (Is this true among people of your acquaintance?) If income level and degree of industrialization and urbanization differ on a geographic basis or if they vary significantly among ethnic groups, then those areas or groups will likely show population growth rates that differ from the average.

Table 6–2 shows rates of natural increase and per capita gross national product, expressed in U.S. dollars, for the 9 republics of this region. Rank the countries to identify the "top third" in the region in GNP per capita (the top three). Indicate these countries on the map in Figure 6–6 by drawing in a pattern of small, open circles within their boundaries. Now rank the countries by natural increase rates to identify the three with the highest such rates and then indicate those on the same map (Figure 6–6) by drawing in a pattern of solid dots. Compare the two map patterns.

**TABLE 6–2** Natural increase rates and per capita gross national product.

| Country | Natural Increase | Per Capita GNP |
|---|---|---|
| Armenia | 0.4 | $ 460 |
| Azerbaijan | 0.9 | 480 |
| Georgia | 0.2 | 970 |
| Kazakhstan | 0.4 | 1340 |
| Kyrgyzstan | 1.5 | 380 |
| Russia | –0.6 | 2260 |
| Tajikistan | 1.6 | 370 |
| Turkmenistan | 1.5 | * |
| Uzbekistan | 1.7 | 950 |

*Not available, but probably relatively high due to extensive oil and gas reserves. An alternative source estimates a GNP per capita as high as $3000.

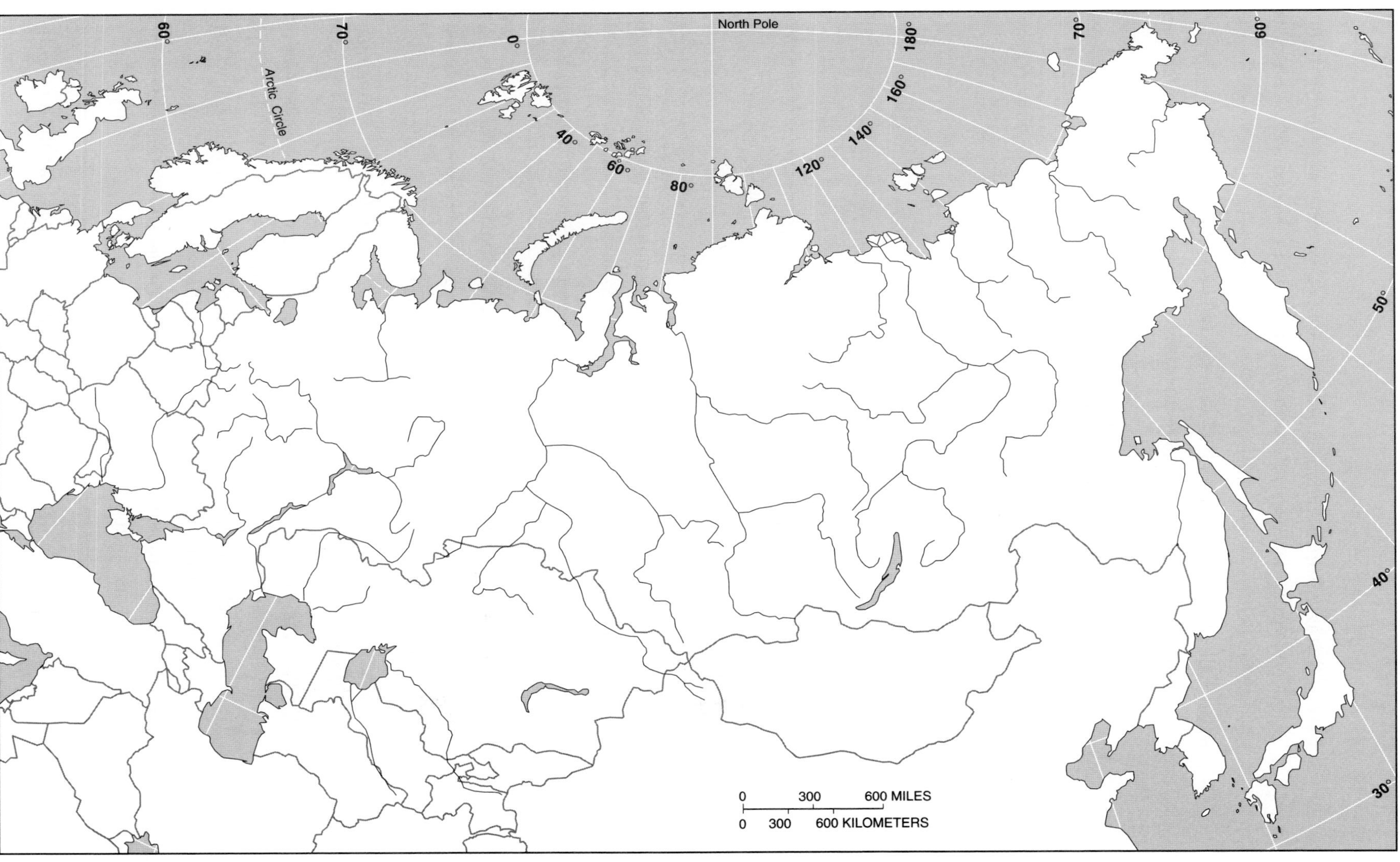


**FIGURE 6–6**
Natural increase rates and per capita gross national products.

**6–12** Is there a *positive spatial correlation* of these two factors? That is, do the countries ranked top in GNP per capita tend to be among the countries ranked top by natural increase rate? ________________

**6–13** Is there a *negative spatial correlation* between the two factors mapped? That is, where one characteristic is high or concentrated, another characteristic is low or absent. ________________

If this is true, the two map patterns would *not* overlap.

**6–14** Which republic has the highest rate of natural increase *and* one of the lowest per capita GNPs, indicating an underdeveloped economy? ________________

________________________________

## CHECK UP

**6–15** The six countries that border Russia on its western frontiers are

________________________________

________________________________

**6–16** The seven countries in Asia that share common land borders with Russia are ________________

________________________________

**6–17** The east–west–trending range of high mountains that is part of the traditional boundary between Europe and Asia is ________________

**6–18** The north–south–trending mountain range that is part of the traditional boundary between Europe and Asia is ________________

**6–19** Russia's major Pacific port, at the eastern terminus of the Trans-Siberian Railroad is ________________

**6–20** The most prosperous Central Asian republic is ________________

**6–21** Circle which of these ethnic groups is *not* mostly Muslim: Armenians, Uzbeks, Kazakhs, Azerbaijanis.

**6–22** Which physical geographic feature provided a relatively easy path for invading German armies to penetrate deep into Russia-Soviet Union? (Circle the correct answer.): Siberian Plain, Trans-Caucasus, Caspian Lowland, North European Plain, Norwegian Coastal Plain.

**6–23** Which two European countries west of the former USSR were once part of the old Russian empire?

________________________________

**6–24** Which of the former USSR's international neighbors, sharing a common border with it, are predominantly Islamic countries? ________________

________________________________

________________________________

**6–25** Which three of the former USSR's 15-member republics, now part of the Europe region, had been independent states before the USSR annexed them in 1940? ______

______

**6–26** Circle the name of the large lake (not part of the "world-ocean") now in serious danger of drying up due to heavy use of its supplying rivers for irrigation: Black Sea, White Sea, Aral Sea, Sea of Okhotsk.

**6–27** International trade using Russia's major Black Sea ports has to pass through narrow straits controlled by which foreign country?

______

**6–28** The Gulf of Finland freezes over during the winter, seriously handicapping the usability of which major western port of Russia? ______

______

**6–29** Moscow is approximately at the center of Russia's "development triangle." Circle the three (approximate) points of this triangle: Vladivostok–St. Petersburg–Murmansk; Volgograd–Yekaterinburg–Novosibirsk; St. Petersburg–Rostov-on-Don–Novosibirsk.

**6–30** The development by the United States of submarines capable of cruising under the floating Arctic icepack meant that which coast of the former USSR was suddenly vulnerable to attack? (Circle the correct answer.): Black Sea; Arctic Ocean; Caspian Sea; Sea of Japan.

# GEOCONCEPTS

## National Minorities and the Potential for Further Fragmentation

Two geographic scales are involved in the definition of a national minority. A minority obviously must be an ethnic or racial group that forms less than half of the total population of a country. A *national minority,* however, forms the majority within a particular part, province, or other subdivision of that country. Furthermore, the ethnic, racial, linguistic, or cultural minority/local majority also must possess a distinctive identity and a determination to achieve independent rule or at least local self-rule. Frequently, a national minority also will have had a period of independence in its past. Almost all the ethnic member republics of the former USSR were independent, self-governing states at some time in their history, and they now have reasserted their independence.

National minorities can be a definite threat to the stability and territorial integrity of a state, as the tragic recent history of Yugoslavia shows (Chapter 5). Unfortunately for the stability of this region, the fragmentation of the USSR into its 15 ethnic republics does *not* eliminate the threat of further political splintering. National minorities definitely exist in most of the new states. Ethnic Russians, for example, are important minorities within Kazakhstan; Armenians and Azeris (Azerbaijanis) are explosively restive minorities in each other's countries. The proper name of Russia is the Russian Federation, for this gigantic state is a chaotic patchwork quilt of no fewer than 23 republics *within* the Russian Federation, along with 11 "autonomous regions." Several of the Russian Federation's republics have declared their ultimate goal of complete independence. Georgians meanwhile, are fighting a bitter civil war along ethnic lines, Armenia and Azerbaijan are at war sporadically, and there are numerous potential border disputes among the successor states to the USSR.

**6–31** Some other countries have national minorities. Name Canada's national minority and the province in which it is a local majority ______________

______________________________________________

**6–32** The United Kingdom has several national minorities within its present borders; they are ______________________________________

______________________________________________

## Continentality

*Continentality* refers to the effect on climate of very large landmasses (continents) compared to ocean-dominated (maritime) climates. A large land area normally heats faster in summer than a large water body at the same latitude (distance from the Equator). Large land areas also usually cool faster in winter than water bodies at the same latitude. This is so because water stores more heat than land, and the sea acts as a moderating influence on temperature. People visit seashore resorts in summer to enjoy cooling breezes caused by the temperature differences between the ocean and land. Coastal cities usually are a little warmer in winter than cities further inland because the oceans generally are warmer than landmasses in winter.

The larger the landmass, and the farther it is from the Equator, and the farther it is inland from the seacoast, the more extreme will be the contrast between coastal and interior temperatures. These conditions certainly apply to Russia. Russia is very strongly dominated by continental-type climatic conditions. Not only does the country occupy the northernmost part of Earth's largest continent, but the Baltic Sea freezes over in winter, meaning that its effect on winter temperatures behaves more like a land area than an ocean area. The icepack that covers most of the Arctic Ocean melts away along Russia's northern coasts only during a very brief summer. The effects of continentality cause winter temperatures for most places in Russia to be much colder than places at similar latitudes on Europe's west coast. Continentality also means that locations far inland in Russia get warmer in summer than places at similar latitudes on Europe's ocean-dominated west coast.

# SPECIAL CHALLENGE

On the outline map showing northern Europe and Russia in Figure 6–7, locate the following places and record their January and July temperature averages (to review latitude and longitude, see Chapter 2):

- Moscow—55° 50' N (a minute ['] of latitude is one-sixtieth of one degree), 12.6°F (–10.7°C) January, 64.4°F (18°C) July.
- Glasgow, UK—55° 53' N, 38.6°F (3.6°C) January, 58°F (14.4°C) July. Note that Glasgow, actually slightly closer to the North Pole than Moscow, is much warmer in the winter and cooler in summer than Moscow.
- Zlatoust—55° 10' N, 3.6°F (–15.7°C) in January and 60.8°F (16°C) in July. Its location in the Ural Mountains almost due east of Moscow is almost 1000 ft higher, which helps explain why it is colder than Moscow in both winter and summer.

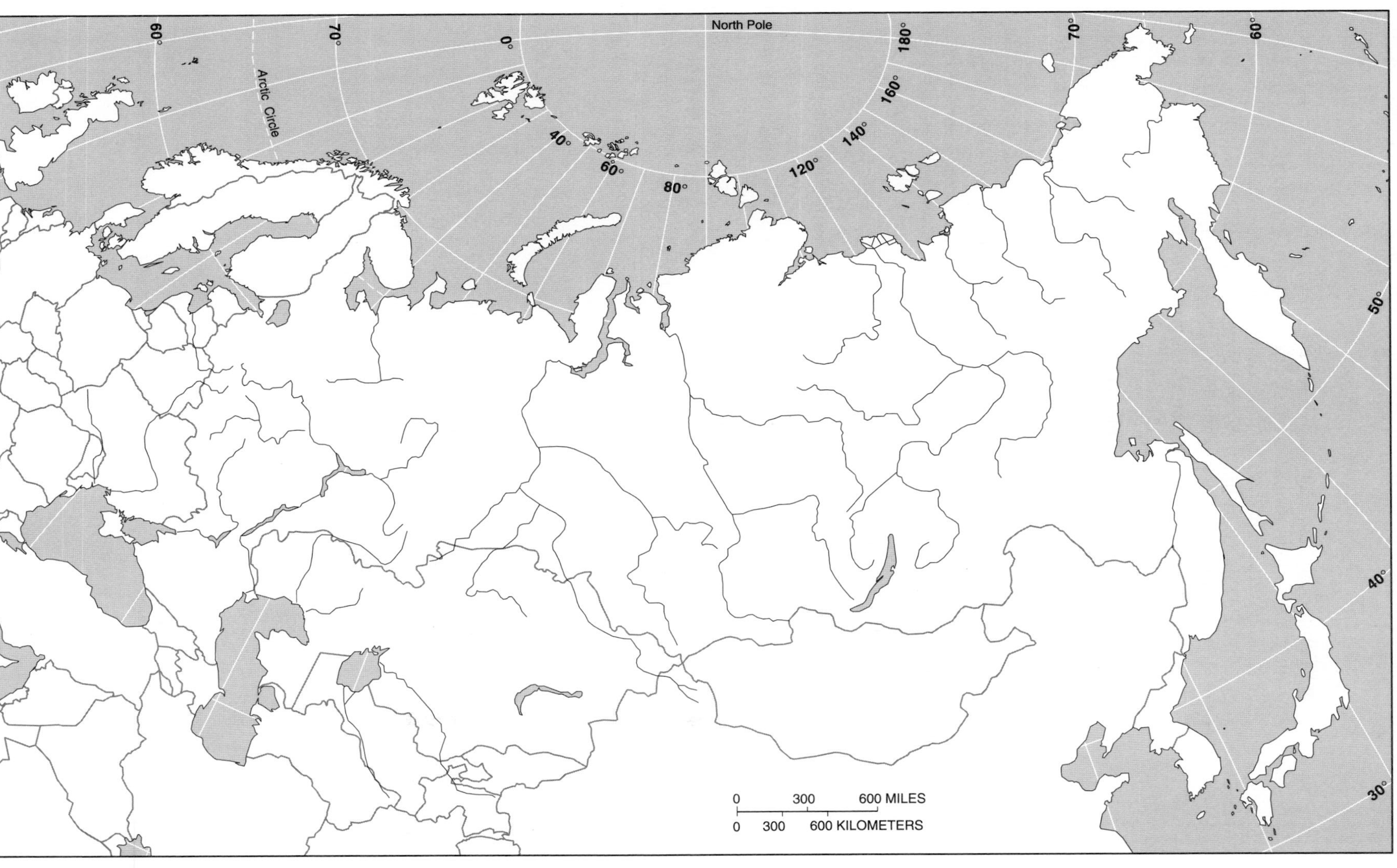


**FIGURE 6–7**
January and July temperature averages.

- Nizhne-Kolymsk—68° 32' N, –40°F (–40°C) in January and 59.9°F (15.5°C) in July. This site is almost on the Arctic seacoast. Compare it to Verkhoyansk, at a similar latitude (67° 33' N), but colder in winter (–58.2°F [–50.1°C] in January; 59.9° F [15.5°C] in July). As Verkhoyansk is only 312 ft (95.1 m) higher than Nizhne-Kolymsk's 16 ft (4.9 m) above sea level, temperature differences again are attributable mostly to continentality, even though the Arctic is frozen at Nizhne-Kolymsk much of the year.

Here are data for two places in North America. Station A's temperatures average 46.3°F (7.9°C) in January and 58.2°F (14.5°C) in July; Station B's average 25.3°F (–3.7°C) in January and 75.5°F (24.2°C) for July. Both stations are close to 42° N latitude.

**6–33** Given what you know of continentality, which is Chicago and which is Brookings, a town on the coast of Oregon? ______________________

## REGIONAL WATCHLIST

The number of independent states and their boundaries can be expected to change within Russia, Transcaucasia, and Central Asia. Whereas such changes on the political map certainly are *possible* in all regions, they are highly *probable* in this region. Look for continued guerilla warfare in Chechnia and Dagestan as Russia attempts to prevent further moves toward independence among its many ethnic groups. Reorientations away from Russia and toward the Middle East may characterize the Muslim-culture republics. This indeed is a region undergoing transformation and, possibly, further disintegration.

# 7

# Australia, New Zealand, and Pacific Islands

## INTRODUCTION

The location farthest away from any place on Earth is the *antipodes,* the exact opposite location on the globe. Because Melbourne, Australia, is about 10,500 air miles (16,898 km) from London, it is nearly opposite the United Kingdom, so it is the antipodes of the UK. Thus, as a British colony, Australia lay at the end of the world's longest imperial lifeline. This great distance from the mother country was an important consideration in the settlement and economic development of Australia and its not-very-close neighbor, New Zealand, 1200 miles (1931 km) away.

The other geographic fact that was important in British attitudes toward their continent-scale colony at the antipodes was that Australia had a very small native population. It is estimated that, at the time of European discovery, only about 300,000 aborigines lived in all of Australia. This would give Australia a population density of about one person to 10 square miles (25.9 $km^2$) at that time. By comparison, New Jersey, the United States' most densely settled state, currently has a population density of over 1000 people per square mile (2.59 $km^2$), or 10,000 times that of Australia before European colonization. No wonder the British regarded Australia as nearly empty, an almost unoccupied land just waiting for colonization.

## PHYSICAL GEOGRAPHY

As the smallest continent, Australia is about three times the size of Greenland, the world's largest island, and about three-quarters the size of Europe. Australia remains by far the most lightly populated of the inhabited continents. (Antarctica has no really permanent population, only small groups of scientists who are present temporarily.)

Sheer distance always has been a factor in Australia's relations with the rest of the world. The most distant of the United Kingdom's former colonies, Australia's nearest neighbor is Indonesia, a country as different from Australia as it possibly could be in cultural terms.

Australia has a fairly compact form, with a much more regular coastline than that of Europe. Geologically stable for millions of years, Australia's only recent mountain-building activities have been along its east coast and a few isolated locations in the interior. New Zealand and most of the Pacific islands are much more recent geologically and generally more rugged.

Australia has a desert heart—in fact, the only parts *not* desert are the mild, summer-dry/winter-rain climates of the southwestern coast and parts of the southern coast, the temperate climates of the southeastern coast, and the tropical, seasonally dry climates of the far north. New Zealand has a cooler, rainy climate very similar to that of northwestern Europe. The many little islands of the Pacific lie mostly in the tropics; Hawaii's tropical climate is typical of most.

On the outline map provided in Figure 7–1 locate and label the following:

| | | | |
|---|---|---|---|
| South Pacific Ocean | | New Zealand, North Island: | 39° S, 176° E |
| Indian Ocean | | New Zealand, South Island: | 44° S, 170° E |
| Tasman Sea: | 29° S, 155° E | Great Barrier Reef: | 18° S, 147° E |
| Coral Sea: | 15° S, 155° E | New Caledonia: | 22° S, 166° E |
| Solomon Islands: | 8° S, 160° E | Great Australian Bight: | 33°S, 130° E |
| Arafura Sea: | 10° S, 135° E | Southern Alps (New Zealand) | 44° S, 170° E |
| Cook Strait; | 41° S, 174° E | Gulf of Carpentaria: | 15° S, 138° E |
| Timor Sea: | 13° S, 125° E | Nullarbor ("No Trees") Plain: | 32° S, 129° E |
| Australian Alps: | 37° S, 149° E | Flinders Ranges: | 32° S, 137° E |
| Bass Strait: | 40° S, 147° E | Great Dividing Range: | 37° S, 148° E |
| Torres Strait: | 10° S, 141° E | Great Sandy Desert: | 22° S, 125° E |
| Cape York: | 11° S, 143° E | Great Victoria Desert: | 30° S, 128° E |
| Tasmania: | 42° S, 147° E | | |

## OBJECTIVES AND STUDY HINTS

Several key geographic questions arise in a study of Australia, New Zealand, and the Pacific islands. Why did Europeans (and, on some islands, Americans) choose to establish colonies in these very different varieties of physical environments? What kinds of colonies were established in terms of ethnic makeup, political status, and economic base? Finally, how do these very different societies and environments function in today's highly competitive global economy?

Every piece of land in this region was a colony to one degree or another, and some in effect remain colonies today. There were three motives for acquiring colonies: (1) strategic/military, (2) complementary production, and (3) "mirror image."

The *strategic/military motive* is exemplified by Gibraltar; a fortress and naval base there control ship movements between the Atlantic Ocean and the Mediterranean Sea. In the South Pacific, islands functioned as naval bases in supplying ships with fresh water, stockpiled fuel, provisions, and ammunition.

The second motive, that of *complementary production,* usually involved tropical commodities. The European colonizing powers did not possess any truly tropical areas in their home territories. They could not grow cane sugar, natural rubber, cocoa, coconut, palm oil, or most spices at home. Tropical colonies were complementary to the European colonizers in that they could produce agricultural commodities that those countries needed. The complementary colonial economies functioned well when they had large populations already in place. Local inhabitants ("natives" to the colonial powers) formed the labor force to grow crops and to mine minerals. Not incidentally, they also formed a large market for the manufactured goods exported by the colonial power. Fiji and New Caledonia are good examples of this variety of colony.

But how would Australia be classified in the colonial era? Although northern Australia is tropical and now produces significant sugar cane and tropical fruit, this area and its potential for tropical crop production was virtually ignored in the colonial era. Its location, relatively remote from major sealanes linking Europe with other important colonies, made it unsuitable as a strategic base. It was

**FIGURE 7–1**
Australia and New Zealand.

the southern and southeastern coasts of Australia—where the climate was temperate midlatitude to subtropical—that attracted early European interests. And what kind of colony was New Zealand? Like the southern fringes of Australia, but even more so, New Zealand's climate was roughly comparable to that of Europe, not tropical. New Zealand, like the major British settlements in Australia, could not produce anything agriculturally that the United Kingdom did not already produce at home, and it too had little significance as a strategic base due to location. These colonies then were not complementary as were the colonized Pacific islands.

New Zealand and most of the coastal Australian colonies are, in fact, examples of a third type of colony. These new lands could become replicas of the home country—*a mirror image* of the home society and cultural landscape. They were reasonably similar to the home country in physical environment. They were not of particular strategic value in themselves, and their indigenous populations were relatively small and lived at a low level of technology. These lands could look like, function economically like, and feel like home.

**7–1** What other British colonies were essentially mirror-image colonies?

---

The mirror-image colonies had some economic problems peculiar to this look-alike relationship with the home country. The complementary-type colonies had a natural basis of trade with the parent country—production of commodities that the colonial power lacked at home. But the very similarity in physical environments between mirror-image colonies and home country meant that these mirror images produced approximately the same agricultural products. If the basis of trade was not complementary, then it had to be colonial production of minerals needed at home or production of considerably cheaper supplies of the same farm and ranch products available in the colonial power's domestic territory. The cost advantage of Australia and New Zealand in exporting agricultural goods to the United Kingdom was, and still is, complicated by the great distances involved. North and South America and Africa have the advantage of shorter distances, and therefore lower transportation costs, to British markets.

In studying Australia, New Zealand, and the Pacific islands, key considerations should be cultural contrasts with nearest neighbors; cultural similarities between Australia and New Zealand with faraway Britain, the United States, and Canada; and trade handicaps and opportunities with the rest of the world.

## POLITICAL GEOGRAPHY

The mirror-image colonies, largely populated by immigrant Europeans in early colonial days, literally were created in the image of the parent. Early on, the mirror images were granted a degree of self-rule and freedom to regulate themselves. An early mirror-image colony that was not given enough control over its own affairs soon rebelled and became the United States of America. Perhaps learning from this experience, the British allowed their other mirror-image colonies to evolve toward independence, rather than have to fight for it. Today, Australia and New Zealand are self-governing members of the Commonwealth of Nations.

On the outline map of the region in Figure 7–2, locate and label the place-names in Table 7–1.

**TABLE 7–1** Australia and New Zealand states/territories, capitals, and important cities.

| State/Territory | Capital | Other Important Cities |
|---|---|---|
| *Australia* | | |
| New South Wales | Sydney | |
| Northern Territory | Darwin | |
| Queensland | Brisbane | |
| South Australia | Adelaide | |
| Tasmania | Hobart | |
| Victoria | Melbourne | |
| Western Australia | Perth | |
| (Federal Capital) | Canberra | |
| *New Zealand* | Wellington | Auckland<br>Christchurch |

**FIGURE 7–2**
Australia and New Zealand: states and territories, capitals, and other important cities.

## The Pacific Islands

The Pacific islands include some of the last colonies to become independent and some of the few to remain colonial. Interestingly, these islands also were among the last areas of the tropical world to come under the colonial domination of the United Kingdom, France, Germany, the United States, and Japan. Many were unknown to people of other regions until the late eighteenth century, and many went unclaimed by any colonial power until the 1840s. On the outline map in Figure 7–3, locate and label all placenames shown in Table 7–2.

**TABLE 7–2** Countries, territories, and capitals of the Pacific Islands.

| Countries/Territories | Former Colonial Powers | Capitals |
|---|---|---|
| *Independent Countries* | | |
| Fiji | UK | Suva |
| Kiribati | UK | Tarawa |
| Marshall Islands | Germany; Japan; U.S. | Majuro |
| Federated States of Micronesia | U.S. | Palikir |
| Nauru | UK | Yaren |
| Palau (Belau) | Germany; Japan; U.S. | Koror |
| Samoa | Germany; NZ | Apia |
| Solomon Islands | UK | Honiara |
| Tonga | UK | Nuku Alofa |
| Tuvalu | UK | Funafuti |
| Vanuatu | UK; France | Port-Vila |
| *French Territories (overseas departments of France)* | | |
| French Polynesia | France | Papeete |
| New Caledonia | France | Noumea |
| *Japanese Territory* | | |
| Okinawa | Japan | Naha |
| *U.S. Territories* | | |
| American Samoa | U.S. | Pago Pago |
| Guam | U.S. | Agana |
| Commonwealth of the Northern Marianas | U.S. | Saipan |

*The U.S. administers many tiny, either sparsely populated or uninhabited Pacific Islands including Wake, Midway, and Johnston Islands, Kingman Reef, Howland, Jarvis, Baker, and Palmyra Islands.

Australia and New Zealand have a three-way mutual defense agreement with the United States. U.S. military power has replaced the UK as the primary shield for New Zealand and Australia against external threat.

## ECONOMIC GEOGRAPHY

Australia and New Zealand both have experienced diversification and maturation of their economies. Compared to the situation in the 1950s, the proportion of total exports accounted for by agricultural products has declined sharply for Australia and decreased somewhat for New Zealand. Australia has become a major exporter of iron ore, coal, and bauxite (aluminum ore) in addition to its traditional exports of wheat, wool, beef, and cane sugar from its tropical northeast. New Zealand remains the world's largest exporter of lamb, mutton, and dairy products and is second to Australia in wool exports.

There has been a reorientation of Australia's and New Zealand's trade relationships. In an atlas, look at a map of the Pacific basin and its coasts. Australia and New Zealand both export high-protein foods and fine-quality wool. Australia also exports metal ores and coal. The region's nearest neighbors, Southeast Asia and the Pacific island groups, are mostly too poor to buy from Australia and New Zealand.

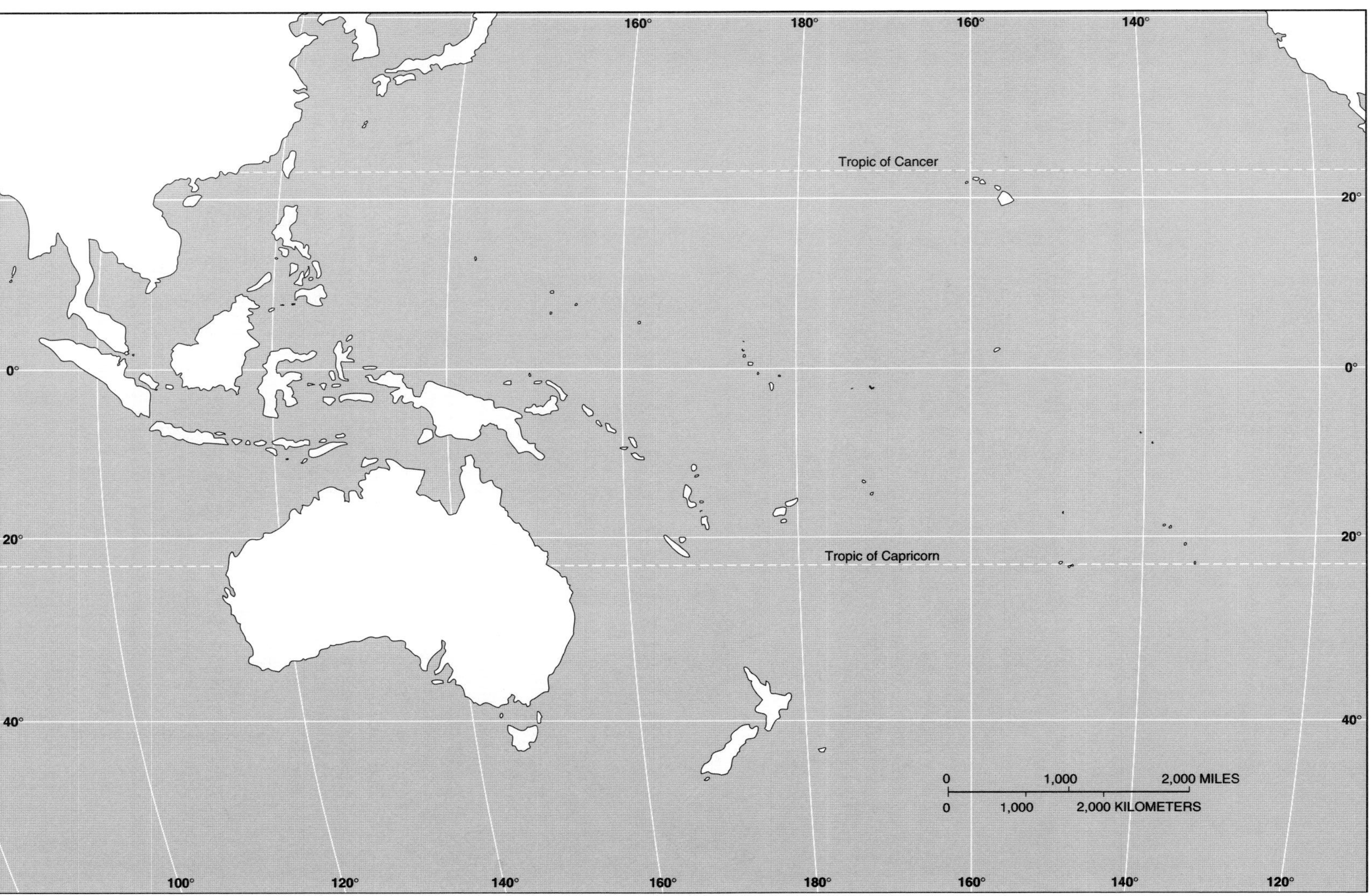


**FIGURE 7–3**
Countries and territories of the Pacific.

7–2 Who is the closest logical trading partner of Australia and New Zealand? (It is a highly industrialized country that needs industrial raw materials and that is upgrading its diet to include more animal products as living standards rise.) ______

______

Tourism is an important industry in most of the Pacific islands. It is notable how many Americans' images of "Paradise" find a perfect match with South Pacific island landscapes—sunny beaches fronting on peaceful lagoons of clear blue water, backed by palm trees and other lush tropical vegetation. However, as strikingly beautiful as the islands may be, they fall a little short of paradise for their inhabitants. Rapidly increasing populations must depend largely on tropical crop exports, all of which face heavy competition in world markets—copra (dried coconut meat), sugar, cocoa, coffee, and bananas. Tourism, too, is very competitive. Many islands offer essentially the same leisure amenities, but international tourism patterns can shift abruptly as new resorts develop, transport technology advances, and high prices or civil unrest discourage visitors at specific places.

7–3 What other area of many islands in a tropical sea has similar economic reliance on tourism, sugar, coffee, bananas, and the like? ______

______

## CULTURAL GEOGRAPHY AND DEMOGRAPHICS

On the outline map of the region in Figure 7–4, draw a line enclosing the large island of New Guinea (6° S, 140° E), the Admiralty Islands (2° S, 147° E), the Bismarck Archipelago (3° S, 151° E), the Solomon Islands (7° S, 148° E), and Fiji (19° S, 175° E). This is *Melanesia,* the *dark islands,* populated by dark-skinned peoples related to Australia's Aborigines. Draw another line around the islands north of Melanesia and east of the Philippines. These include the Caroline Islands, Marianas, and Marshall Islands; they are *Micronesia,* or the *tiny islands,* and are inhabited by people with light brown skins. Then, draw a line from New Zealand, to Hawaii, to Easter Island, and back to New Zealand. This huge triangular ocean and island region is known as *Polynesia,* or *many islands.* Polynesian people are brown-skinned; "Polynesians" are as much a language group as a distinctive racial group.

About 15% of New Zealand's population is Maori, a Polynesian people. Aside from the Maori and Australian Aborigines, the populations of New Zealand and the "island continent" (Australia) are growing at the low rates typical of highly industrialized and highly urbanized societies.

On the outline map of Australia in Figure 7–5, draw a line from Gladstone, Queensland, inland 200 miles (322 km). Then go southward, paralleling the seacoast, to Canberra. Then continue the line northwestward, again paralleling the seacoast, to Port Pirie on South Australia's coast. Start another line in western Australia at the southwestern tip, West Cape Howe, then northward, paralleling the seacoast about 100 miles (161 km) inland to a point about 200 miles (322 km) north of Perth, again bringing the line to the coast. You have separated the nearly empty interior from the much more densely settled coastal strips.

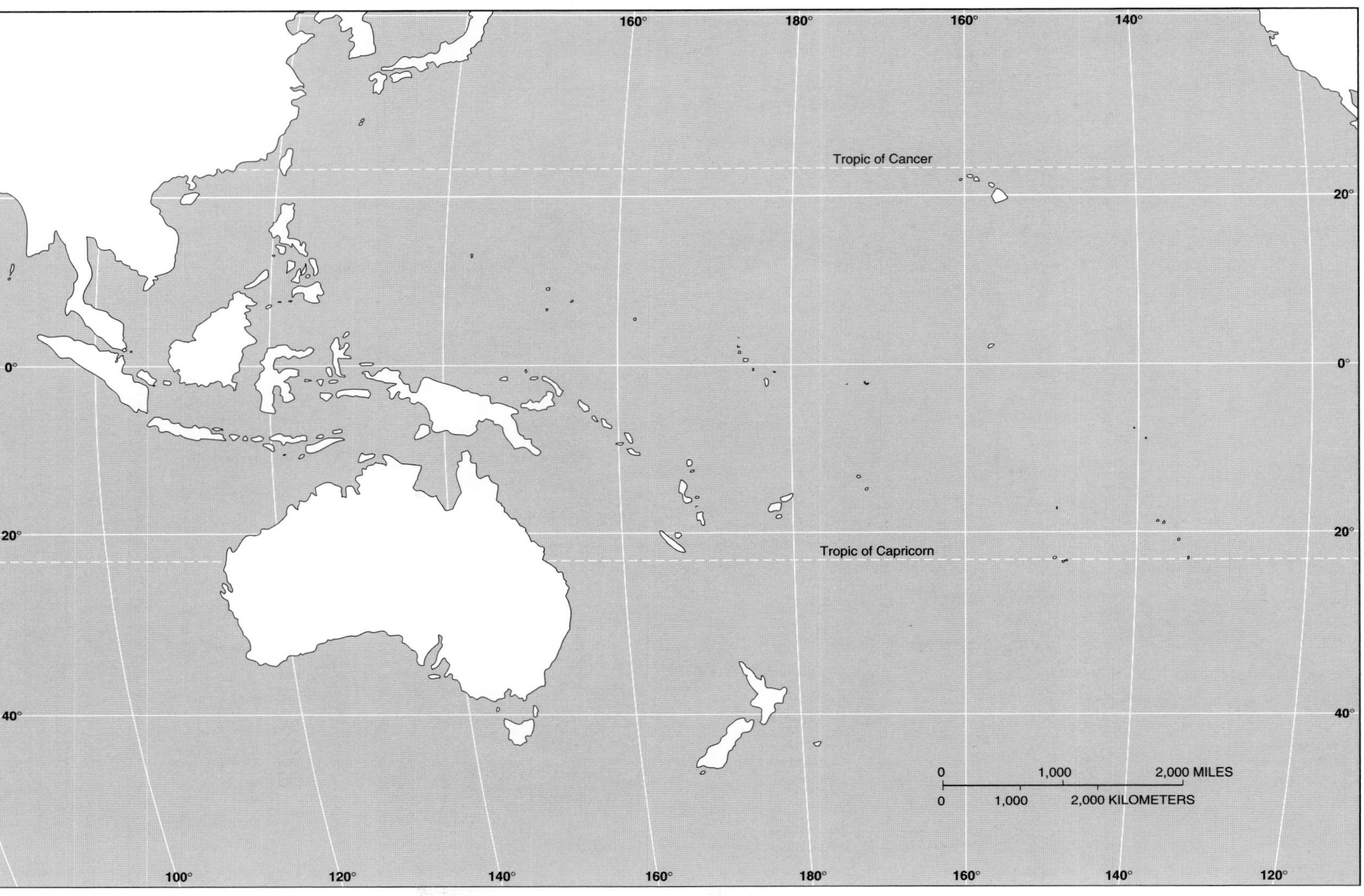


**FIGURE 7–4**
Polynesia, Melanesia, and Micronesia.

**FIGURE 7–5**
Australia.

## CHECK UP

**7–4** Australia's only island state is ______________. New Zealand's capital, ______________, is located on ______________ island.

**7–5** At about 42° S, Wellington is the world's southernmost national capital. How many other national capitals lie south of 30° S? ______________

**7–6** The only large part of Australia not a state of that federation, ______________ territory, is not a state because it has so few inhabitants.

**7–7** What other large federal nation also has territories for the same reason very light population? ______________

**7–8** About when would it become spring in Sydney? (Remember which hemisphere Sydney is in.) ______________________________

**7–9** Seabathers at Perth's City Beach would be swimming in which ocean?

______________________________

**7–10** If a cruise liner sailed from Sydney heading for the nearest tropical Polynesian island, it would go to which of the following islands? (Circle the correct answer.): Tasmania, Fiji, Port Moresby, Samoa Islands, Yap Islands.

**7–11** Australia's geographically closest neighbors that are *not* part of the developing world would be ______________________________

**7–12** Australia and New Zealand's own antipodes would be which of the following areas? (Circle the correct answer.): Japan, Southeast Asia, Hawaii, South Africa, northwestern Europe.

**7–13** Logically, most of Australia's domestic production of cane sugar and tropical fruits would come from which of the following? (Circle the correct answer.): Central Australia, South Australia, Queensland, Tasmania.

**7–14** Referring to Figure 7–4, label whether the following islands belong to Melanesia, Micronesia, or Polynesia. Marquesas Islands: 9° S, 141° W; ______________; Palau: 7° N, 143° E; ______________; Guadalcanal; 10° S, 158° E; ______________; Guam: 14° S, 143° E; ______________; Tahiti: 17° S, 149° W; ______________; Bougainville: 6° S, 155° E; ______________.

# GEOCONCEPTS

## Choosing a Capital

In the colonial era, each of the Australian colonies of the UK had its own colonial administrative center. Each was a seaport for convenience of communication with London. In time, each colonial capital became the largest city in its colony, and each large city fostered a local railway network focused on itself. The two largest colonial capitals (state capitals after independence) were Sydney and Melbourne. Either could have become the national capital in terms of being a large city with excellent transport and communications facilities. Nearly the same size (Sydney is slightly larger), the two were strong rivals. But Victoria State would never agree to Sydney as the new nation's capital, and New South Wales could not accept Melbourne as the national seat of government. Federal states whose member states or provinces have rivalries or strong cultural differences commonly choose a neutral national capital location to avoid favoring one contender over another. So, Canberra became the Australian capital territory. It is located in the Dividing Range of mountains, roughly between Sydney and Melbourne.

**7–15** Actually, Canberra is a little closer to one of the two state capitals than the other. Look at the map. Was Sydney, the larger city, able to influence Canberra's location to make it more accessible from Sydney?

______________________________

**7–16** Name two other federal capitals located in relatively small cities between contending regions. ______

## Urbanization

Australians are a highly urbanized people. Whereas about three-quarters of U.S. citizens are classified as urban rather than rural, the comparable figure for Australia is 85%. Over half of Australians live in cities exceeding 100,000 population. Whereas popular images of Australia focus on the ranchers and farmers of a rugged frontier, Australians are actually more likely to live in big cities and their suburbs than are people living in the United States.

The 1987 map of world urbanization in Figure 7–6 shows relatively wealthy Australia and New Zealand as heavily urban. Relatively poor Southeast Asia is less urban.

**7–17** Which were the most urban countries in East Asia? ______

**7–18** Were these also relatively poor countries? ______

**7–19** Which were the least urbanized countries in Europe? ______

**7–20** Were they relatively rich or poor compared to the European average? (Consult Chapter 5 on Europe.) ______

**7–21** Were there any exceptions on the map to the generally positive correlation between percentage urban and income level? ______

**7–22** How might this be explained? ______

In a world of rapid urbanization, the 1987 data used in Figure 7–6 might be expected to be out of date. Table 7–3 categorizes countries by percentage of the population urban, using the same percentage groupings as Figure 7–6, but with 2000 data. Using a blank outline map, Figure 7–7, construct a 2000 map to compare with the 1987 data map, Figure 7–6.

**7–23** As a generalization, which major world region, as identified in this workbook, is the least urban? ______

**7–24** Which Scandinavian country appears to have become less urbanized by the year 2000? ______

**7–25** Has there been any change, 1987 to 2000, in South American countries in the top category (most urbanized)? ______

**7–26** Which of these large countries (in territory) has/have become more urbanized from 1987 to 2000. (Circle the correct answer.): China; Canada; Russia; USA ______

**7–27** Which mainland Southeast Asian country apparently has become more urbanized, 1987 to 2000: ______

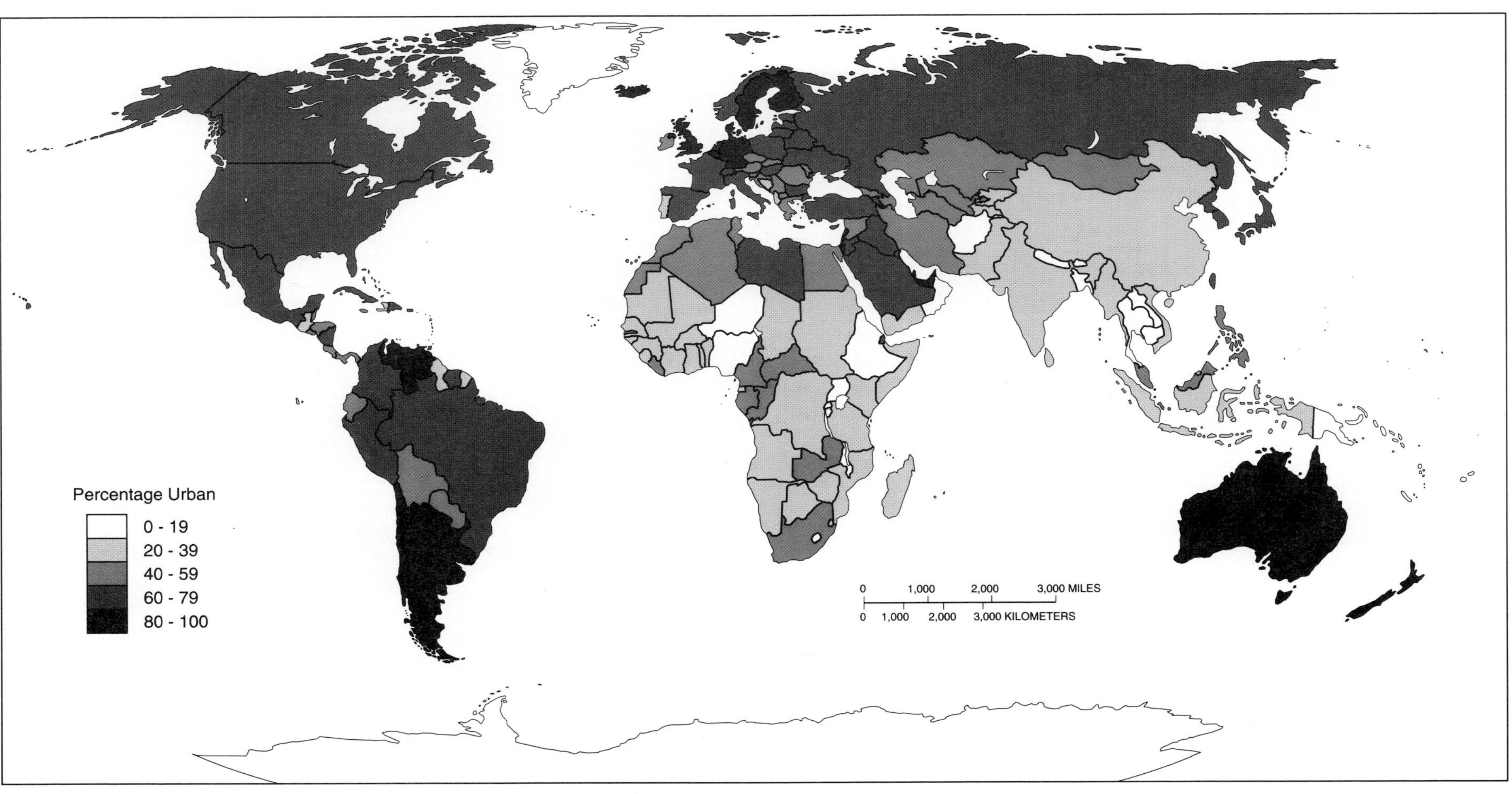


**FIGURE 7–6**
World urbanization, 1987. (Source: Data from Population Reference Bureau, 1987)

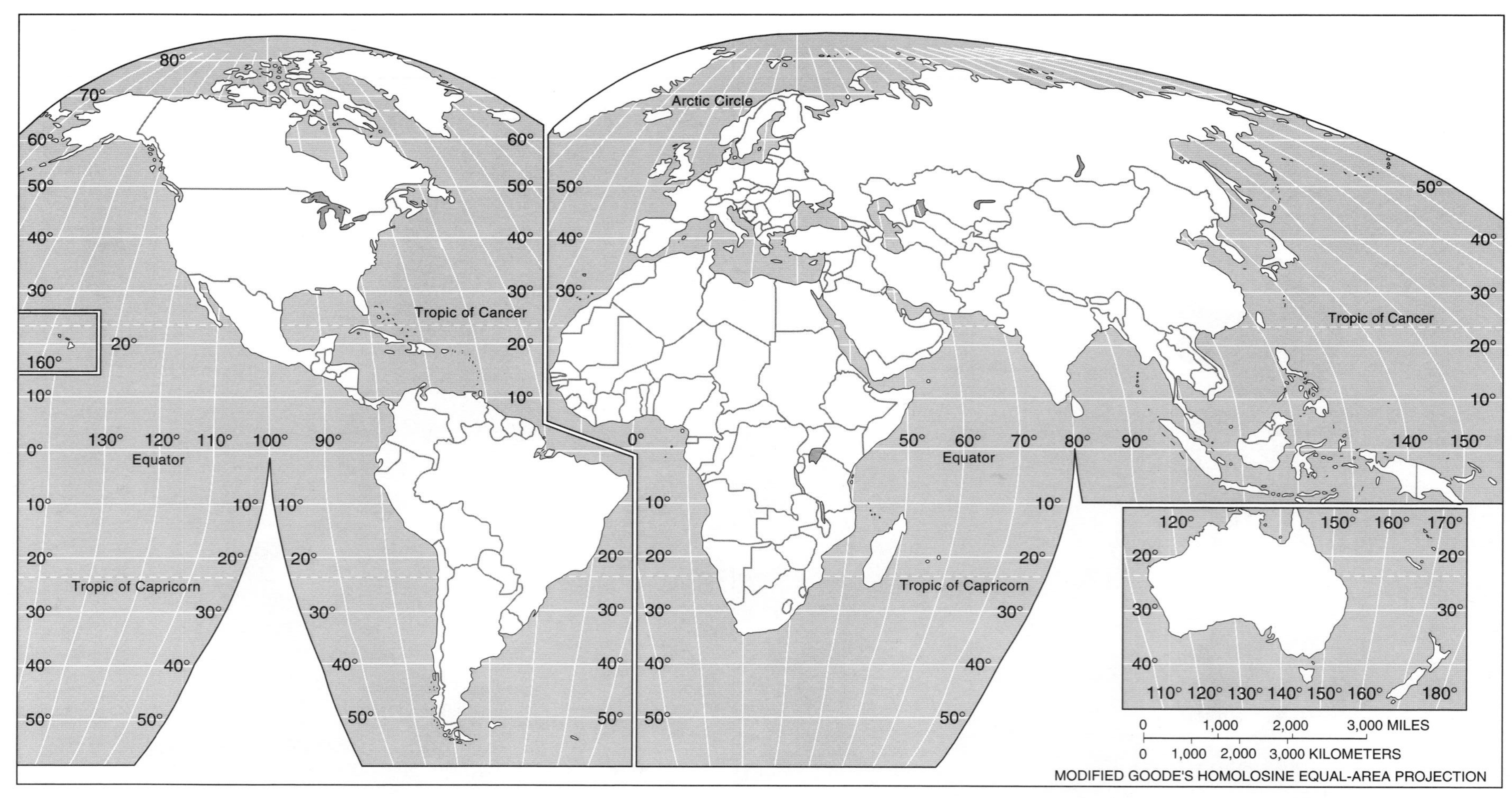


**FIGURE 7–7**
World urbanization, 2000.

**TABLE 7–3** Percentage of total population living in urban centers, 2000.

| | | |
|---|---|---|
| **19% or Less** | | |
| Burkina Faso | Lesotho | Rwanda |
| Niger | Bhutan | Uganda |
| Burundi | Nepal | Laos |
| Eritrea | Cambodia | Papua New Guinea |
| Ethiopia | | |
| **20% to 39%** | | |
| Sudan | Kenya | Democratic Republic of Congo |
| Benin | Madagascar | Equatorial Guinea |
| Gambia | Malawi | Namibia |
| Ghana | Mozambique | Swaziland |
| Guinea | Somalia | Guatemala |
| Guinea-Bissau | Tanzania | Tonga |
| Mali | Zambia | Chad |
| Nigeria | Zimbabwe | Vietnam |
| Sierra Leone | Angola | China |
| Togo | Central African Republic | Thailand |
| Haiti | Pakistan | India |
| Guyana | Sri Lanka | Kyrgyzstan |
| Yemen | Tajikistan | Indonesia |
| Afghanistan | Uzbekistan | Myanmar |
| Bangladesh | | |
| **40 to 59%** | | |
| Algeria | Belize | Mongolia |
| Egypt | Costa Rica | Ireland |
| Morocco | El Salvador | Moldova |
| Cape Verde | Honduras | Portugal |
| Cote D'Ivoire | Panama | Slovenia |
| Liberia | Jamaica | Yugoslavia |
| Mauretania | Paraguay | Fiji |
| Senegal | Azerbaijan | Slovakia |
| Cameroon | Georgia | Albania |
| Congo | Syria | Bosnia-Herzegovina |
| Botswana | Kazakhstan | Croatia |
| South Africa | Turkmenistan | Greece |
| Philippines | Malaysia | Macedonia |
| North Korea | Romania | New Caledonia |
| **60 to 79%** | | |
| Tunisia | Norway | Surinam |
| Gabon | Austria | Armenia |
| Canada | France | Cyprus |
| USA | Netherlands | Iraq |
| Mexico | Switzerland | Jordan |
| Nicaragua | Belarus | Lithuania |
| Cuba | Oman | Bulgaria |
| Dominican Republic | Turkey | Czech Republic |
| Puerto Rico | Iran | Hungary |
| Bolivia | Brunei | Poland |
| Brazil | Japan | Russia |
| Columbia | South Korea | Ukraine |
| Peru | Taiwan | Spain |
| Equador | Estonia | Latvia |
| French Guiana | Finland | |
| **80% and Above** | | |
| All the rest | | |

## SPECIAL CHALLENGE

Third-World developing countries tend to have faster-growing populations than do highly industrialized, highly urbanized societies. (See the geoconcept, "The Population Explosion," in Chapter 9.) Australia and New Zealand are growing at relatively modest rates. Most of this region's Southeast Asian neighbors, already densely populated, are growing at much faster rates (Table 7–4).

**TABLE 7–4** Population and increase rate comparison.

| Country | Present Annual Rate of Population Growth (Natural Increase) | Estimated Additional Population by 2010 (Millions) |
|---|---|---|
| Australia | 0.6% | 3.6 |
| New Zealand | 0.8% | 0.6 |
| Indonesia | 1.6% | 61.2 |
| Malaysia | 2.1% | 13.7 |
| Philippines | 2.3% | 37.0 |
| Thailand | 1.0% | 10.1 |
| Myanmar (Burma) | 2.0% | 19.1 |
| Laos | 2.6% | 3.2 |

**7–28** Should Australia and New Zealand greatly increase their immigration quotas from Southeast Asia? Why or why not? ____________________

**7–29** Would Australia's admission of the entire net gain in population from Southeast Asian countries during 1997 to 2010 solve the population-pressure problems of Southeast Asian states? ____________________

________________________________________

**7–30** What would happen to the ethnic/racial composition of Australia and New Zealand if they absorbed Southeast Asia's net gains in population?

________________________________________

________________________________________

**7–31** How might such huge increases in total population affect employment and ratios of farmland to population in Australia and New Zealand?

________________________________________

________________________________________

## REGIONAL WATCHLIST

Australia seems likely to further deemphasize its largely symbolic ties with the United Kingdom. Although this will not significantly change the political geography of this region, it may reflect an increasingly obvious, fundamental economic reorientation. Australia in particular, but also the rest of this region, is conscious of an emerging Pacific basin economy, replacing its former imperial ties with the UK. Industrial raw materials, food, energy, manufactured products, capital, and tourists now travel in ever-increasing volumes among the countries of this region and those of East Asia, Southeast Asia, the United States, Canada, and Latin America.

# 8

# East Asia

## INTRODUCTION

Asia is the giant of continents, in both territory and population. Its billions of people and splendid variety of cultures, countries, and economies can be better understood by dividing it into three Asian cultural regions (not counting Siberia, or Asiatic Russia, which was covered in Chapter 6): East Asia, South Asia, and Southeast Asia.

The *East Asia region,* as used in this workbook, includes the People's Republic of China (PRC), the Republic of China (island of Taiwan), Mongolia, North Korea (Democratic People's Republic of Korea), South Korea (Republic of Korea), Japan, and the colonial relics of Hong Kong and Macau, now merged with China. Outline the East Asia cultural region on the outline map in Figure 8–1.

In size of territory, the People's Republic of China is third-largest in the world, and the one-time Portuguese colony of Macau, about six square miles (15.5 $km^2$) in area, was one of the globe's smallest political units. Highly industrialized Japan is the region's wealthiest state and the only nation in Asia to have achieved levels of technology and prosperity fully comparable to western Europe and the United States. The People's Republic of China, by contrast, has a per capita gross national product (each person's average share of the money value of all goods and services produced in the nation) that is less than 2% that of Japan's. South Korea is one of the world's fastest-expanding economies, and landlocked Mongolia is only in the earliest stages of industrialization. Hong Kong prospers at about two-thirds the per capita GNP of Japan, but the merger with the People's Republic in 1997 produces an uncertain future for this great trading and manufacturing center. Both Hong Kong and Macau are "special administrative regions" (SARs) of the People's Republic of China.

One of East Asia's unifying features is the tremendous influence of China on its neighbors. Most East Asians read a written language based on Chinese ideographs (symbols of objects, concepts, and emotions). Basic Chinese technologies in metalworking, pottery, weapons, transport, architecture, and agriculture were adapted by most regional neighbors many centuries ago.

For many reasons, we look across the Pacific to East Asia with strong interest. We long have been fascinated by the "mysterious Orient"—the ancient civilization of China, its industrious people, and its huge potential as a trading partner. Japan has become one of the largest merchandise suppliers to the United States, building the world's largest trade surplus, to the occasional annoyance of its best customer.

On the outline map of East Asia in Figure 8–1, locate and label the nations and capitals shown in Table 8–1.

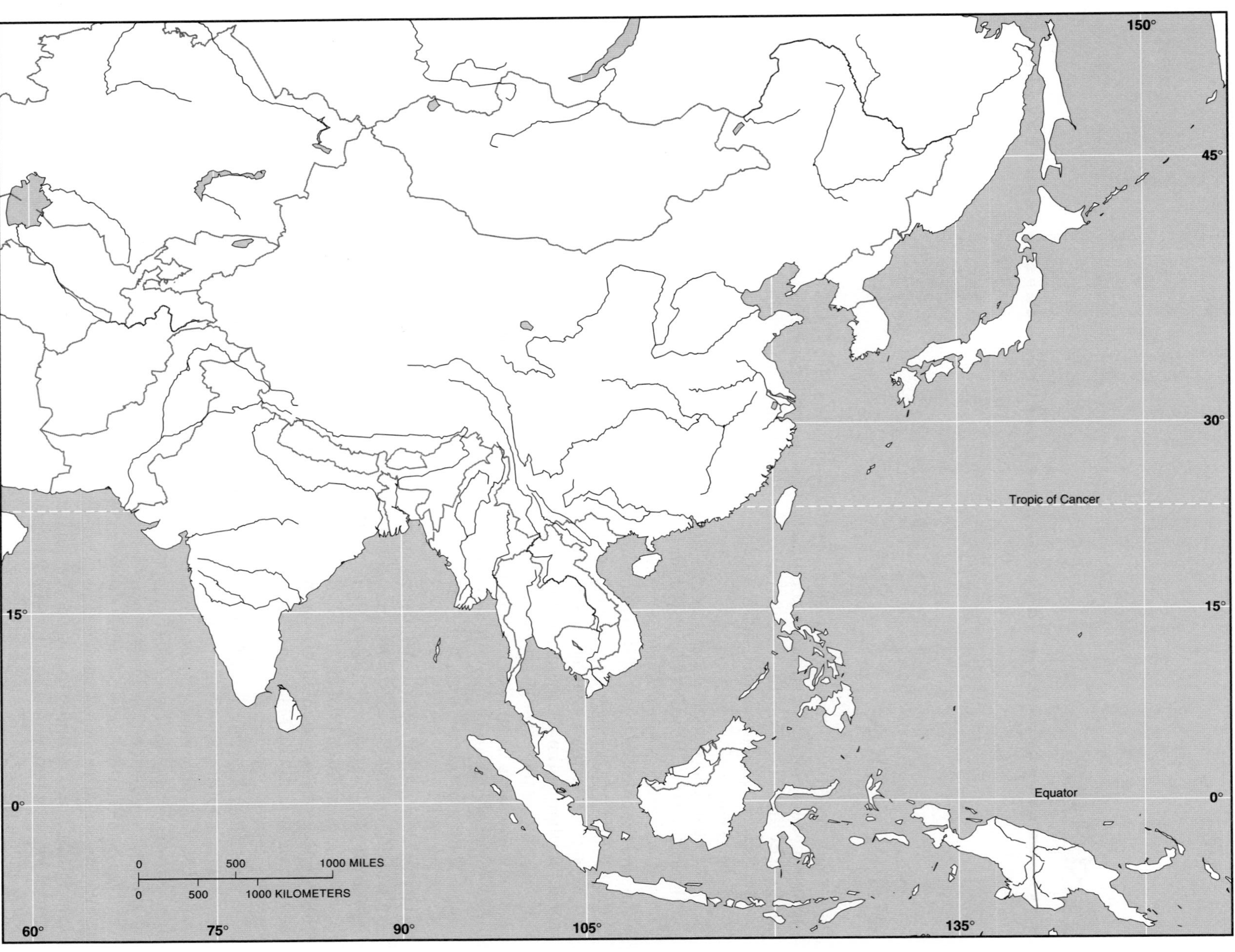


**FIGURE 8–1**
East Asian cultural region.

**TABLE 8–1** East Asian nations and capitals.

| Country | Capital |
|---|---|
| People's Republic of China | Beijing (Peking) |
| Republic of China (Taiwan) | Taipei |
| Japan | Tokyo |
| North Korea (Democratic People's Republic of Korea) | Pyongyang |
| South Korea (Republic of Korea) | Seoul |
| Mongolia | Ulaanbaatar (Ulan Bator) |

## PHYSICAL GEOGRAPHY

This major region's northern boundary is with Siberia; its southern lands border a tropical sea. One of the world's driest deserts lies in western China, while heavy summer rains help produce crops of bananas and pineapples in Taiwan, and southern China grows two crops of rice a year, along with tea and subtropical fruits.

The boundaries of mainland China (PRC) with India, Nepal, Bhutan, Pakistan, Afghanistan, Russia, Kazakhstan, Kyrgyzstan, and Tajikstan lie along some of the most rugged mountain ranges in the world. Geographers used to talk about "natural boundaries," by which they meant political boundaries that coincided with an obvious physical feature that served as a natural barrier to cultural and economic interaction. States that occupy islands or island groups, such as Japan, New Zealand, or Cuba, are said to have natural boundaries. Rugged mountains, which once restricted movement across them and were unproductive areas with few inhabitants, made relatively stable, trouble-free borders, as in the Pyrenees Mountains along the French–Spanish border. Borders that crossed flat lowlands, such as the mutual borders of Germany/Poland and Cambodia/Vietnam, seem to be unusually troublesome, with frequent realignments resulting from invasion and war.

However, natural boundaries of high mountains are not necessarily trouble-free. China and India both claim as their own some territories along their rugged borders, to the east of Bhutan and in India's northernmost territory of Kashmir. On the other hand, Japan's natural boundaries as an island nation have been relatively stable over the centuries.

On the outline map of East Asia in Figure 8–2, locate and label the following:

**Pacific Ocean**

| | | | |
|---|---|---|---|
| Sea of Japan: | 40° N, 135° E | Himalaya Mountains: | 28° N, 90° E |
| Yellow Sea: | 37° N, 123° E | Karakoram Range: | 35° N, 76° E |
| East China Sea: | 30° N, 125° E | North China Plain: | 33° N, 118° E |
| South China Sea: | 20° N, 115° E | Szechwan (Sichuan) Basin: | 30° N, 105° E |
| Taiwan (Formosa): | 24° N, 121° E | Manchurian Plain: | 45° N, 125° E |
| Tarim Basin: | 40° N, 84° E | Greater Khingan Mountain Range: | 46° N, 120° E |
| Japanese islands of Honshu, Hokkaido, Kyushu, Shikoku: | 36° N, 138° E | Huang Ho (Yellow River): | 35° N, 115° E |
| Shandong (Shantung) Peninsula: | 37° N, 122° E | Xiz River: | 23° N, 112° E |
| Korea Strait: | 34° N, 128° E | Chang Jiang (Yangtze River): | 31° N, 110° E |
| Formosa Strait: | 24° N, 119° E | Kanto (Kwanto) Plain: | 36° N, 140° E |
| Korean Peninsula: | 38° N, 127° E | | |

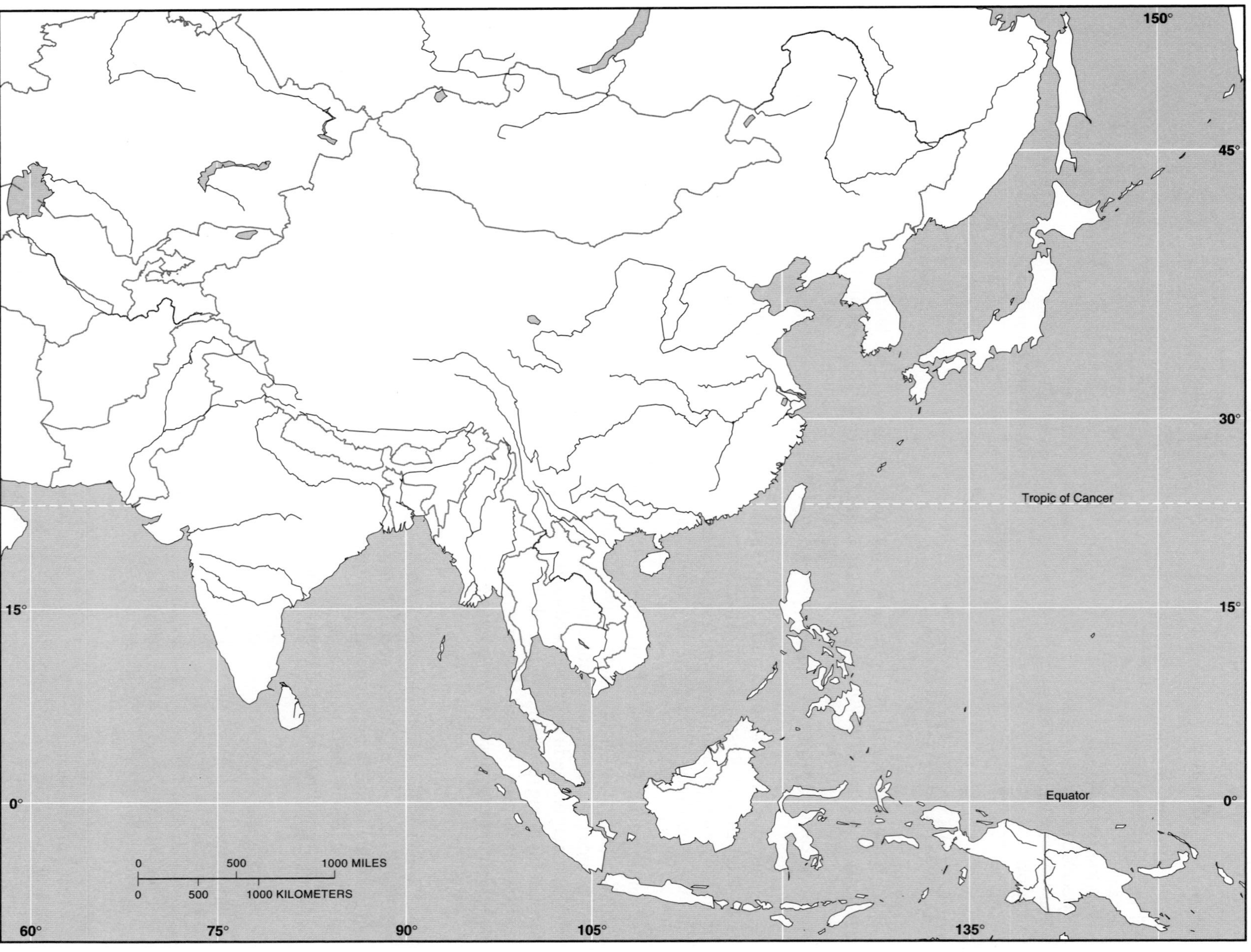


**FIGURE 8–2**
East Asia.

| | | | |
|---|---|---|---|
| (Japan's) Inland Sea: | 34° N, 133° E | Hainan Dao (Island): | 19° N, 110° E |
| Sea of Okhotsk: | 47° N, 146° E | Okinawa Island: | 27° N, 128° E |
| Tyan Shan Mountains (Tien Shen): | 36° N, 118° E | Gobi Desert: | 44° N, 110° E |
| | | Wuhan Basin: | 31° N, 114° E |

# OBJECTIVES AND STUDY HINTS

There are several interesting geographic interactions to watch in the East Asia region. The distribution of population appears to be strongly influenced by topography and climate. Until this century, East Asia's people were mostly farmers, sensitive to the agricultural usability of the land. Flat land, or at least land in reasonably gentle slope, was most desirable as cropland. To produce crops, there had to be an adequate water supply—precipitating from the sky and/or supplied by irrigation canals. As a result, most Chinese live in eastern China, on the North China Plain, the lower Chang-Jiang Basin, the well-watered South China Hills, the interior Wuhan and Szechwan basins, and the Manchurian Plain. Similar to the United States, China had a "wild west"—western lands that were drier and more sparsely settled than in the east. China's ethnic minorities—Mongols, Tibetans, Turkic peoples, and Vietnamese—tended to live in these southern and western border areas. Because of a shortage of water and level land, western China's mountains and deserts are used to graze animals rather than plant crops.

Other areas of eastern Asia have similar problems of topography and climate. The western portion of the Korean peninsula is lower and less rugged than the eastern mountain spine, as is reflected in population distribution. Taiwan also has a chain of steep mountains along its east coast and gentler gradients to the west. About 80% of Japan is in steep, forest-covered mountains causing most Japanese to live in the pockets of flat land around the coast where rivers exiting to the sea have built small plains. All of Japan's major cities are located close to the sea, allowing most of Japan's industry to enjoy the advantages of easy access to ocean shipping, both to import raw materials and to export manufactured goods. Note that the largest coastal plain, the Kanto Plain, is the site of Tokyo, Japan's largest city and one of the largest on Earth. Another geographic theme to watch in studying East Asia's place geography is that of the changing directional flows of new technology and new ideas. For most of this region's history, new ideas and technologies flowed outward from the center, eastern China, toward the periphery, such as Japan and Korea. Eastern China contains the world's oldest continuous civilization. At one time, Chinese technologies were superior to, or at least equal to, any in the world.

In modern times, however, the flow has reversed. Japan certainly is the regional leader in developing and applying new technology. Japan was the first Asian nation to industrialize, and rapidly industrializing South Korea, Taiwan, and Hong Kong all enjoy higher levels of modernization and personal incomes than the PRC. Once the cultural source for East Asia, China now is in the position of slowly adapting to the new world of high-technology, most of which was developed elsewhere.

## Bringing Chinese Placenames into the 21st Century

Is it Peking or Beijing? It's Beijing if we want to use the modern, official Chinese pronunciation. Older books and atlases use the traditional forms of Chinese placenames, such as Peking, Canton, and Chungking. But the names of these cities now are given as Beijing, Guangzhou, and Chongqing, respectively. The new official pronunciation more accurately reflects the northern dialect of spoken Chinese.

## POLITICAL GEOGRAPHY

The People's Republic of China is the world's largest *unitary state.* Although its territory is divided into provinces, these provinces do not constitute another level of government with legislative powers. The world's other large countries, faced with administering huge territories with varied physical and cultural geography and strong regional interests, have organized as federal states: The USSR, Canada, the United States, Australia, India, and Brazil all chose the federal system. China's reliance on the highly centralized unitary state, with all legislative, executive, and judicial decisions and policy determined at Beijing, may reflect the reaction to a recent history of central government weakness and internal chaos.

There seems to have been a repeating cycle in Chinese history. A powerful central government takes firm control, but then gradually weakens. This allows the rise of provincial rulers (warlords) who were not responsive to direction from the national capital. Typically, this situation was followed by a new resurgence of power at the national government level. In the Chinese experience, decentralization of power to the provinces was a result of disintegration of the national government. It is no wonder that the Chinese are uneasy about granting the provinces some powers of local government.

The PRC's real territorial grievances lie on the Indian and Russian borders. In the Chinese view, the Russian empire took Chinese territory in the late nineteenth century, taking advantage of temporary Chinese weakness.

On the outline map in Figure 8–3, shade the area of Russia east of the northward-flowing Amur River. This, to the Chinese, is an *irredenta,* an "unredeemed" territory; it *was* Chinese historically, it *should be* Chinese now, and the implication is that it *will be* Chinese territory in the future. Shade a zone about 100 to 200 miles (161 to 322 km) deep into Russian territory along the border from the Amur Bend at Khabarosk westward to the Mongolian border as another, more vaguely defined *irredenta.* (Use the map scale to estimate how wide a strip would represent 100 to 200 miles, or 161 to 322 km.)

Shade all of Mongolia, in a contrasting color or pattern, as a former *tributary state* or satellite of China. Tributary states were those not directly ruled by China, but that paid tribute money to the Chinese emperor to acknowledge Chinese power and protection. Draw a series of short, dashed lines across the Chinese–Indian border from Pakistan to Nepal and from Bhutan to Burma to indicate that these borders are not settled to both India's and China's satisfaction.

For centuries, Mongolia has alternated from Chinese domination to Russian/Soviet domination; it is now an independent country, but heavily dependent on Russian aid and trade. Ironically, the Mongols once ruled from the coasts of China all the way west into Hungary and Poland. Mongol horsemen once terrorized both the Chinese and the Russians; their ferocity encouraged the building of China's Great Wall.

Korea is one nation, but two states. North and South Korea reflect the Cold War tensions between the United States and the former USSR. Korea is the Asian counterpart of once-divided Germany in Europe—a result of western and Soviet occupation zones at the end of World War II. As with Germany, Korean reunification seems inevitable, eventually.

## ECONOMIC GEOGRAPHY

Using the outline map of the region in Figure 8–4, construct an income map using GNP per capita data from the Appendix. Use three colors. For GNP per capitas over $20,000, use a very dark color. This is a rich country comparable with the

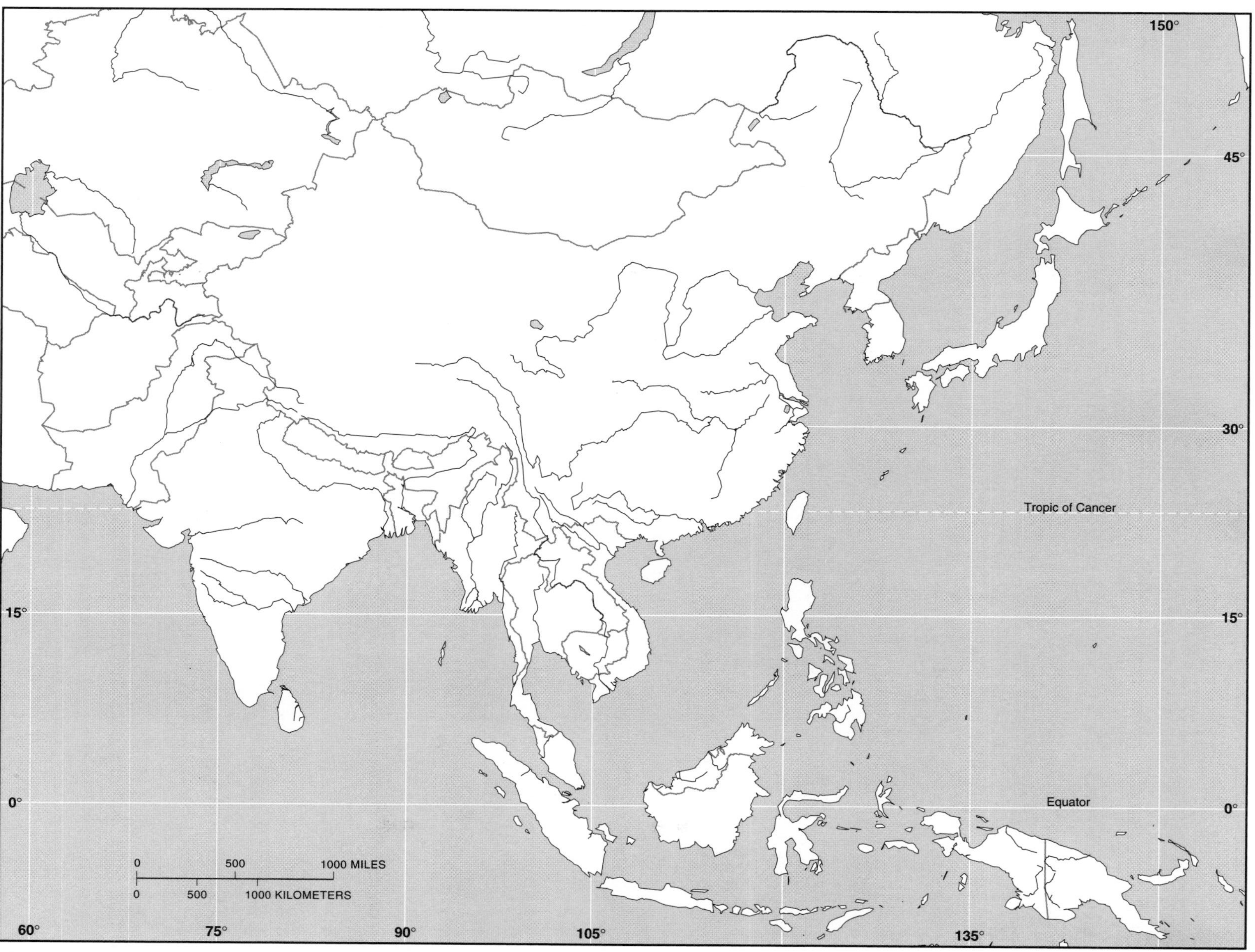


**FIGURE 8–3**
Potential border disputes.

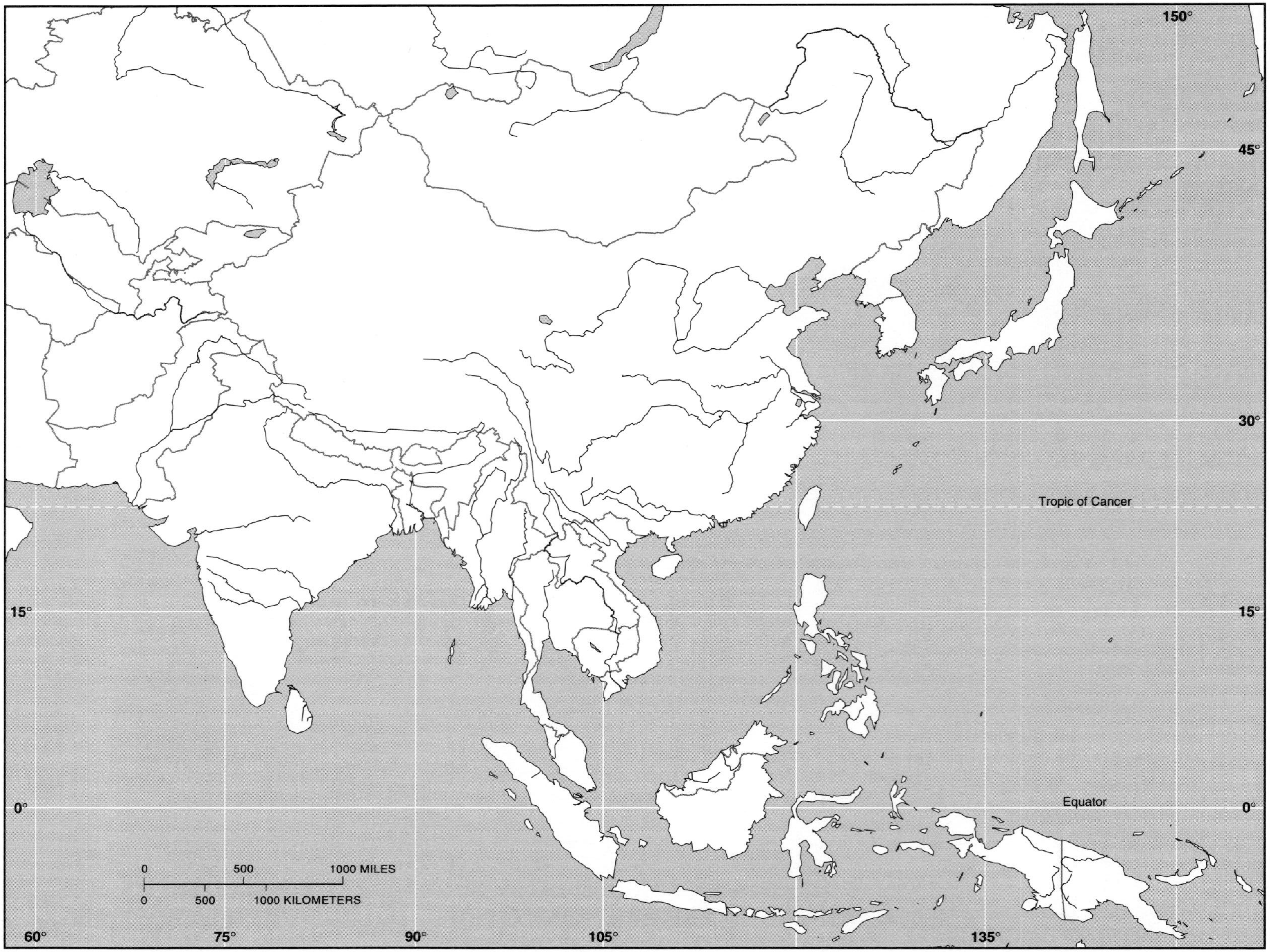

**FIGURE 8–4**
GNP per capita per year.

United States, Canada, or Western Europe. Using a lighter color, indicate which political unit(s) fall into the $5000–$19,999 category. Use the lightest color to designate countries in the $1000–$4,999 group. Leave white those national economies below $1000 GNP per capital per year.

**8–1** List some nations in other regions that have rich economies, some that have middle-class economies, and some that have poor economies.

______

______

______

## CULTURAL GEOGRAPHY AND DEMOGRAPHICS

It will be demonstrated in Chapter 9 that strong links exist between degree of industrialization and rate of population increase. Again referring to data tables in the Appendix, use the outline map (Figure 8–5) to record average annual rates of population growth for the recent past. Relatively rich countries tend to have low population increase rates, whereas poor countries usually are characterized by higher rates of population increase.

**8–2** Which East Asian country has the lowest rate of population increase?

______

**8–3** Is it also the richest country in the region? (Check the maps you made in the Economic Geography section.) ______

______

Judging by level of GNP per capita, the PRC's population increase rates should be similar to those of other poor countries. Compare the PRC's rates of population increase to those of countries of similar GNP per capita, like India, Central African Republic, Pakistan, and Ghana.

**8–4** Is the PRC's population growth rate more typical of wealthy or poor countries? Does the growth rate suggest that the Beijing government has worked effectively to influence the birth rate? ______

______

______

**8–5** If low rates of population increase seem to be a consequence of industrialization (see Chapter 9), will increased industrialization be a product of lowered population growth rates? ______

______

## CHECK UP

**8–6** Two large islands lie off the coast of China and were considered part of the old Chinese empire. Which is administered by the PRC and which claimed independent status as the Republic of China? ______

______

**8–7** Which is the region's only landlocked state? ______

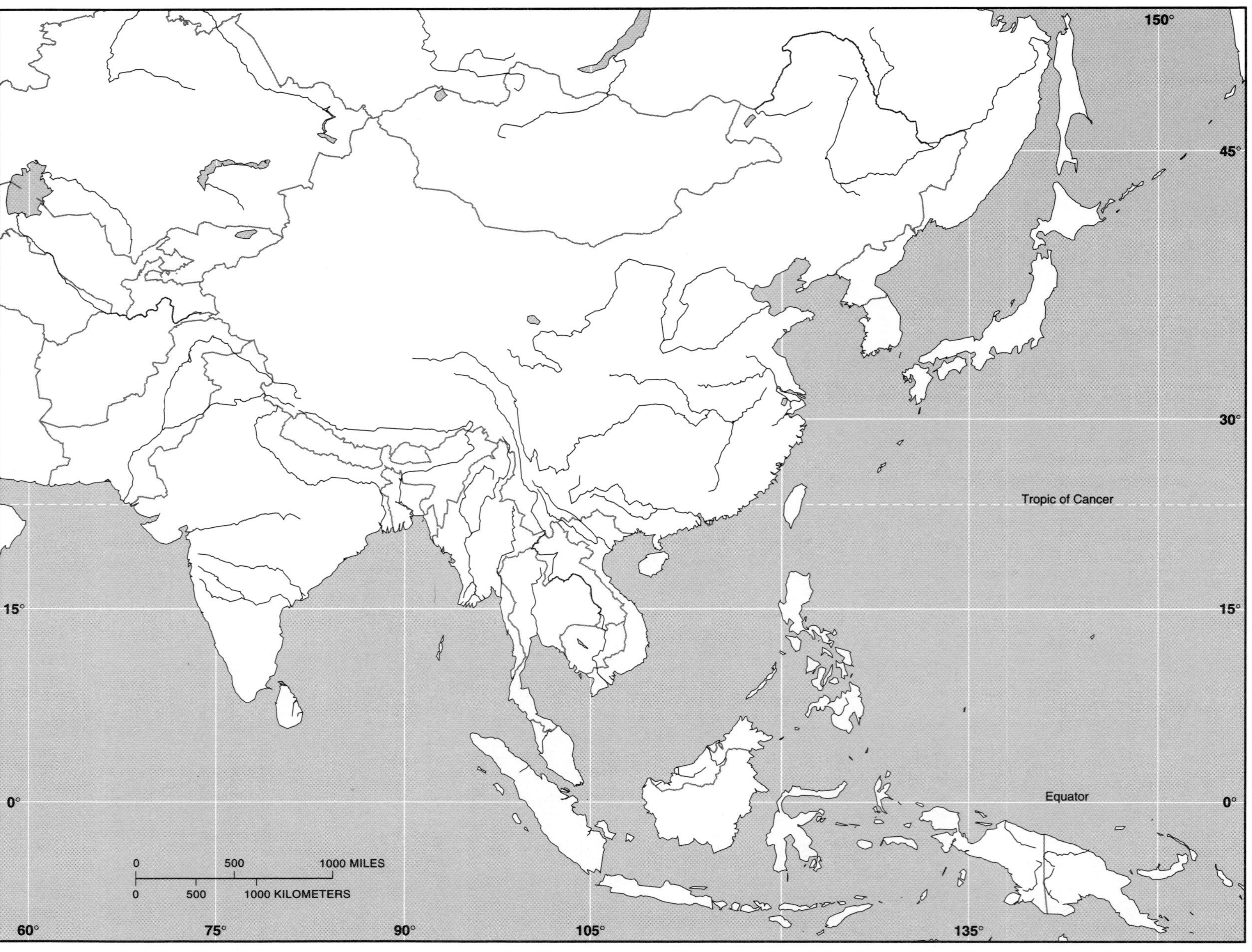


**FIGURE 8–5**
Annual rates of population growth.

**8–8** Which two of this region's countries have extensive land borders with Russia? ____________

**8–9** Russia's major port and naval base at Vladivostok is located on land once claimed by which East Asian country? ____________

**8–10** Name the four main islands of Japan: ____________

**8–11** Which three seas border the PRC? ____________

**8–12** Which two political units in East Asia are relics of European colonial interests? ____________

**8–13** Other than those two non self-governing territories, only one of the region's national capitals is also a seaport: ____________

**8–14** Which three countries are officially Communist states? ____________

**8–15** Which is the region's largest state in territory? ____________

**8–16** Which is the region's largest state in population? ____________

**8–17** If large oil and gas deposits are discovered under the seabed of the South China Sea, which countries would try to claim control over at least part of those resources? ____________

**8–18** Which three other major world regions, as defined in this workbook, share a boundary with East Asia? ____________

**8–19** In what areas would you expect to find the highest population densities? (Circle the correct answer.): Mongolia, Western China, Eastern China.

**8–20** Which regional country's territory extends farthest south? ____________
Farthest north? ____________
Farthest east? ____________

**8–21** Japan's latitudinal position is comparable to that of the east coast of the United States from northern New York state to southern Georgia. Which Japanese island would have a climate roughly similar to that of New England? ____________

**8–22** Which foreign country controls island territories closest to Japan?

____________

**8–23** So-called *buffer states* are relatively small states that separate two states that have been, or may become, hostile toward one another. Which East Asian state would best fit the role of buffer? ____________

**8–24** Which nation was divided into two mutually hostile states by post–World War II/Cold War politics? ____________

# GEOCONCEPTS

## Diffusion of Technology

Ideas, designs, tools, technologies—all are essential to civilization itself. Civilizations often are characterized as much by their material culture, or things, as by their values and beliefs. Who could separate cars, TVs, and air conditioners from any study of contemporary American civilization? All aspects of culture, both material (computers, cars) and nonmaterial (language, religion), have a geographic expression. Believers in a religion are concentrated in some places and scarce in others; the number of computers in ratio to human population is high in some places and low in others.

Another geographic aspect of any product of human creativity, any idea, or any technology is the *origin-area* of that idea or mechanism, and there is a pattern in time and space of the spread of that idea or technology. Ideas are first discussed and things are invented or first assembled at a particular place. This origin-point or zone is called the *cultural hearth.* Ideas, tools, designs, systems, and machines are all carried away from the hearth and eventually accepted or used by other people in other places.

Seldom is distance the only influence on the pace of the outward spread of ideas or technologies. The outward-spreading use or adaptation of anything new is not necessarily like tossing a rock into a pool of water and watching waves move evenly outward from the point of impact. Whereas the waves move in a uniform pattern, the geographic pattern of transmission-acceptance, or *diffusion,* occurs at unequal rates. Diffusion rates are influenced by cultural barriers at least as much as by physical barriers. Large, stormy oceans may not retard outward diffusion of a new thing or concept as effectively as a neighboring society that is just not interested. Understanding why some peoples and cultures do not readily accept a new idea is not always easy, but diffusion patterns can be mapped and confirm that sheer distance does not explain the rate of diffusion over space and through time.

## Diffusion Rates of the Industrial Revolution

The basic ideas of the Industrial Revolution came together in northwestern Europe, northeastern United States, and southeastern Canada in the late 1700s through the 1860s. The practical application of science to problems of production, transport, and communication; the use of enormous quantities of inanimate energy; the factory system; and the assembly line are the foundations of the modern world. The society or country that moves through the complex process of industrialization is rewarded with huge gains in standard of living.

Look at the thematic map of diffusion of the Industrial Revolution in Figure 8–6. Note that Japan's industrialization clearly was underway by 1900, yet southern Ireland, right next to the hearth or centers of origin of modern industrialization, was a relative latecomer. And why did the south coast of Australia accept the new ideas of industrialization so much sooner than north African states, which have just now begun to industrialize? Language, religion, and race seem to have little, if anything, to do with it. Some neighbors close to the hearth were slow to pick up and apply principles of industrialization, whereas some far-distant countries were faster. No one has demonstrated any significant regional, national, or racial variations in basic intelligence, so that geographic variable appears not to exist.

One geographic variable we might consider is colonialism. Look at the thematic map of colonial empires of the past in Figure 8–7. Of all the East Asian nations, Japan's modern history shows that it alone was never a satellite or dependent of another country before its emergence as a modern industrial state.

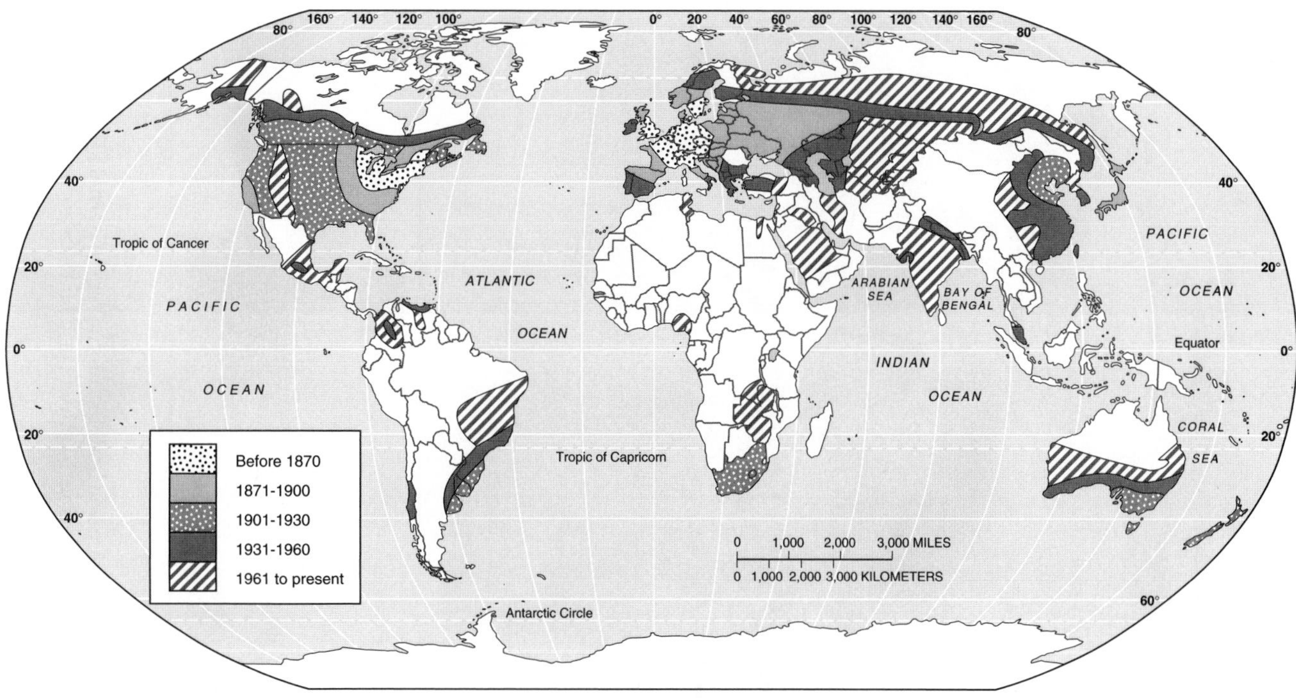


**FIGURE 8–6**
The diffusion of the Industrial Revolution.

**8–25** Was Ireland a colonial territory subject to foreign rule until the twentieth century? ______

______

**8–26** Were the North African states? ______

**8–27** What do your answers suggest about the relationship of foreign domination and likelihood of industrialization? ______

______

**8–28** Which of these areas has yet to become part of the industrialized world according to Figure 8–6? (Circle the correct answer.): eastern U.S.; western Europe; western China; eastern China

**8–29** Which of these regions is *not* considered a "cultural hearth" of the industrial revolution? (Circle the correct answer.): England, France, Germany, Belgium, Netherlands; U.S. states: MA, MI, OH, PA, IN, NJ, NY, RI, CN; Sudan, Ethiopia, Eritera, Somalia.

**8–30** The part of East Asia which began to industralize before 1870 was (Circle the correct answer.): Japan; eastern China; south Korea; (none).

**8–31** The East Asian country which appears to be the least industralized is (Circle the correct answer.): China; North Korea; Japan; Mongolia.

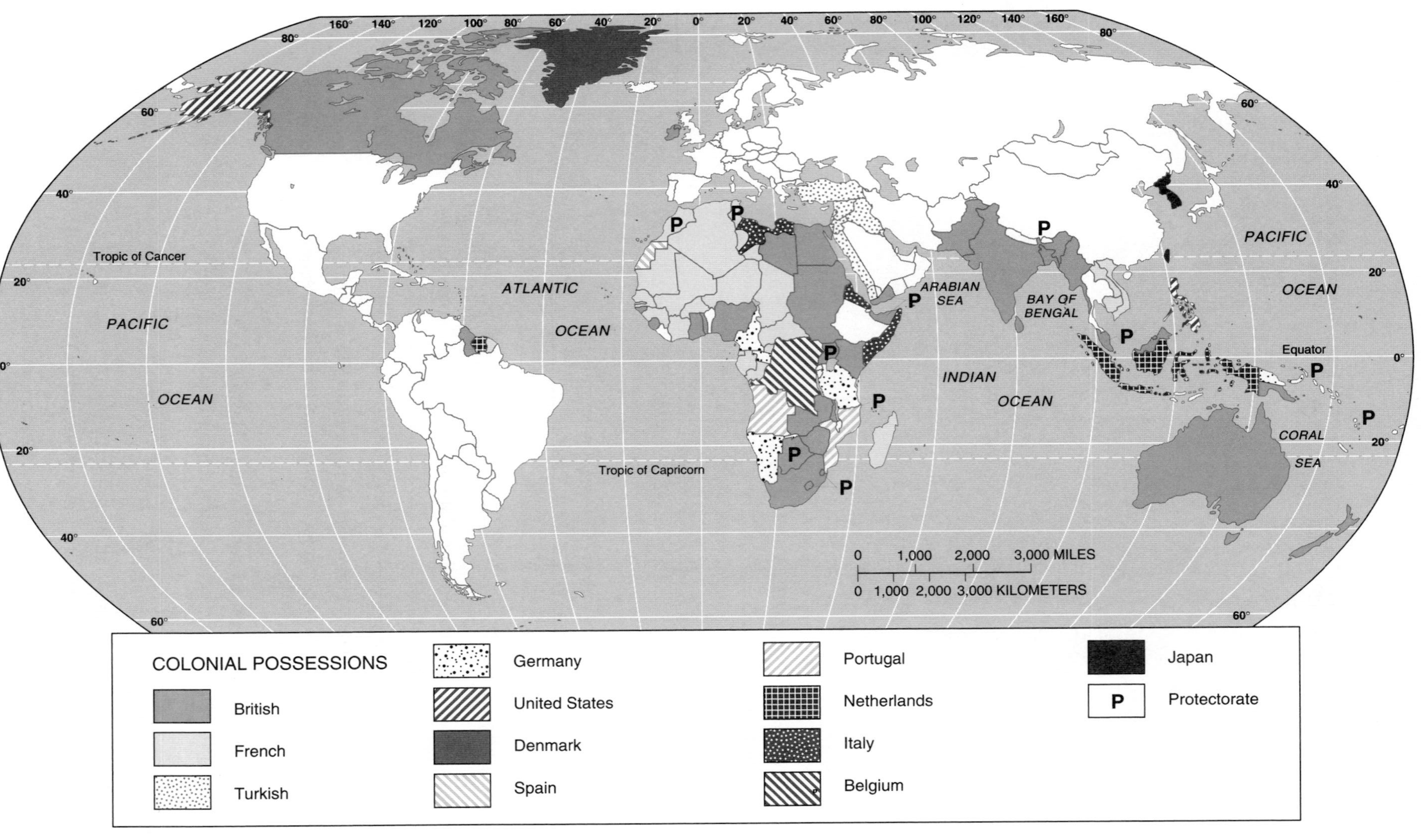


**FIGURE 8–7**
Colonial empires/protectorates, ca. 1914. A political unit attains the status of a true state when it is free to determine both its internal and external affairs. Units that controlled only the former were called *protectorates* rather than colonies. They were often sought by colonial powers to protect crucial transportation routes. Borders shown are current; patterns show 1914 colonial control.

## HONG KONG'S CHANGING STATUS

Hong Kong and Macau (Macao on some older maps) were relics of a past era of weak central government in China. They were the last of the treaty ports—small territories sold or leased to foreigners who would enforce their own laws and maintain order to the benefit of trade. When Europeans held a clear edge in the technologies of weaponry, transport, and communications, this technological advantage coincided with a cycle of disintegration of central authority in China. Chunks of territory were conquered, claimed, or informally administered by foreign powers. Coastal ports in particular were grabbed by European countries as treaty ports.

China allowed Hong Kong (British) and Macau (Portuguese) to survive into the late twentieth century because these colonial relics benefited the Chinese economy as much as the British or Portuguese economies. Hong Kong was a handy trading partner and intermediary in the days when the United States did not recognize nor officially trade with mainland China. Hong Kong was the UK's most populous and most important remaining colony, and as such received large foreign investments in industrial development. See Figure 8–8. Hong Kong was an internationally important banking and financial center. The UK–China Treaty

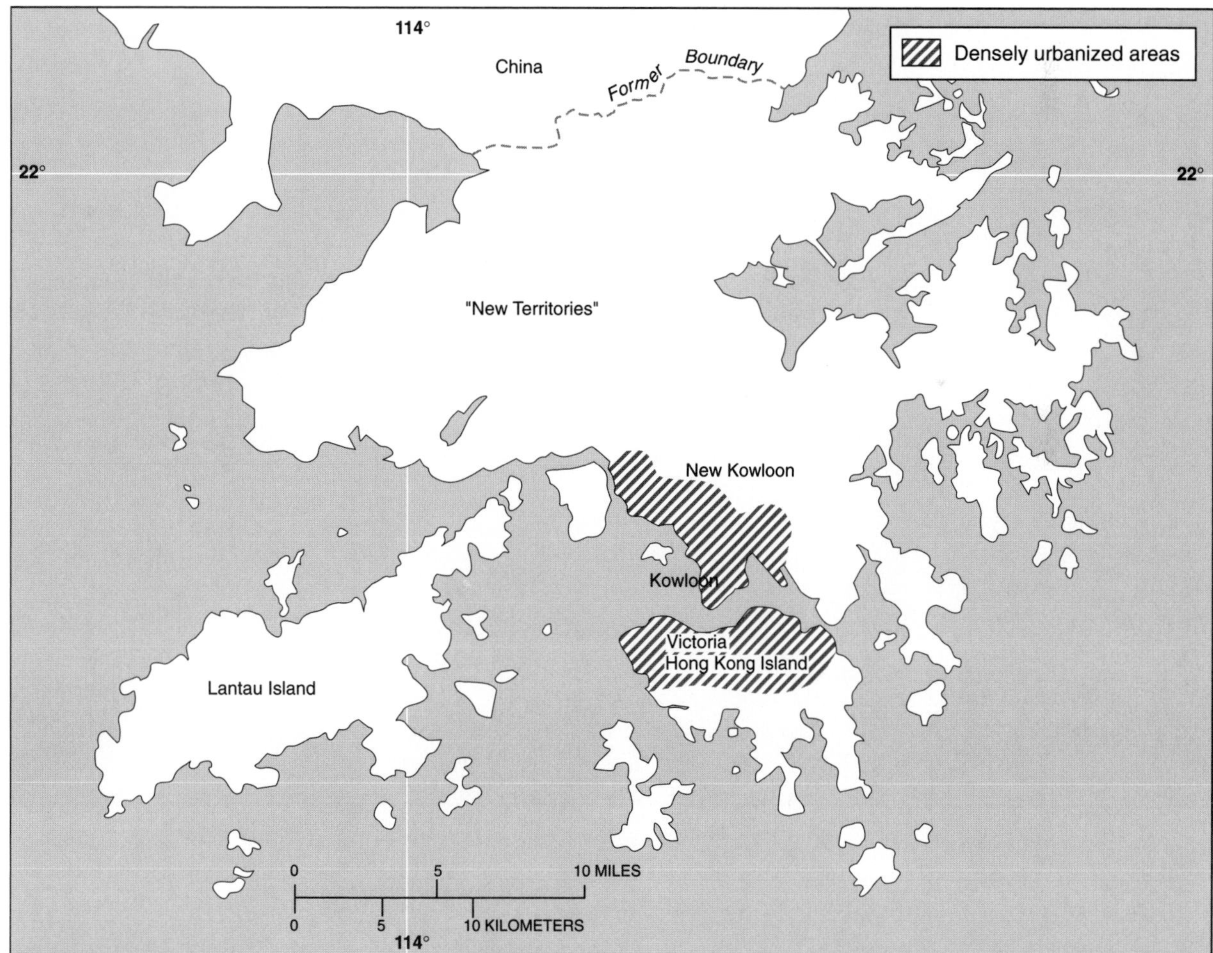


**FIGURE 8–8**
Hong Kong was one of the last European colonies in Asia. In 1997, the United Kingdom returned Hong Kong to China.

returning the 410 square mile colony to China in 1997 includes a Chinese pledge not to overturn Hong Kong's capitalist economic system. Macau reverted to Chinese rule in 1999.

**8–32** If the Chinese takeover leads to increased economic activity, where is the expansion of the built-up area most likely? ______________________

______________________________________________

## REGIONAL WATCHLIST

The relationship of mainland China (People's Republic of China or PRC) with the Republic of China (Nationalist China) on the island of Taiwan could well dominate future political news from this region. And, sooner or later, the non-Chinese ethnic groups on China's land frontiers, especially in the sparsely settled western interior and Tibet, could openly aspire to some real autonomy. If the Russian Federation does not respond effectively to secessionist pressures along its frontiers with China, China could take advantage of this fragmentation to reassert control in territories once Chinese. Look for talks on possible reunification of the two Koreas.

# 9

# South Asia

## INTRODUCTION

South Asia is the major world region south of China and the newly independent states, west of Southeast Asia, and east of the area known as the Middle East. South Asia includes India, Pakistan, Bangladesh, Afghanistan, Nepal, Bhutan, and Sri Lanka (the island of Ceylon). At well over one billion people, India is the largest country in South Asia, both in population size and territory. Just as China is the dominant culture in East Asia, India has been the star of South Asia.

Two of the world's great religions, Hinduism and Buddhism, originated in India. A third, Islam, claims the vast majority of the residents of Pakistan, Afghanistan, and Bangladesh as believers. The countries of Bhutan, Nepal, and Sri Lanka are divided among the religions. These religious differences, often producing persecution and bloodshed, led to the split of the British Indian Empire into Hindu and Muslim states.

South Asia was almost entirely controlled by one colonial power in the age of European colonial expansion. India, Pakistan, and Bangladesh all were once part of the British-controlled Indian Empire. Ceylon, now an independent country known as Sri Lanka, also was a British colony. Nepal and Bhutan were British-dominated buffers along the border with Tibet. Afghanistan was a buffer between British India and an expanding Russian empire. The Soviet occupation of Afghanistan during 1978–1988 was simply the latest expression of many attempts by Russia to dominate this rugged, preindustrial land.

On the outline map of South Asia in Figure 9–1, locate and label the countries and capitals listed in Table 9–1.

**TABLE 9–1** South Asian countries, capitals, and major cities.

| Country | Capital | Major Cities |
|---|---|---|
| Afghanistan | Kabul | |
| Bangladesh | Dhaka (Dacca) | |
| Bhutan | Thimphu | |
| India | New Delhi | Mumbai (Bombay), Calcutta, Kanpur |
| Nepal | Kathmandu | |
| Pakistan | Islamabad | Lahore, Karachi |
| Sri Lanka | Colombo | |

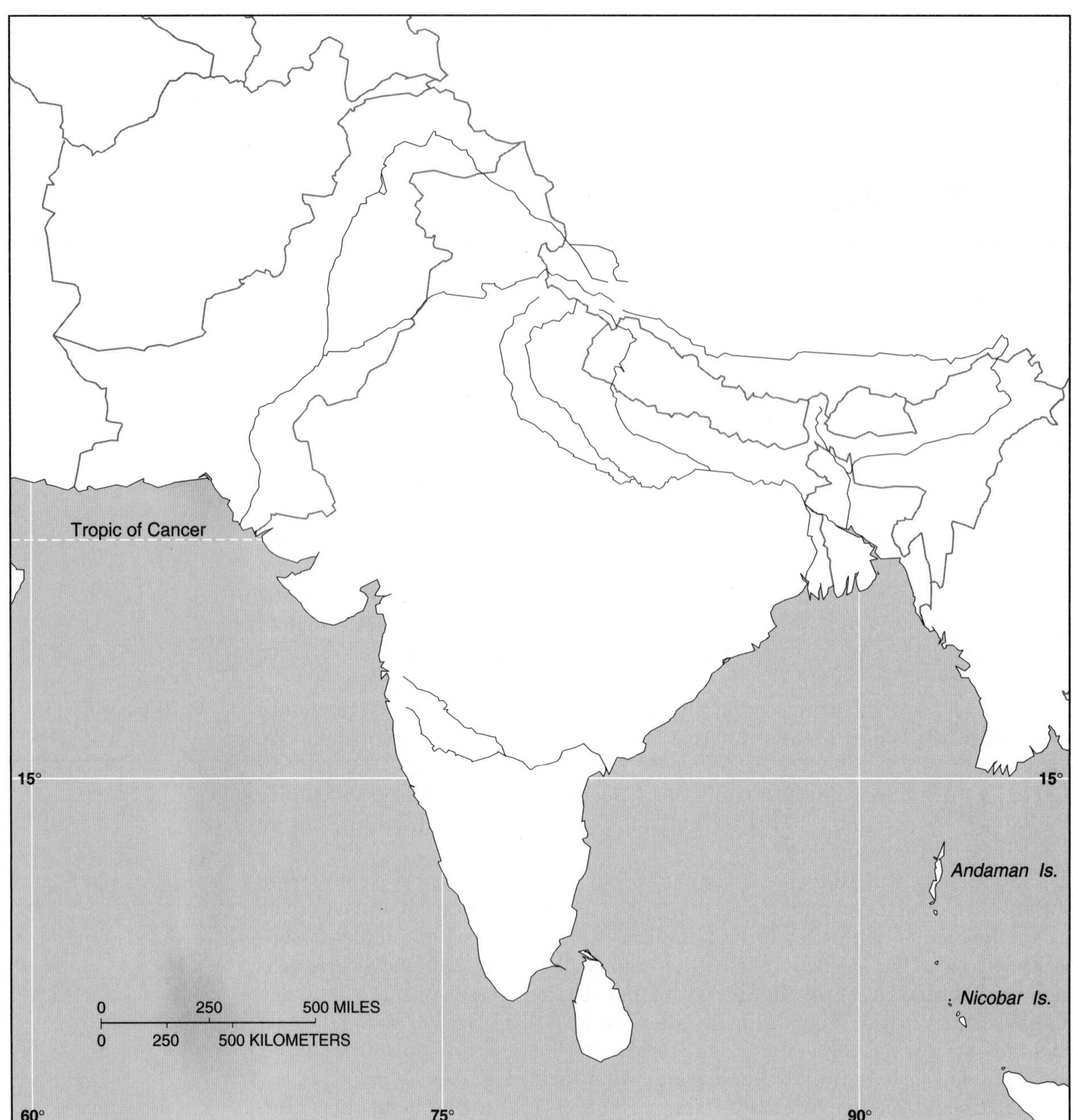


**FIGURE 9–1**
South Asian countries, capitals, and major cities.

## PHYSICAL GEOGRAPHY

India and its flanking Islamic neighbors of Pakistan and Bangladesh often are referred to as a subcontinent, reflecting the large scale of this region. Many superlatives are used in describing the physical geography of this region. Of the 16 famous mountain peaks in the world that soar over 25,000 feet (7,620 m) above sea level, 12 are within South Asia or are directly on its borders with China. Nepal, Bhutan, and Afghanistan lie mostly in the high mountain ranges that rise steeply above two enormous lowland plains. These huge lowlands, covered in river-deposited mud and silt, stretch to the sea in opposite directions from the Himalayan foothills of Kashmir and northern Pakistan. The larger of the two plains is that of the Ganges and Brahmaputra Rivers. It is 1200 miles (1931 km) long and is the home of hundreds of millions of Indians and all Bangladeshis. The other plain, half as long, is that of the Indus River and its tributaries. This plain is split between India and Pakistan.

In the most general view of South Asia, there are three broad features: a great arc of rugged mountains (Bhutan, Nepal, Afghanistan, part of Pakistan, and the northern borders of India); the two great connected lowlands that form the heart of India, all of Bangladesh, and part of Pakistan; and the huge triangular plateau of Deccan in southern India, a detached piece of which forms the island of Sri Lanka. The Thar or Great Indian Desert lies across India's northwestern border with Pakistan, with both countries trying to irrigate portions of this otherwise desolate land. For much of India, however, the problem is not as much a shortage of water as it is the timing of its availability.

On the outline map of South Asia in Figure 9–2, locate and label the following:

| | | | |
|---|---|---|---|
| Indian Ocean | | Palk Strait: | 10° N, 80° E |
| Bay of Bengal: | 18° N, 90° E | Karakorum Mountains: | 35° N, 78° E |
| Arabian Sea: | 20° N, 65° E | Sulaiman Range: | 30° N, 69° E |
| Laccadive Sea: | 10° N, 74° E | Himalaya Mountains: | 30° N, 80° E |
| Gulf of Mannar: | 7° N, 78° E | Andaman Islands : | 13° N, 94° E |
| Gulf of Khambhat (Cambay): | 21° N, 74° E | Indus River: | 28° N, 68° E |
| Gulf of Kutch: | 22° N, 68° E | Ganges River: | 26° N, 80° E |
| Cape Comorin: | 8° N, 77° E | | |

## The Monsoon

Continentality, a geographic concept explained in Chapter 6, helps account for the wet and dry seasons typical of much of South Asia. Many of Earth's seasonal atmospheric pressure cells are a direct result of surface temperatures beneath them. High temperatures on the surface tend to produce heated air, which expands and rises upward; this upward movement of air from the surface causes a low-pressure area. Low temperatures on the surface usually mean that cold, dense air above the surface is settling down, not lifting up, thus producing a high-pressure area. Winds flow out of high-pressure areas into low-pressure zones.

The enormous landmass of Eurasia is colder in winter than the tropical Indian Ocean and most Pacific waters. Therefore, atmospheric pressure is higher over the heart of the landmass, and lower over the warm seas to the south and southeast. In winter, cold winds bearing little moisture flow outward from the continent, giving most of South Asia an offshore wind. For most of the South Asia region, as in much of China and Siberia, winter skies are sparkling blue, with little precipitation.

In summer, the surface temperature contrast is reversed. The land heats faster than the surrounding seas, providing a low atmospheric-pressure area over the land, while the now-relatively cooler seas have a high-pressure area above them. In summer, warm, moist air flows in over the coasts toward the low-pressure centers over southwest Asia. When these warm winds, laden with water vapor, rise over the hills and then mountains of South Asia, heavy downpours drop from dark, low clouds. With few exceptions, summer is the rainy season, and winter is much drier. These strong, persistent, and seasonal winds are known as *monsoons* in this region. The timing of the seasonal reversal of wind

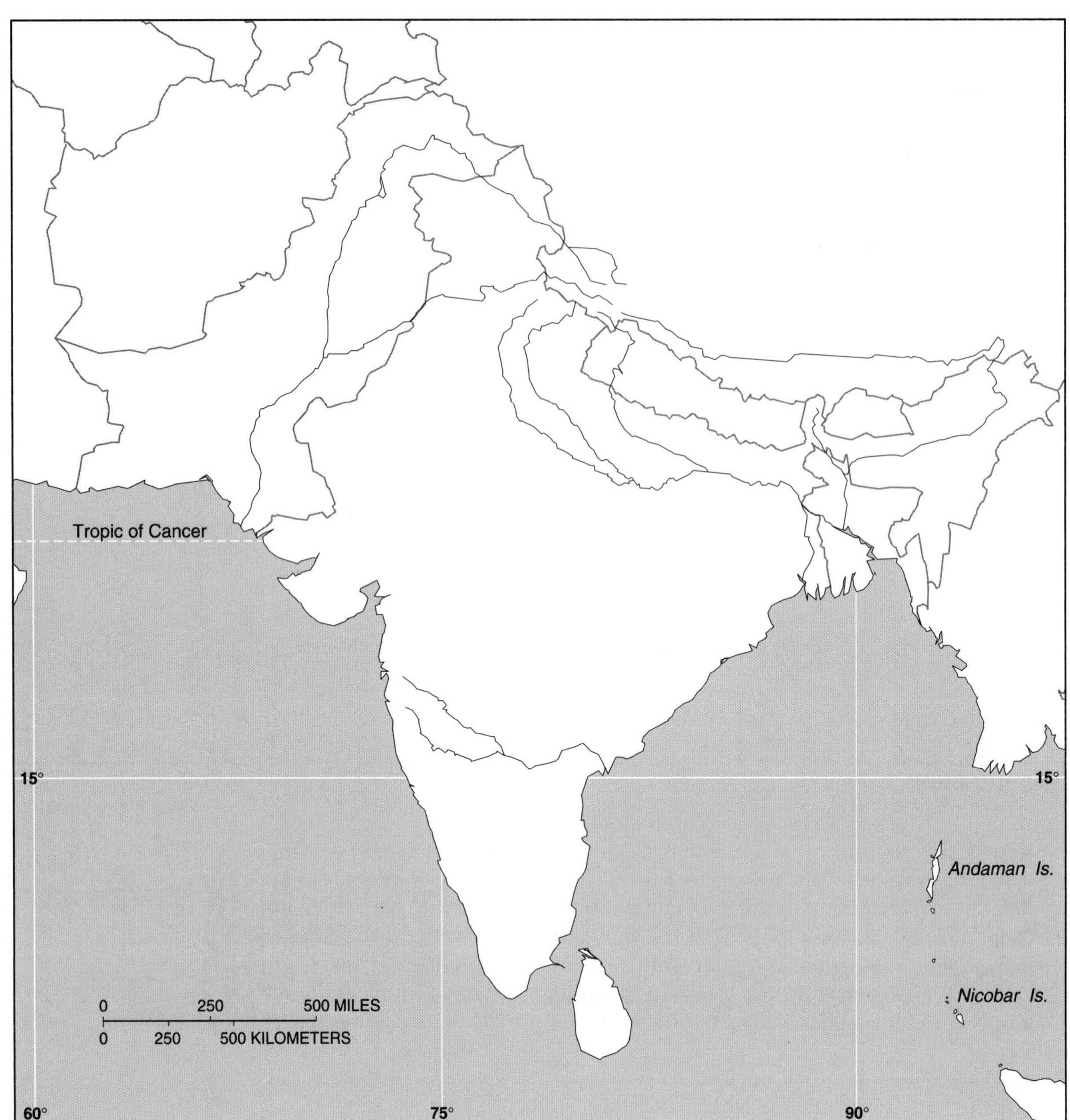


**FIGURE 9–2**
Major physical features of South Asia.

direction over South Asia is critical. Farmers must plant crops before the rainy summer monsoon arrives, but not too much before, or the seedlings will die of lack of moisture.

The summer monsoon winds, when forced to rise against the flanks of high mountains like the Himalayas, cool quickly and condense much of their load of water vapor. Cherrapunji, India, in the Himalayan foothills, averages 426 inches (1082 cm) of rain a year. Its July average is 99.52 inches (252.8 cm), whereas December averages only 0.2 inch (0.5 cm)!

Draw precipitation graphs for three cities to illustrate this remarkable difference between winter's dryness and summer's monsoon wetness. On the outline map of South Asia in Figure 9–3, mark and label the cities of Mumbai,

Calcutta, and Kathmandu. In open space near each city, draw a 1/2-inch horizontal line to be the base for each graph. Choose a vertical scale to represent precipitation averages for the wettest month and driest month. (For example, if 1/8 inch [0.3 cm] represents 1 inch [2.5 cm] of precipitation, then 25 inches [63.5 cm] of precipitation in a month would be represented by a vertical line 3-1/8 inches [7.9 cm] high.) On the left end of each line, draw a vertical line of proper length to represent July's rainfall, as shown in Table 9–2. On the right end of each line, graph winter's precipitation from data in the table. Can you see why the annual monsoon has such great impact on South Asia?

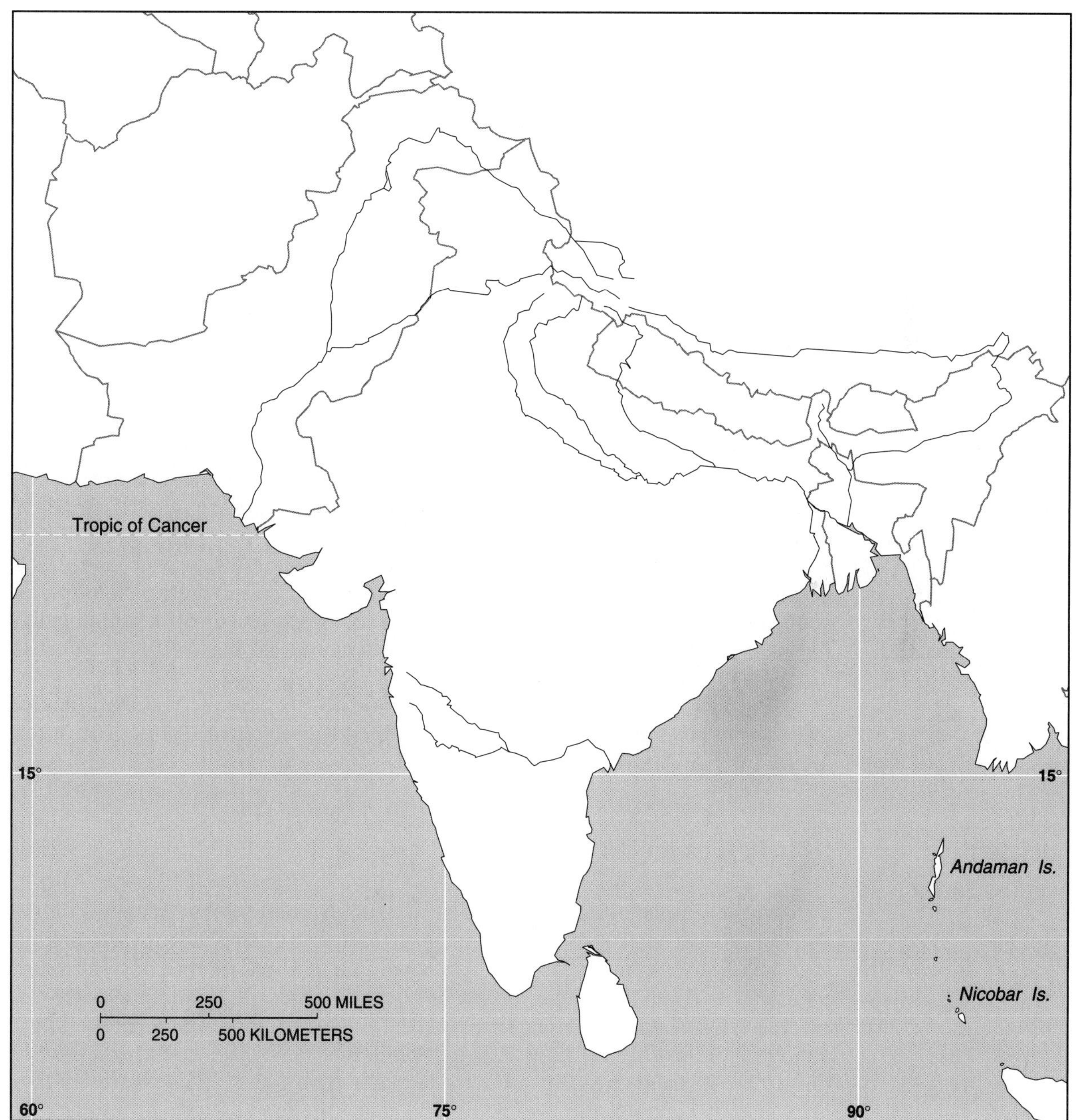


**FIGURE 9–3**
Precipitation differences between dry winter months and the summer monsoon.

**TABLE 9–2** Precipitation data.

| Place | Wettest Month and Amount | Driest Month and Amount |
|---|---|---|
| Mumbai, India | July, 25.17 inches (63.9 cm) | March, 0.02 inch (0.05 cm) |
| Calcutta, India | July, 12.89 inches (32.7 cm) | Dec., 0.21 inch (0.53 cm) |
| Kathmandu, Nepal | July, 14.91 inches (37.9 cm) | Nov., 0.31 inch (0.79 cm) |

## OBJECTIVES AND STUDY HINTS

As mentioned in the Introduction, religion is a key factor in studying the region of South Asia. Religious differences and intolerance required the division of Britain's Indian Empire into predominantly Hindu and Muslim states. India and Pakistan have fought several wars, and the potential for further conflict between them remains high.

It is useful to look at India as a huge country struggling to cope with deep-rooted religious conflicts, literally hundreds of languages, and strong ethnic and racial tensions. It is more relevant to compare India with all of Europe than with any other single country! Only since 1947 has India been unified, with one elected federal government.

For a country that is officially pacifist and neutral, India's power and influence have been felt by neighboring countries. India maintains a large, well-equipped army and has regional superiority in naval and air forces. It has fought border skirmishes with China as well as several wars with Pakistan. As mentioned earlier, Bhutan is almost a province of India. Nepal is strongly influenced by its gigantic neighbor. Besides giving Nepal economic aid, India allows its port of Calcutta to function as a port for Nepal as well. India actively encouraged former East Pakistan to break away from (West) Pakistan and become independent Bangladesh.

India must be the center of attention in any geographic look at South Asia because of its internal tensions among religions, linguistic, and ethnic groups, its continuing dispute with Pakistan over control of mostly Muslim but Indian-ruled Kashmir (the northernmost state of India), and its role as "big brother" to most of its smaller neighbors.

Recently, the Indian state of Kashmir has tried to achieve greater autonomy (self-rule in local matters). This may be the first step in withdrawing from India in order to join Pakistan.

## POLITICAL GEOGRAPHY

European opinions of India were once so strongly positive that Britain, France, and Portugal all competed for colonies and trading privileges. Europeans looking at India three or four centuries ago were astonished by a picture of wealth and plenty. At the time, India was the world's largest source of diamonds. Sapphires, ivory, spices, beautiful tropical hardwoods, silk and cotton textiles of the finest quality in the world, pearls, brass, and excellent steel swords and knives—all of these poured from India in seeming torrents. The lavish lifestyles of Indian princes and maharajahs living in opulent palaces made a powerful impression on European visitors. Obviously, there were many poor Indians at the time, too, but this detracted little from the desirability of trading with India.

Trade was the magnet that drew Europeans to the South Asia region. Attempts to extend and protect trade led to military conquest and territorial acquisition in India. The need to defend their growing Indian empire led the British to extend their imperial control to the frontiers of Afghanistan and China and to conquer Burma (now Myanmar). India became the brightest star of the British Empire, and Britain's global strategy centered around holding on to this glittering asset.

On the world outline map in Figure 9–4, draw a line southward from the UK through the Atlantic and around the southern tip of Africa. Then bring the line northeastward through the Indian Ocean to either Mumbai (west coast) or Calcutta (east coast) on the Indian coastline. This line was the most direct route between Britain and India before the Suez Canal was built in Egypt. Notice how the southern tip of Africa occupied a key position on this imperial route. The British took Capetown from its Dutch founders in 1795 as a convenient supply point on the UK–India route, just as the Dutch had needed Capetown as a base on their way to the Dutch East Indies (Indonesia). As part of its naval strategy, Britain created fortified naval bases to control natural sea routes between various oceans and seas; Capetown overlooked the passage between the south Atlantic and Indian oceans.

On the same outline map, draw in the Suez Canal. This canal, which opened in 1869, greatly shortened sea voyages between the UK and India and gave the British new strategic considerations. Draw another line on the map between the UK and India, this time using the Suez Canal. Draw it approximately south from the isle of Britain, turning eastward at the southern tip of Spain, through the Strait of Gibraltar and into the Mediterranean Sea. Continue the line through the Suez Canal and into the Gulf of Suez, the Red Sea, the Gulf of Aden, the Arabian Sea, and to Mumbai. From there, extend the line around India and up the Bay of Bengal to Calcutta.

On Figure 9–4, locate and label the following colonies, bases, and protectorates. All were used by the British to defend their sea routes to and from India: Gibraltar (still a British colony); Malta; Egypt (in the late 19th century, a theoretically independent country but dominated by the British until 1953); Aden, at the southwest tip of the Arabian peninsula; and finally, India itself, with Sri Lanka helping to safeguard routes to India's east coast.

## ECONOMIC GEOGRAPHY

South Asia provides a good opportunity to examine the problem of cheap labor. In increasingly competitive world markets, developing countries like those of South Asia may seem to have a huge advantage—a large supply of people willing to work for very low wages. The countries of South Asia have per capita GNPs in the range of $210 to $810. This means that South Asia is the major world region with the most *uniformly* low incomes on a national average basis. Other world regions, composed mostly of less-industrialized countries, include at least a few countries that are climbing the ladder of industrialization toward higher incomes, important exceptions to the prevalence of low incomes.

Do low wages guarantee the employer profitability? No. Profit and wage level do not have any predictable relationship. It doesn't do anyone any good to be paying a worker ten cents an hour if they are adding only nine cents worth of value to the company's product or service per hour. On the other hand, no employer would hesitate to pay a worker $100 an hour if that person were adding

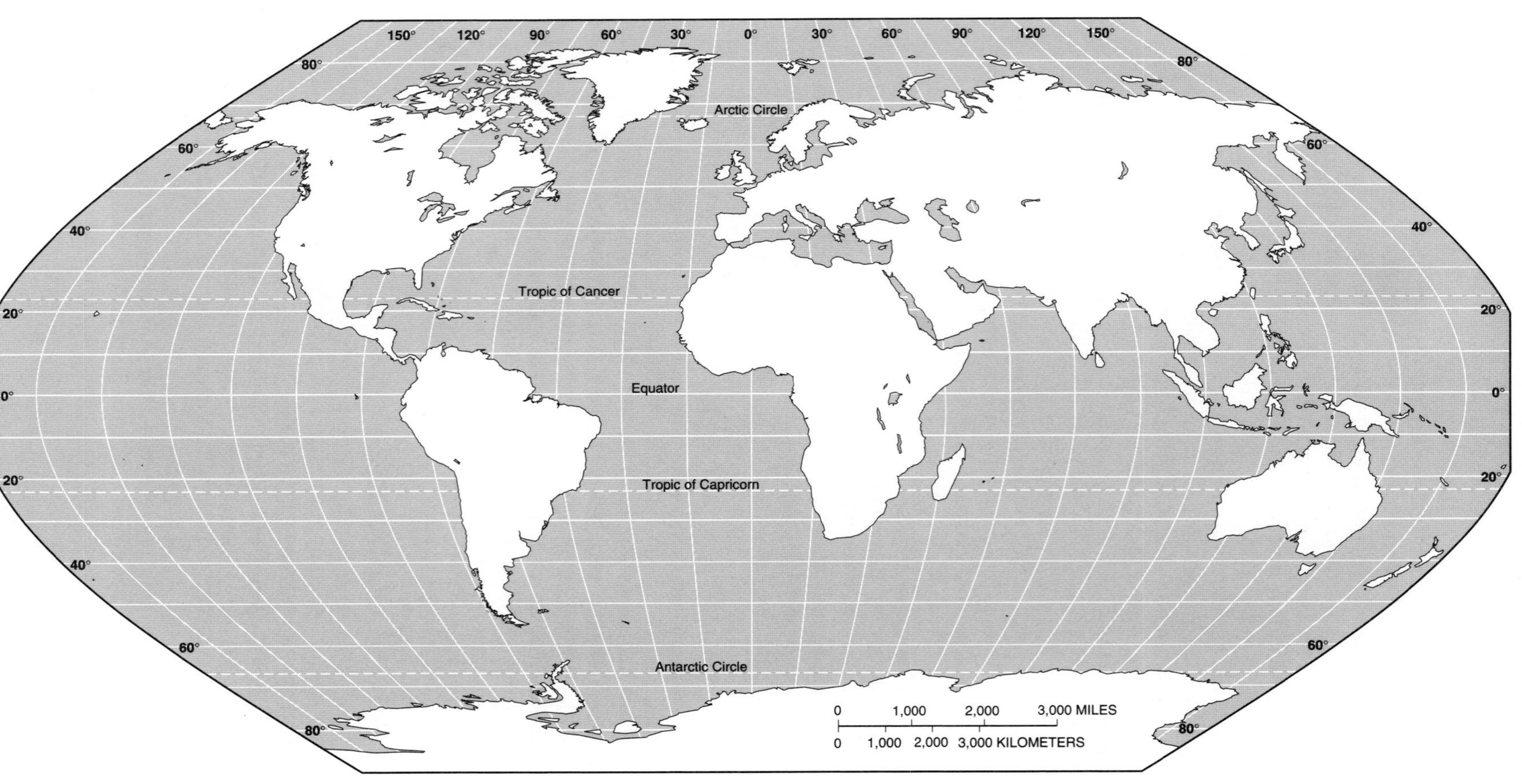


**FIGURE 9–4**
Sea routes between the UK and India.

$200 worth of value hourly to the line of production or services provided. It is this high ratio of *value added* to the goods or services through the productivity of the worker that brings prosperity to the enterprise and the country.

In the following, which person and associated technology and capital equipment would be able to process medical insurance claims faster and more accurately? Person A has a fourth-grade education, a pencil, and paper. Person B is a college graduate with a major in business administration and a minor in computer programming, an air-conditioned office equipped with a computer terminal connected to a variety of databases and a modem, and a FAX machine, electronic calculator, and telephone. Hardly a fair competition, is it? Even though Person B almost certainly is earning more per hour than A, B's much larger output of correctly processed claims more than justifies a higher wage. Of course, the technology and capital equipment being used by B, plus the advanced education, represent some costs to the enterprise and to the whole society. Education and capital investment more than pay for themselves in the long run, but these costs must be met up front. This is the dilemma of poor, developing countries—how to pay for education and imported technologies out of a very tight national budget.

On the world outline map in Figure 9–5, make a map showing postsecondary (beyond high school) education enrollments for the selected countries listed in Table 9–3 (enter the actual number for each country).

**TABLE 9–3** National per capita GNP vs. post-secondary education.

| Low-Income Countries (Under $810 per Capita GNP) | Percent of Age-Group Enrolled in Post-secondary Education |
|---|---|
| Bangladesh | 5% |
| Nepal | 5% |
| China (PRC) | 2% |
| India | 5% |
| Sri Lanka | 4% |
| Afghanistan | 1% |
| **Middle-Income Countries (Over $810, Less Than $10,000 per Capita (GNP)** | |
| Mexico | 16% |
| Poland | 17% |
| South Korea | 33% |
| **Upper-Income Countries (Over $10,000 GNP per Capita)** | |
| United Kingdom | 22% |
| Australia | 29% |
| France | 30% |
| Sweden | 37% |
| Canada | 55% |
| Japan | 29% |
| United States | 59% |

**9–1** Which country listed has the highest rate of post-secondary enrollment?

______________________________________________

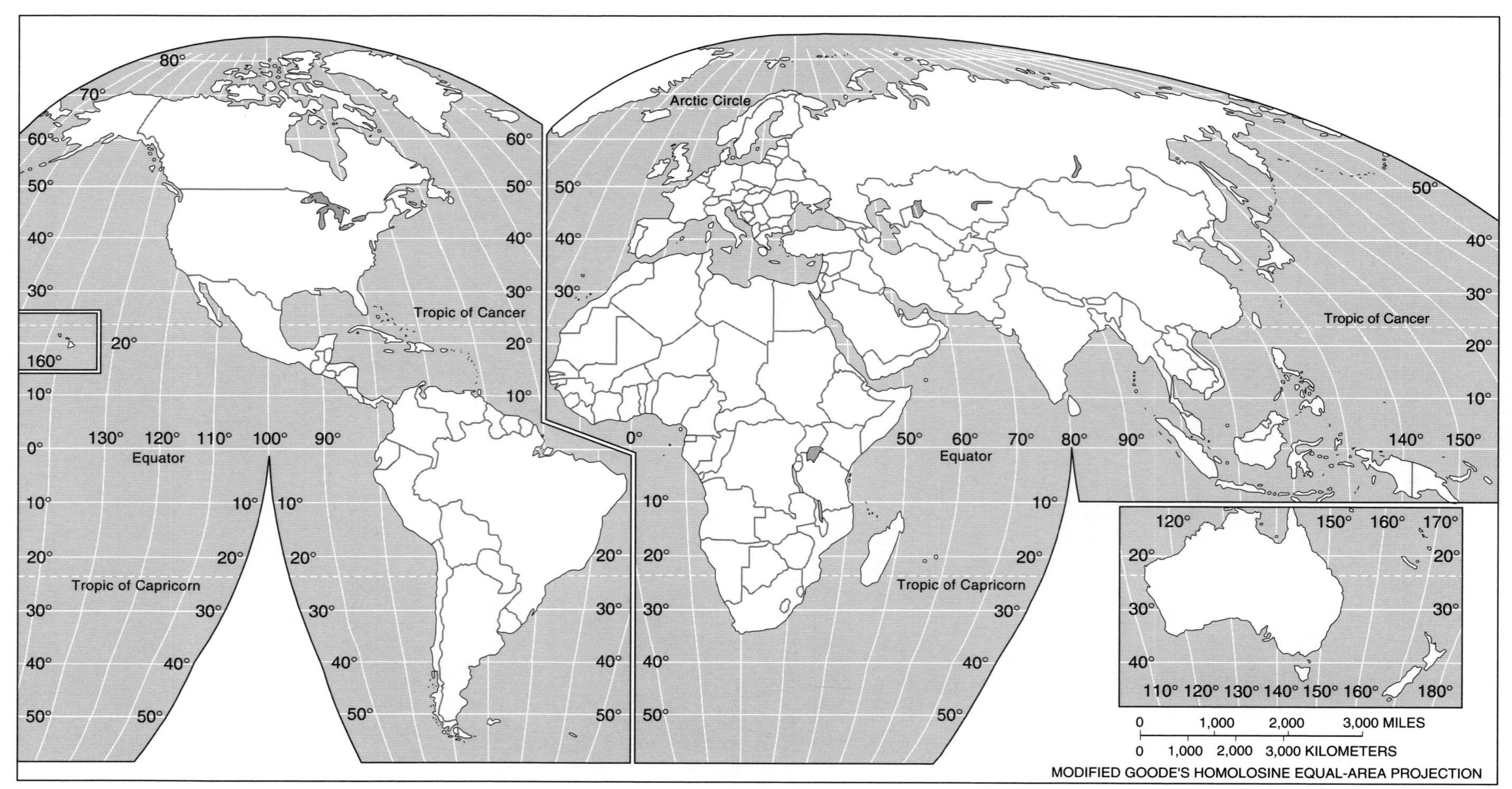


**FIGURE 9–5**
Post-secondary education enrollment and per capita GNP.

**9–2** Which of the "middle income" countries has the highest rate of post-secondary enrollment? ______________________________

**9–3** What does this map *suggest* (but not necessarily prove) about the relationship of higher education to income level?

______________________________

______________________________

## CULTURAL GEOGRAPHY AND DEMOGRAPHICS

The key to South Asian cultural geography is the rainbow of religions and languages. In India, for example, several hundred languages and dialects are spoken, some by only a few thousand people in remote isolated areas. The largest language in number of users is Hindi, but only a third of the Indian people speak it. Imagine trying to communicate effectively with all of India's citizens, even when limiting language choices to the 15 official languages! To add to the confusion, higher education normally is in English. Therefore, most well-educated Indians who run the government and major companies speak English.

While other South Asian nations do not have as complex a variety of languages to contend with, few have anything like language uniformity. Pakistan, for example, has many citizens speaking Urdu, Dari, Punjabi, Baluchi, Brahui, and Kashmiri languages. In many cases, language groups also are ethnic groups—each with its own traditions, culture, and history—and often, each with its own set of long-remembered rivalries and wars with other groups.

Even the powerful unifying force of adherence to the same religion cannot override these ethnic-linguistic divisions. The initial breakup of the Indian Empire on its independence from Britain produced India and Pakistan. But Pakistan was split into West Pakistan (now called Pakistan) and East Pakistan (now called Bangladesh), separated by nearly a thousand miles (1609 km) of Indian territory. Muslim Pakistan was separated from Hindu India in the first place to separate warring Muslims and Hindus, but East Pakistan (Bangladesh) split off in a civil war in 1971, mostly due to language and ethnic differences. Trying to hold together states containing a bewildering variety of languages, religions, and ethnic groups is a real challenge. Watch the news for recurrent examples of civil unrest, revolt, and possible civil war in the following confrontations:

India: Muslims vs. Hindus; Hindus vs. Sikhs
Sri Lanka: Tamils vs. Sinhalese
Nepal: Hindus vs. Buddhists
Pakistan: Indic language groups vs. Iranic language groups

## CHECK UP

**9–4** South Asia's largest country in territory is ______________; in population is ______________.

**9–5** South Asia's three landlocked states are ______________________________

______________________________.

**9–6** Circle the country that does *not* share international boundaries with South Asian countries: Turkmenistan, China, Myanmar (Burma), Laos, Iran.

**9–7** For most of South Asia, which of the following is the rainy season? (Circle the correct answer.): Winter, spring, summer, fall.

**9–8** The South Asian country occupied by the Soviet Army in the 1980s was ______________________.

**9–9** Circle the South Asian country that was never part of the British Empire: India, Pakistan, Sri Lanka, Afghanistan.

**9–10** The only island state in South Asia is ______________________.

**9–11** Which state was born in a 1971 civil war? ______________________

**9–12** Which South Asian state's territory extends farthest east? ______________________
Farthest north? ______________________

**9–13** Which two countries share the Thar Desert? ______________________
______________________

**9–14** Which two countries lie mostly in the Himalayas and their foothills?
______________________

**9–15** Foreign exports to easternmost India most likely would use which port? (Circle the correct answer.): Bombay, Calcutta, Madras.

**9–16** South Asia's two national capitals that also are seaports are ______________________
______________________.

**9–17** Kashmir is a territory disputed by which two South Asian countries?
______________________

**9–18** Which South Asian state is mostly a low-lying, flat river plain and delta?
______________________

**9–19** In correct north–south order, which major landform regions would you pass over by flying from the northernmost part of India to its southern tip? (Circle the correct answer.):
a. plateau—mountains—lowlands
b. mountains—lowlands—plateau
c. lowlands—plateau—mountains

**9–20** Which religion is *not* the majority in at least one South Asian state? (Circle the correct answer.): Hinduism, Buddhism, Christianity, Islam.

**9–21** For most of South Asia, the winter monsoon winds come from which direction? ______________________

**9–22** Enormous cultural variety helps explain why this is South Asia's only federal state. ______________________

# GEOCONCEPTS

## The Population Explosion

The human population of the planet has experienced truly explosive growth over the past few centuries. It took about 600,000 years for the human population to reach 1 billion, around 1830. It took a century to add a second billion, by 1930. Thirty years later, in 1960, the world total was 3 billion. Another 15 years or so added a fourth billion (1975), and the five-billionth person alive on this planet arrived sometime during 1987, only 12 years afterwards. By late 1999, the population total had topped 6 billion.

South Asia, specifically India, has the image of a rapid population growth overwhelming the land's capacity to support more people. But in fact, South Asia does not contain most of the world's fastest-expanding populations. Still, India's huge population of over one billion registers impressive gains even with rates of increase lower than about 30 other countries. India is expected to become the world's largest country in population, surpassing China, around 2030.

On the outline map of the world's political units in Figure 9–6, make a map of natural increase rates. Use solid black to show high rates, a lighter color for moderately high, and a very light color for moderately low. Leave white all countries in the low category (which includes a few negative growth rates). Use the following information:

**High growth rates (3% per year and higher)**

Bhutan
Angola
Niger
Togo
Oman
Eritrea
Democratic Republic of the Congo (Zaire)
Mali
Saudi Arabia
Nicaragua
Liberia
Swaziland
Chad

**Moderately high (2 to 2.9% per year)**

Nepal
Cameroon
Laos
Côte d'Ivoire
Pakistan
Algeria
Egypt
Lesotho
Central African Republic
Ecuador
Philippines
Burkina Faso
Guinea-Bissau
Libya
Paraguay
Tanzania
Dominican Republic
Honduras
Ghana
Zambia
Syria
Afghanistan
Malaysia
Kenya
Rwanda
Gabon
Papua New Guinea
Mexico
El Salvador
Senegal
Guinea
Jordan
Uganda
Somalia
Djibouti
Guatemala
Mauritania
Myanmar (Burma)
Ethiopia
Kuwait
Cambodia
Congo
Burundi
Mozambique
Sudan
Colombia
Brunei
Nigeria
Peru
Gambia
Yemen
Iraq
Madagascar
Bolivia
Benin
Venezuela
Belize
Sierra Leone
United Arab Emirates

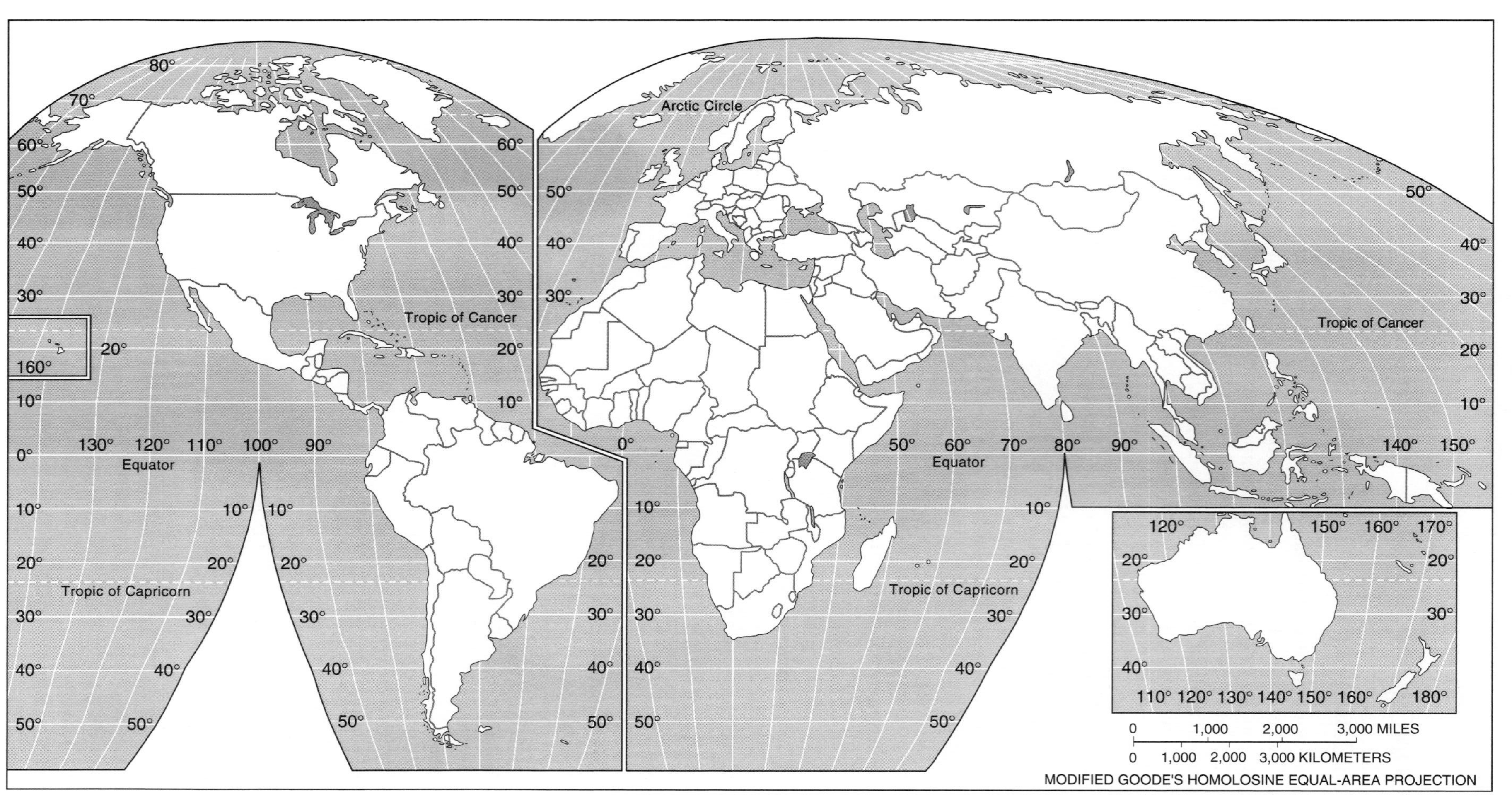


**FIGURE 9–6**
Population growth rates.

**Moderately low (1 to 1.9% per year)**

| | | |
|---|---|---|
| Thailand | Tunisia | Tajikistan |
| Brazil | Lebanon | Kyrgystan |
| India | Costa Rica | Zimbabwe |
| Albania | Bahrain | Turkey |
| Israel | South Africa | Bangladesh |
| Surinam | North Korea | Botswana |
| Chile | Guyana | Uzbekistan |
| Swaziland | Vietnam | Morocco |
| Sri Lanka | Jamaica | Qatar |
| Indonesia | Mongolia | Malawi |
| Haiti | Argentina | Turkmenistan |
| Panama | Iran | Namibia |

**Low (less than 1% per year)—(includes negative growth rates)**

| | | |
|---|---|---|
| All of Europe except Albania | Australia | Martinique |
| Uruguay | Japan | South Korea |
| Trinidad and Tobago | Cuba | Taiwan |
| Cyprus | Armenia | Russia |
| Singapore | China | Puerto Rico |
| Canada | United States | Georgia |
| Barbados | Azerbaijan | Dominica |
| | Kazakhstan | New Zealand |

**9–23** Why is Saudi Arabia an apparent exception to the generalization that high-growth-rate countries tend to be poor countries? ________________

________________________________________

**9–24** How many South Asian countries are in each category? ________________

**9–25** Which category is *most typical* of each of these world regions? (Circle the correct answer.):

Southeast Asia

Africa South of the Sahara

Middle East–North Africa

Latin America

**9–26** What is the only South Asian country in the "high growth rate" category?

________________________________________

**9–27** The three South Asian countries in the "moderately high" growth category are: ________________________________________

________________________________________.

## World Languages

Some languages make a greater impact on the world than others. While some languages are in common use only in one part of one country, others are understood by millions of people on each of several continents.

English is a good example of a *world language*—one spoken by many people around the world as their first (native) or second language. Energetic immigration of English speakers to the United States, Canada, Australia, South Africa, and New Zealand, plus onetime British colonial control in India, Pakistan, and many tropical African countries, has spread that language in the past. The present motivation for learning English is the trade and tourism impact of Americans. All communications between nondomestic airliners and airport control towers around the world are in English. One language has to be mutually intelligible to avoid fatal misunderstandings, and English is now recognized as the first among world languages.

## REGIONAL WATCHLIST

South Asia's giant, India, could be headed for serious threats to its political stability and territorial integrity; Kashmir continues to be a troublespot. The potential for the many national minorities to seek more autonomy or even full independence is quite high. The hopelessness of many Indians living in extreme poverty could fuel rebellions and riots on a grand scale when mixed with ethnic and religious tensions that have festered for centuries. Relations between India and Pakistan, never really friendly, repeatedly worsen and ease; Sri Lanka's guerrilla civil war is unlikely to disappear; and Afghanistan seems to have no effective central government at all.

# 10

# Southeast Asia

## INTRODUCTION

Only a half-century ago, Southeast Asia was largely controlled by outside powers—the British, French, and Dutch empires, and the United States (Table 10–1). Only Thailand ruled itself.

Eventual freedom from colonialism did not end the interests of outside powers in this region. Vietnam was the arena of prolonged confrontation between the United States, which supported the Republic of Vietnam (South Vietnam), and the Soviet-supported Democratic Republic of Vietnam (North Vietnam). The U.S. military role in Vietnam, which started with a few advisors, became large-scale in 1964. It reached a peak troop strength in 1969, when gradual withdrawal began. The war ended in 1975 when the Saigon Embassy was evacuated.

Vietnam, with the region's largest standing army, also fought a brief border war with the People's Republic of China, and had intervened in Cambodia and Laos. In the Philippines, the renewal of U.S. leases on important military bases was refused in a highly controversial decision. Meanwhile Indonesia struggles with the attempted independence of East Timor, a onetime Portuguese colony, and tries to integrate culturally different West Irian (Western New Guinea) into Indonesian political and economic life.

Economic development is taking place at an accelerating rate in Malaysia and Thailand, and Vietnam and Myanmar (Burma) seek to end their respective economic isolations by improving relations with free-market industrial countries. Referring to current events and media analyses to update the following list, create a "political alignment" thematic map for the region (Figure 10–1). Leave neutral countries white, apply a dark color to pro-Western states, and a light color to communist-dominated states.

| Pro-Western | Neutral | Communist-Dominated |
|---|---|---|
| Brunei | Indonesia | Cambodia |
| Malaysia | Myanmar | Laos |
| Papua New Guinea | | Vietnam |
| Philippines | | |
| Singapore | | |
| Thailand | | |

**TABLE 10–1** Southeast Asian nations formerly controlled by colonial powers.

| Country | Colonizer |
|---|---|
| Brunei | UK |
| Cambodia | France |
| Indonesia | Netherlands |
| Laos | France |
| Malaysia | UK |
| Myanmar (Burma) | UK |
| Papua New Guinea | UK and Germany; then Australia under UN mandate until 1975 |
| Philippines | U.S. |
| Singapore | UK |
| Thailand | (None) |
| Vietnam | France |

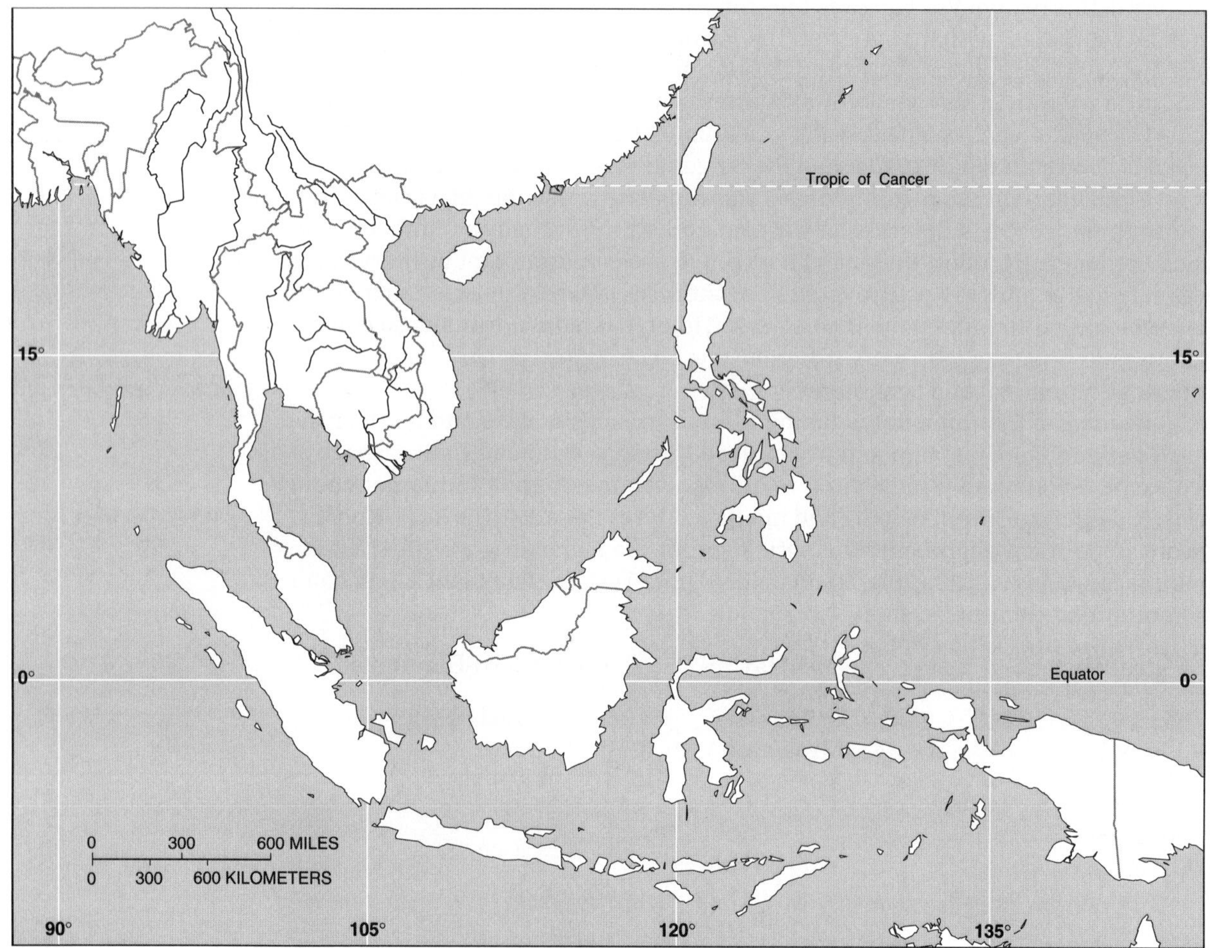


**FIGURE 10–1**
Southeast Asian political alignments.

## PHYSICAL GEOGRAPHY

Southeast Asia has a complex physical geography. It consists of the southeasternmost peninsulas and offshore islands of Earth's largest continent, Asia. Southeast Asia's easternmost islands lie to the north of the world's smallest continent, Australia. New Guinea and Kalimantan (Borneo) are the world's second- and third-largest islands. The planet's eleventh-longest river, the Mekong, flows through Laos, Thailand, Cambodia, and Vietnam. The South China Sea, which lies within this region, is the world's largest sea; it is a partly enclosed portion of the largest ocean, the Pacific.

The intermingling of land and water is an outstanding characteristic of this region. This collection of peninsulas and archipelagoes leads to frequent comparisons with the Mediterranean—an "Asian Mediterranean" of mountainous coasts and islands, set in a warm sea.

On the accompanying outline map in Figure 10–2, locate and label the following physical features:

**Bodies of Water**

| | | | | | |
|---|---|---|---|---|---|
| Indian Ocean | | Java Sea: | 5° S, 112° E | Timor Sea: | 10° S, 129° E |
| Pacific Ocean | | Coral Sea: | 12° S, 148° E | Arafura Sea: | 8° S, 135° E |
| South China Sea: | 20° N, 115° E | Torres Strait: | 10° S, 143° E | Gulf of Tonkin: | 18° N, 107° E |
| Isthmus of Kra: | 10° N, 98° E | Sunda Strait: | 6° S, 106° E | Strait of Malacca: | 3° N, 100° E |
| Philippine Sea: | 14° N, 133° E | Lombok Strait: | 8° S, 116° E | Singapore Strait: | 2° N, 104° E |
| Sulu Sea: | 8° N, 120° E | Cam Ranh Bay: | 16° N, 107° E | Makassar Strait: | 0° , 118° E |
| Celebes Sea: | 3° N, 123° E | Luzon Strait: | 21° N, 121° E | Manila Bay: | 14° N, 121° E |
| Banda Sea: | 6° S, 127° E | Gulf of Thailand: | 11° N, 102° E | | |
| Flores Sea: | 6° S, 120° E | Bay of Bengal: | 19° N, 90° E | | |

**Islands**

| | | | | | |
|---|---|---|---|---|---|
| Borneo (Kalimantan):* | 1° S, 114° E | Celebes (Sulawesi): | 2° S, 120° E | Timor: | 8° S, 125° E |
| Sumatra: | 0° S, 102° E | Luzon: | 15° N, 122° E | Samar: | 12° N, 125° E |
| New Guinea (Irian): | 5° S, 141° E | Mindanao: | 7° N, 125° E | Andaman Islands: | 13° N, 93° E |
| Jawa (Djawa): | 7° S, 110° E | Palawan: | 10° N, 118° E | Nicobar Islands: | 8° N, 94° E |

## OBJECTIVES AND STUDY HINTS

Now that you are familiar with the basic physical frame of Southeast Asia, you are ready to study the political, economic, and cultural geography of this fascinatingly complex world region. It is important to *link* various geographic facts to one another. This makes it easier to remember geographic information, and evaluating the significance of apparent linkages or relationships contributes to your increased awareness of the true significance of geography.

The geoconcepts illustrated by these Southeast Asian examples are core and periphery. The *core* is the center, the heartland of the country; it concentrates the population, economic activity, and power. The *periphery* is the edge, or the frontier—often less populated and sometimes less integrated into the state's political and economic life. You should think about applying these concepts to other geographic phenomena in other world regions.

---

*Traditional names, as found in most atlases, are listed first. More recent or locally recognized names are in parentheses.

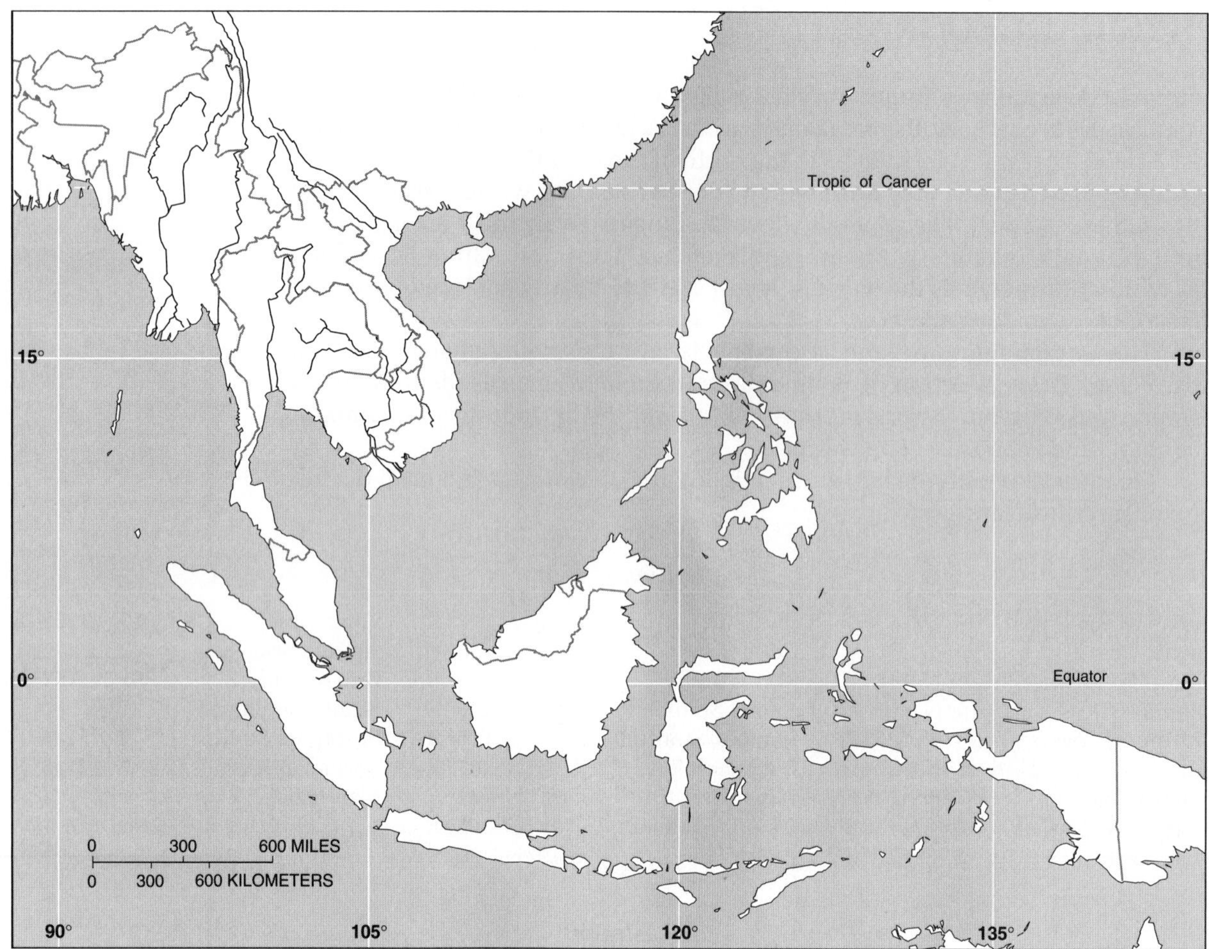


**FIGURE 10–2**
Physical features of Southeast Asia.

Also, as you enter geographic information on place maps, consider the consequences of these location factors: Is there any relationship between the facts that the region's only *landlocked* state is also one of its poorest? What advantages and disadvantages lie in the facts that two states consist of island archipelagoes and a third has both mainland and island territories? Would Singapore's strategic location advantages be affected by the construction of a canal through the Isthmus of Kra? How is Japan's import of Persian Gulf (Arabian) oil affected by Indonesian restrictions on the size of oil tankers that can use the Sunda and Malacca straits?

## POLITICAL GEOGRAPHY

Southeast Asia's political geography is nearly as complicated as its physical geography. As you learned in the introduction, this region is comprised of 11 national units. Locate each of these countries and their capitals (shown in Table 10–2) on Figure 10–3.

Indochina, as an *historic* region, refers to the one-time French colonies of Vietnam, Laos, and Cambodia. The term *Indochina* reflects the two long-standing cultural influences on this region from its two huge neighboring centers of ancient civilizations—India and China. As a *physical* geographic region, Indochina also includes these three states plus Thailand, Myanmar (Burma), and mainland Malaysia. On a world map, note how Southeast Asia can be seen as a "land between"—a region caught between two neighboring, strongly expansionist, richly developed cultures.

**TABLE 10–2** Southeast Asian nations and capitals.

| Country | Capital |
|---|---|
| Myanmar (Burma) | Rangoon (Rangon) |
| Thailand | Bangkok (Krungthep) |
| Laos | Vientiane (Viangchan) |
| Vietnam | Hanoi |
| Cambodia (Kampuchea) | Phnom Penh |
| Malaysia | Kuala Lumpur |
| Indonesia | Jakarta (Djakarta) |
| Singapore | Singapore |
| Brunei | Bandar Seri Begawan |
| Philippines | Manila |
| Papua New Guinea | Port Moresby |

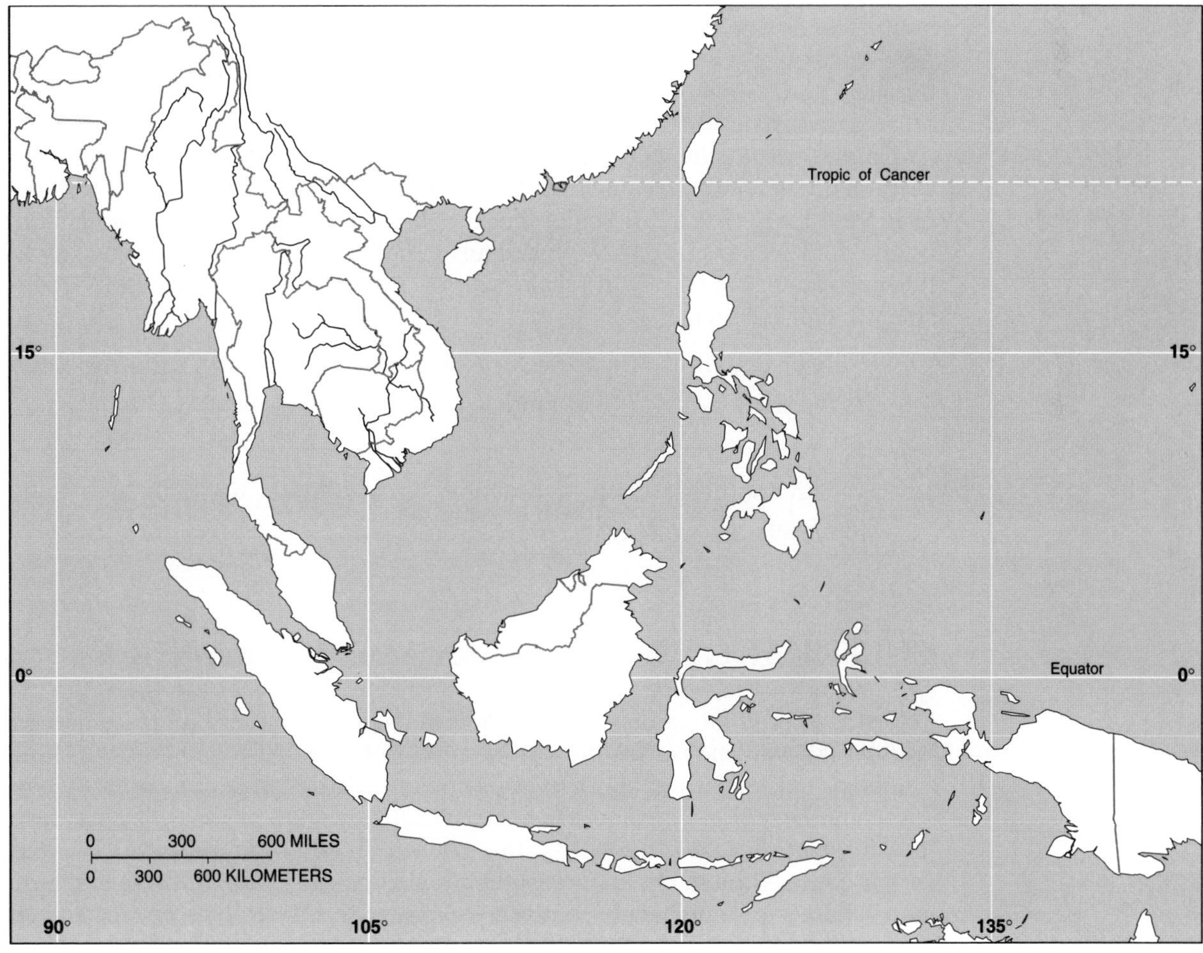


**FIGURE 10–3**
Southeast Asian countries and capitals

As previously mentioned, Southeast Asia was largely controlled by outside powers, but starting in the mid-1940s these onetime colonies were granted independence. Indonesia, a onetime colony of the Netherlands known as the Dutch East Indies, achieved independence in 1945. Cambodia and Laos became independent in 1953, and Vietnam achieved independence from France in 1954, though a communist government in the north fought a noncommunist south until withdrawal of U.S. forces in 1975 led to unification in 1976.

Malaysia's mainland portion was freed from British control in 1957. In 1963, two former British colonies on the north coast of Borneo, Sarawak and Sabah, joined Malaysia, along with the city-state of Singapore. Singapore later withdrew as an independent state in 1965. Oil-rich Brunei, on the northern Borneo coast, remained a British protectorate until 1984. Thailand was never a European colony.

The Philippines, conquered from Spain in 1898 by the United States, were granted independence in 1946. Myanmar (Burma) achieved freedom from the British Empire in 1948. The western portion of New Guinea, called West Irian by the Indonesians, was not transferred to Indonesia by the Dutch until 1962. Papua New Guinea was an Australian mandate (protected territory) until 1975.

## ECONOMIC GEOGRAPHY

As a region, Southeast Asia contains widely varied levels of economic development. The city-state of Singapore is the wealthiest political unit in Southeast Asia. Historically an *entrepot* (a trading center for an extensive, often international, region), Singapore has become a major industrial center. Cambodia, at the other extreme, is one of the world's poorest states. Referring to data in the Appendix, create a thematic map (Figure 10–4) showing per capita GNP in Southeast Asia. Use four colors—one for GNPs over $8000; another for GNPs ranging from $1500 to $7999; a third color for those between $600 and $1499, and blank (white) for GNPs below $599. When data are not available, place that country in the lowest category. On a world scale, countries with $6000 or more might be considered high-income.

**10–1** Why do you think Singapore is richer than most European countries, such as the UK, France, or Germany? ____________________

________________________________________

**10–2** Which Southeast Asian states would rank as lower middle-income ($600 to $7999) on a worldwide comparison? ____________________

________________________________________

## CULTURAL GEOGRAPHY AND DEMOGRAPHICS

Referring to population data in the Appendix, map the natural increase rate for Southeast Asian states. Use the outline map in Figure 10–5 and color to show the following categories: blank (white)—below 1% per year; one color—1% to 1.5%; another color—1.5% to 2%; a third color—over 2% per year. Geographers look for *spatial correlations*—situations in which two or more factors seem to vary in some relationship to one another. For example, if two different thematic maps show the same general patterns—areas with high values on one match areas with high values on the other—this would be a *positive correlation*. If, on the other hand, high values or concentrations on one map seem to be linked with low values on another map (and low values on the first map with high

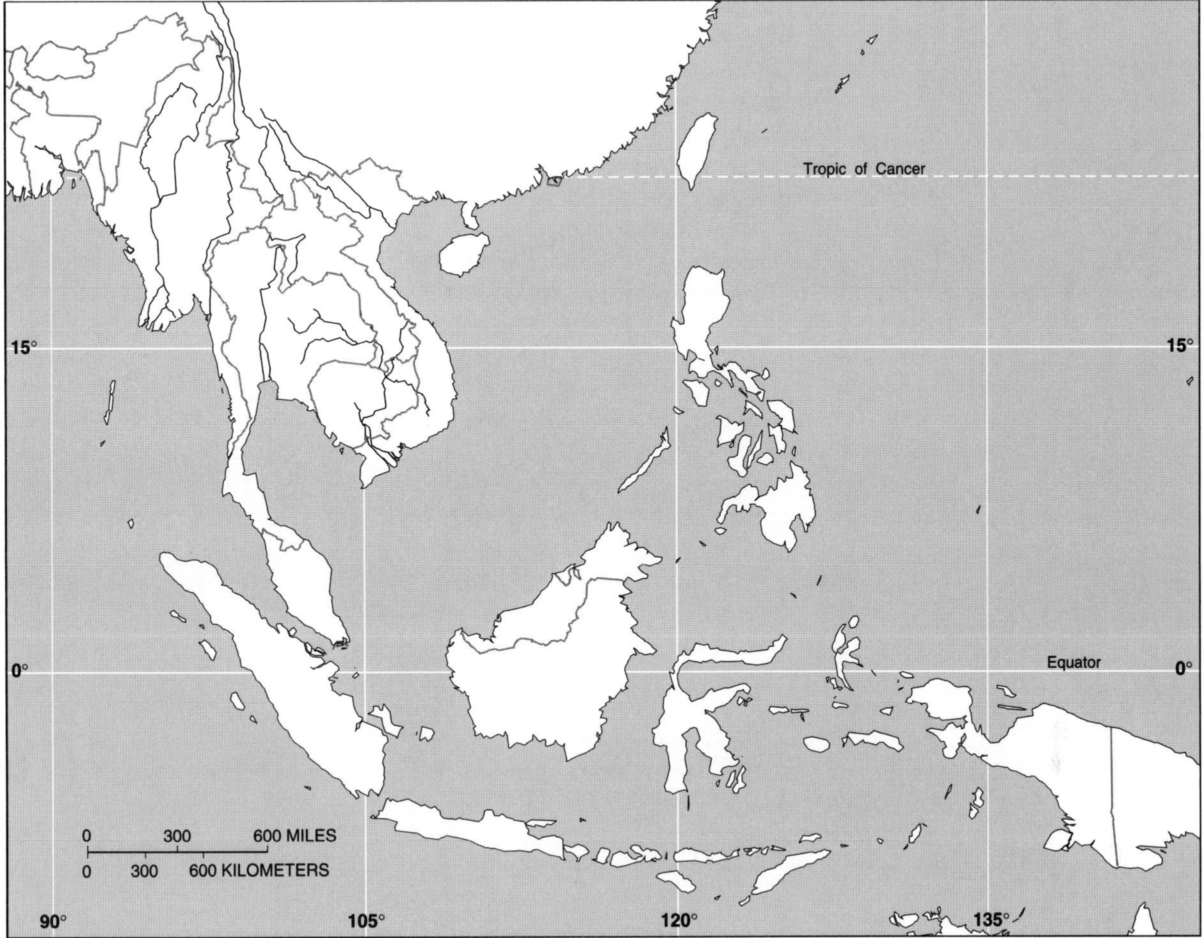


**FIGURE 10–4**
Per capita GNP in Southeast Asia.

values on the second map), this would be a *negative correlation.* Is there a positive or negative correlation (or no correlation at all) between your maps of per capita GNP and projected natural increase rate? (Note that any spatial correlation suggests a *possible* cause-and-effect relationship, but *does not prove* that one exists).

Culturally, Southeast Asia traditionally has been viewed as a *shatterbelt*—a zone of complex mixing and occasional confrontation of different ethnic groups and cultures. A prime example is the Philippines. These islands form the third-largest English-speaking nation in the world (although many Filipinos also speak Tagalog or Filipino). They are the region's only predominantly Christian society. Long-time colonial control by Spain, which actively sought converts to Christianity, and 47 years of American occupation, which created an English-language educational system, helped produce this unique culture.

Besides Christianity, there is also Islam, spread by seagoing Arab traders, which has adherents from the coast of Burma (Myanmar) to West Irian. Indonesia is the world's largest Muslim state; Malaysia is predominantly Muslim; and Mindanao, the southernmost large island in the Philippines, is predominantly Muslim. Evidence of Hindu culture can still be seen throughout Southeast Asia, in the form of ancient temples from Indonesia to the gigantic Hindu-influenced temple complex at Angkor Wat in the Mekong Valley.

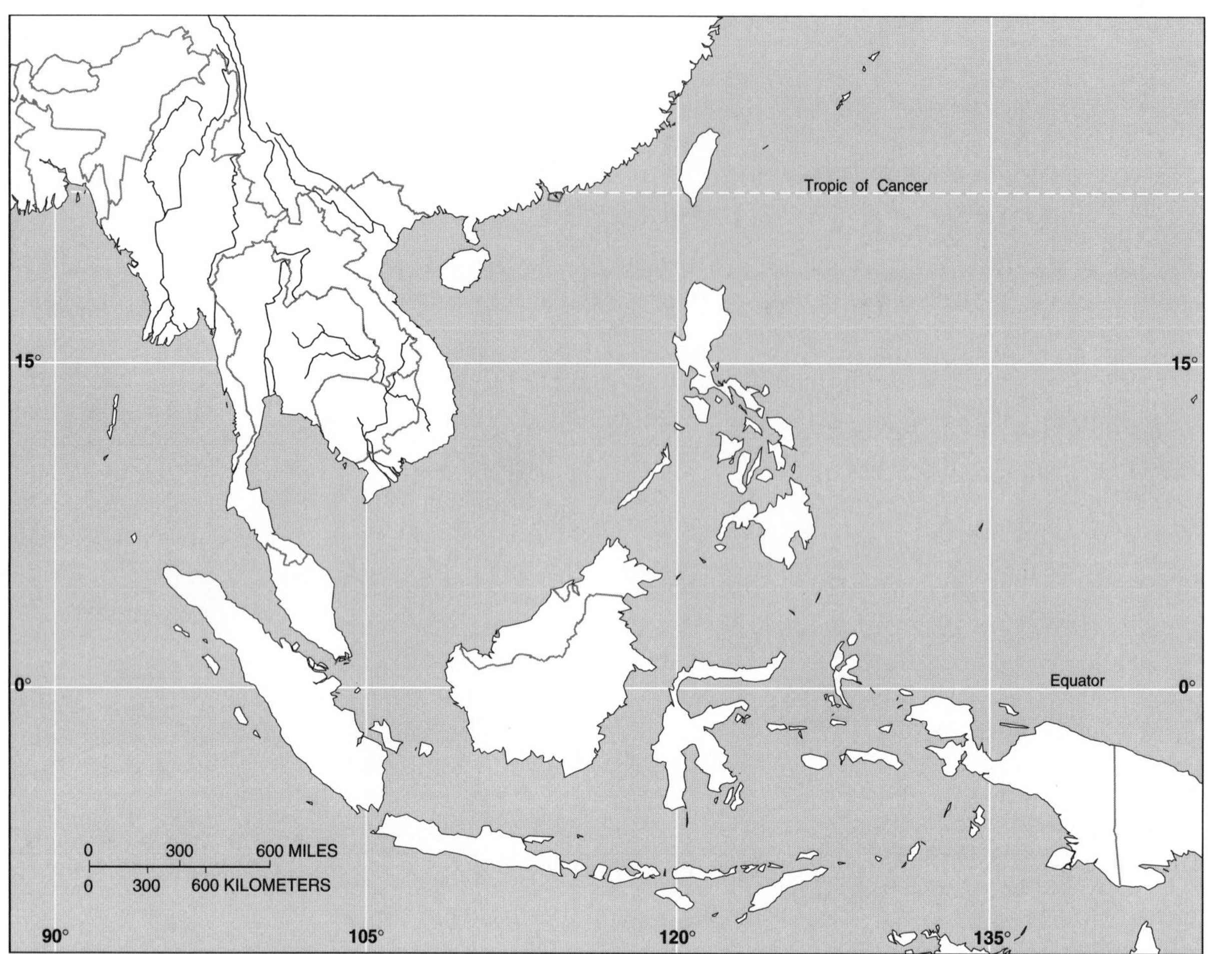


**FIGURE 10–5**
Rates of natural increase in Southeast Asia.

The Indonesian island of Bali retains its Hindu religion today, and a creature from Hindu mythology (Garuda) is portrayed on the national emblem of Indonesia.

Buddhism, a fourth great religious tradition, also is well represented in mainland Southeast Asia from Burma to Vietnam, where it has blended with traditional Chinese beliefs. Thus, with the exception of Judaism, each of the world's major religious traditions is well represented in this cultural shatterbelt.

## CHECK UP

**10–3** The world's largest sea, the ____________________, lies between the People's Republic of China and which three Southeast Asian countries?

____________________________________________

**10–4** Southeast Asia's largest unit in both territory and population is

____________________________________________

**10–5** The Southeast Asian country with the fastest rate of natural increase projected is ____________; the region's slowest natural increase rate is that of ____________.

**10–6** By using *global* comparisons, the Southeast Asian state with a per capita GNP higher than those of several European countries is ____________ ____________.

**10–7** By using *regional* comparisons, the five poorest Southeast Asian states are ____________ ____________.

**10–8** The increasingly prosperous upper-middle-income nation of ____________ ranks second only to ____________ in regional per capita GNP.

**10–9** The last Southeast Asian territory to fly the flag of a European colonial power was ____________.

**10–10** The first of the region's states to achieve independence from a colonial power was ____________, while ____________ was never a colony.

**10–11** ____________ and ____________ are national units composed of many islands, while ____________ includes both mainland and island territories.

**10–12** Which three Southeast Asian countries once were French possessions?

____________

**10–13** Which nation was ruled by the United States for almost half a century? ____________ Which three were once part of the British Empire?

____________

____________

**10–14** Which four of the region's national capitals are *not* also seaports?

____________

**10–15** By using both a world map and the regional map, which straits most likely would be used by oil tankers carrying Persian Gulf oil to Japan?

____________

**10–16** The fall of the British naval base at Singapore was an important event of World War II. In what way was Singapore comparable strategically to Gibraltar? (Use maps in Chapter 5 or 12 to locate Gibraltar and see its significance.) ____________

____________

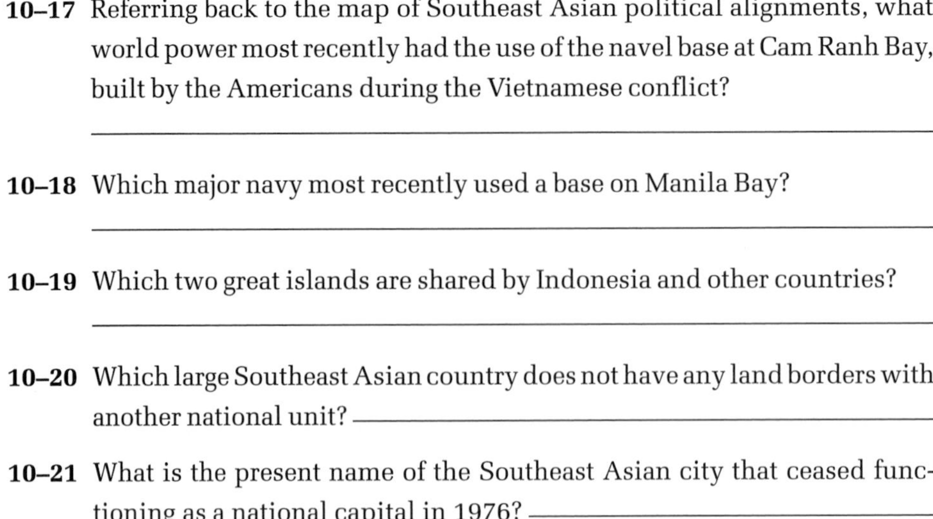

**10–17** Referring back to the map of Southeast Asian political alignments, what world power most recently had the use of the navel base at Cam Ranh Bay, built by the Americans during the Vietnamese conflict?

________________________________

**10–18** Which major navy most recently used a base on Manila Bay?

________________________________

**10–19** Which two great islands are shared by Indonesia and other countries?

________________________________

**10–20** Which large Southeast Asian country does not have any land borders with another national unit? ________________

**10–21** What is the present name of the Southeast Asian city that ceased functioning as a national capital in 1976? ________________

# GEOCONCEPTS

## Core and Periphery: The Political State and Its People

Geographers often apply the concepts of *core* and *periphery* in seeking answers to geographic questions. In political geography, the core of the state is its *heartland*—the concentration of population, economic activity, and political-military power. Economic development and political direction flow outward from the core, which is commonly the historic birthplace of the state and society. The periphery, or the edge, lies at the fringes or frontiers of the state. Quite often, the periphery is less advanced economically, less densely populated, and may be occupied by ethnic/cultural groups that are different from those of the core.

Southeast Asia provides an opportunity to observe core-periphery contrasts in at least three states— Myanmar, Thailand, and Vietnam. Refer to a map showing population distribution in Southeast Asia. Notice that Burma recently changed its official name to Myanmar, partly in an attempt to include a variety of ethnic groups along with the ethnic Burmese. Also note the two concentrations of population—a lower core in the delta of the Irrawaddy River around the capital, Rangoon, and an older, inland core around more centrally located Mandalay. These two cores are occupied primarily by ethnic Burmese, and they produce most of Myanmar's exports of rice and other agricultural commodities. Neighboring Thailand also has a densely occupied core stretching along a productive valley north from Bangkok.

In both Myanmar and Thailand, the periphery consists of lightly populated hilly and mountainous terrain surrounding the denser core. Many of these hill-country people are tribal groups with distinctly different cultures and lifestyles, and they may have little loyalty to the state whose frontiers they happen to occupy. The authority of the central government may be weak in these border zones where people are not effectively integrated into the state's political or economic life. A good example is the area where poorly organized, preindustrial Laos (a state virtually without any core) borders Myanmar and Thailand. It is known as the infamous "golden triangle"—a lawless area noted for illegal opium production.

The population map also shows two very separate cores in Vietnam—a northern core in the Red River Valley focused on Hanoi and a southern core in the

Mekong Delta focused on Ho Chi Minh City (formerly Saigon). Ethnic Vietnamese developed a traditional agricultural society based on rice grown in flat, flooded fields. The mountainous spine of Vietnam was not usable to them, and so they concentrated in the fertile river valleys of the north and south and in the many small, coastal pockets of flat land. This left the mountains to be occupied by hill tribes of different ethnicity, different culture, and different livelihood.

## Dividing the Seabeds for Resource Exploitation

Who owns the resources found on and under the seabeds beyond both the "territorial sea" and the continent shelf? (See the Geoconcept in Chapter 3.) International law is still evolving concerning control of minerals on and under deep seabeds. There is a high potential for conflict in the South China Sea, where Chinese maritime claims, based on historic claims to the Paracel and Spratly islands, overlap claims made by Vietnam, the Philippines, Brunei, Malaysia, and Indonesia. See Figure 10–6. These conflicting claims become important in terms of the granting of leases to explore for subseabed oil and gas.

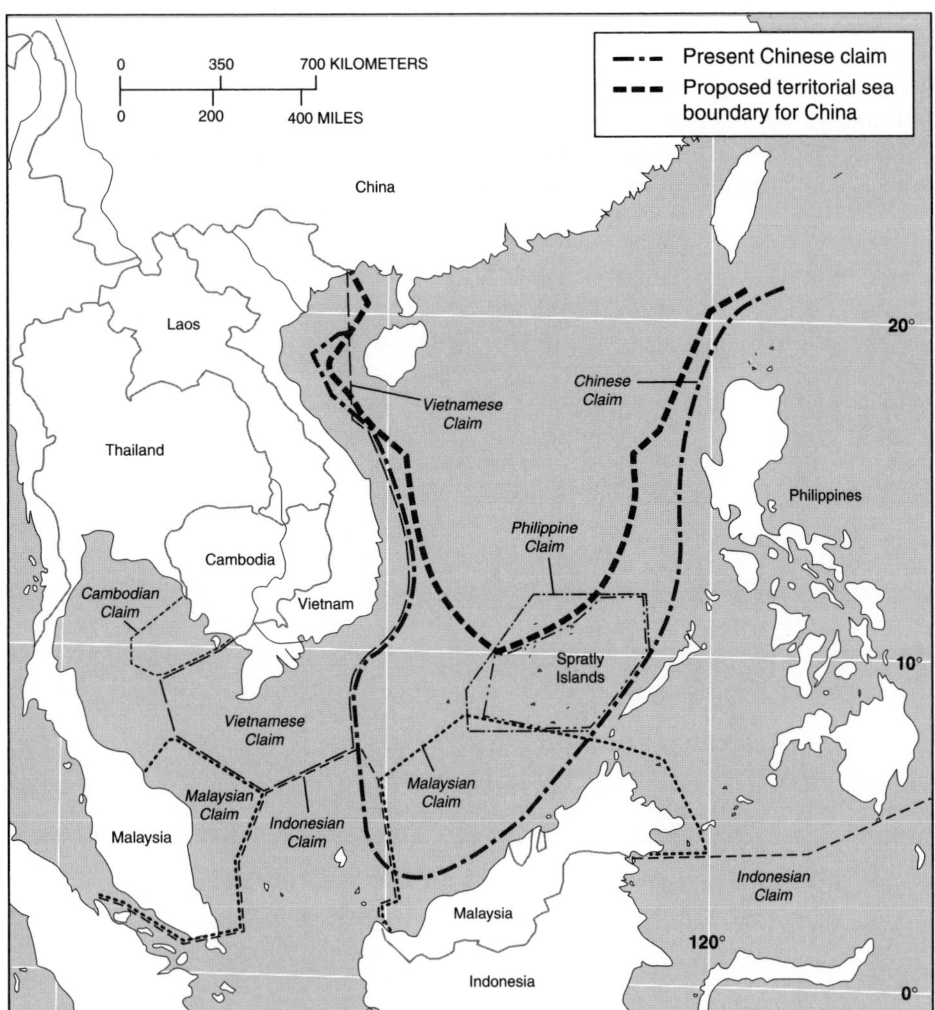


**FIGURE 10–6**
Conflicting maritime claims in the South China Sea.

**10–22** How might Vietnam's status as Southeast Asia's largest military force affect Chinese claims? ______________________________

______________________________

**10–23** Which Northwest European seabed has been completely divided among coastal states for oil and gas production? ______________________________

______________________________

## REGIONAL WATCHLIST

The rising popularity among U.S. investors of "international," "global," "Pacific opportunity," and "emerging nations" mutual funds includes those specializing in one national economy, such as stocks of companies in Malaysia, Thailand, or Singapore. Investor enthusiasm for Southeast Asian economies may be explained by noticing the country-of-origin labels ("Made in ______________") on purchases by American consumers. Where a generation ago popularly priced consumer electronics, toys, and gadgets may have come from Japan if they were of Asian origin, many labels on such imports now boast Southeast Asian sources. On a less positive note, almost all Southeast Asian states have potential national minorities to strain their internal stabilities. Watch for revolts and terrorism in Mindanao as that predominantly Muslim island attempts independence or local autonomy from the mostly Christian Philippines. Indonesia appears to be on the verge of fragmentation on an ethnic basis as dissident groups in Sumatra, Borneo, and Celebes struggle for independence.

# 11

# Africa South of the Sahara

## INTRODUCTION

The continent of Africa is a natural region on the basis of its continental status. But, like the continents of North America and Asia, Africa customarily is subdivided—into North Africa and Africa south of the Sahara—on the basis of its cultural geography. North Africa—the Mediterranean coast and the Sahara desert—has much more in common with the Middle East than with Africa south of the Sahara. North Africa shares both a key physical characteristic (desert terrain) and a major cultural quality (Islam) with the Middle East.

The Sahara can be seen as the largest of a series of deserts that dominate an area extending from the Atlantic coast of Africa eastward to the Red Sea, the Arabian peninsula, much of Israel, Syria, Jordan, Iraq, and Iran, then across mountains northeastward to the deserts of Central Asia, western China, and Mongolia. This great belt of desert and near-desert coincides remarkably with the distribution of Islam. The world's great religions are innovators, teachers, and preservers of culture, and Islam has fostered a distinctive culture.

The religion and culture of Islam have crossed the Sahara into the semidesert southern fringes of the Sahara, so that the cultural divide is not a sharp line but rather a zone of gradual transition. Nonetheless, Africa south of the Sahara has its own internal unifiers and is sufficiently different from North Africa to be a distinctive region.

## PHYSICAL GEOGRAPHY

Africa is the world's second largest continent. Its coastline is relatively smooth and regular without the complex combination of peninsulas, bays, and islands that typifies Europe or Southeast Asia. As a result, natural harbors are in short supply along African coasts. Africa is composed of a series of plateaus and mountain systems with little coastal plain. Africa even lacks the extensive offshore continental shelf typical of northeastern North America, the Arctic coasts of Eurasia, or southeastern South America.

Africa lies almost entirely within the tropics. Bizerte, Tunisia, on Africa's Mediterranean coast, is at 37° N latitude, and Capetown, South Africa, is close to 30° S. Only the extreme northern and southern coasts of Africa have subtropical rather than tropical climates.

**11–1** Which is the only other continent to lie entirely equatorward of 40° latitude north or south? ____________________

Africa is the only continent that is literally splitting. Parts of eastern Africa appear to be slowly pulling away from the main mass of the continent. Look at a detailed map of landforms for Africa. Note the long, stringbean-shaped Lake Nyasa and Lake Tanganyika along the borders of, respectively, Malawi—Tanzania and Zaire—Tanzania. These great lakes occupy parts of a massive series of trenches that begin in Mozambique and extend northward to Uganda, splitting and branching off into Kenya, where Lake Rudolph is a smaller version of Lakes Nyasa and Tanganyika. The Ethiopian plateau also is split by this series of trenches known as the Great Rift Valley. The trenches are characterized by roughly parallel ridges that form their edges. Where the trenches are submerged, the rocky ridges become the mountainous coasts of long, narrow seas or gulfs. You can find a sea and three gulfs along the northeastern flank of Africa that are shaped by the slow pulling apart of the Earth's continental crust.

**11–2** The sea and three gulfs are ______________________________________________

______________________________________________.

**11–3** Another branch of this trenchlike valley moves north from the Gulf of Aqaba to include the Jordan Valley of the Israeli—Jordanian border and which sea of Biblical fame (look at a detailed map of the Middle East)?

______________________________________________

On the outline map of Africa south of the Sahara in Figure 11–1, locate and label the following:

| | | | |
|---|---|---|---|
| Atlantic Ocean | | Bight of Benin: | 5° N, 3° E |
| Indian Ocean | | Bight of Biafra: | 3° N, 8° E |
| Gulf of Guinea: | 3° N, 5° E | Mozambique Channel: | 17° S, 42° E |
| Lake Nyasa: | 12° S, 33° E | Cape of Good Hope: | 35° S, 19° E |
| Cape Agulhas: | 35° S, 20° E | Lake Tanganyika: | 6° S, 29° E |
| Niger River: | 8° N, 7° E | Cap Blanc (Cape Blanc): | 22° N, 18° W |
| Cape Fria: | 18° S, 12° E | Lake Victoria: | 1° S, 33° E |
| Cap Vert: | 15° N, 18° W | Comoro Islands: | 13° S, 43° E |
| Zambezi River: | 16° S, 30° E | Zaire (Congo) River: | 0° , 18° E |
| Cape Palmas: | 4° N, 7°W | | |

## OBJECTIVES AND STUDY HINTS

Africa may be the least understood continent in the perceptions of most Americans and Canadians. This may reflect the recent history of the continent and especially the recent history of Africa south of the Sahara.

Use of Africa's great rivers to sail into the interior was difficult. The Nile was blocked by rapids (called cataracts) and the mammoth vegetation—choked swamplike Sudd region in Sudan. While shifting sandbars and mudflats are typical of the mouths of most West African rivers, the great Zaire (Congo) River tumbles over huge rapids a few miles upstream from its mouth. In addition, pounding surf on beaches backed by mangrove swamp did not encourage landings on many African shores. Either treacherously shallow water or rapids blocked use of the smaller rivers to penetrate the unknown. Until the late nineteenth century, only about 120 years ago, European interest in, knowledge of, and territorial claims to Africa south of the Sahara were all limited to a few coastal trading stations in the area south of the Mediterranean African countries and north of the present-day nation of South Africa.

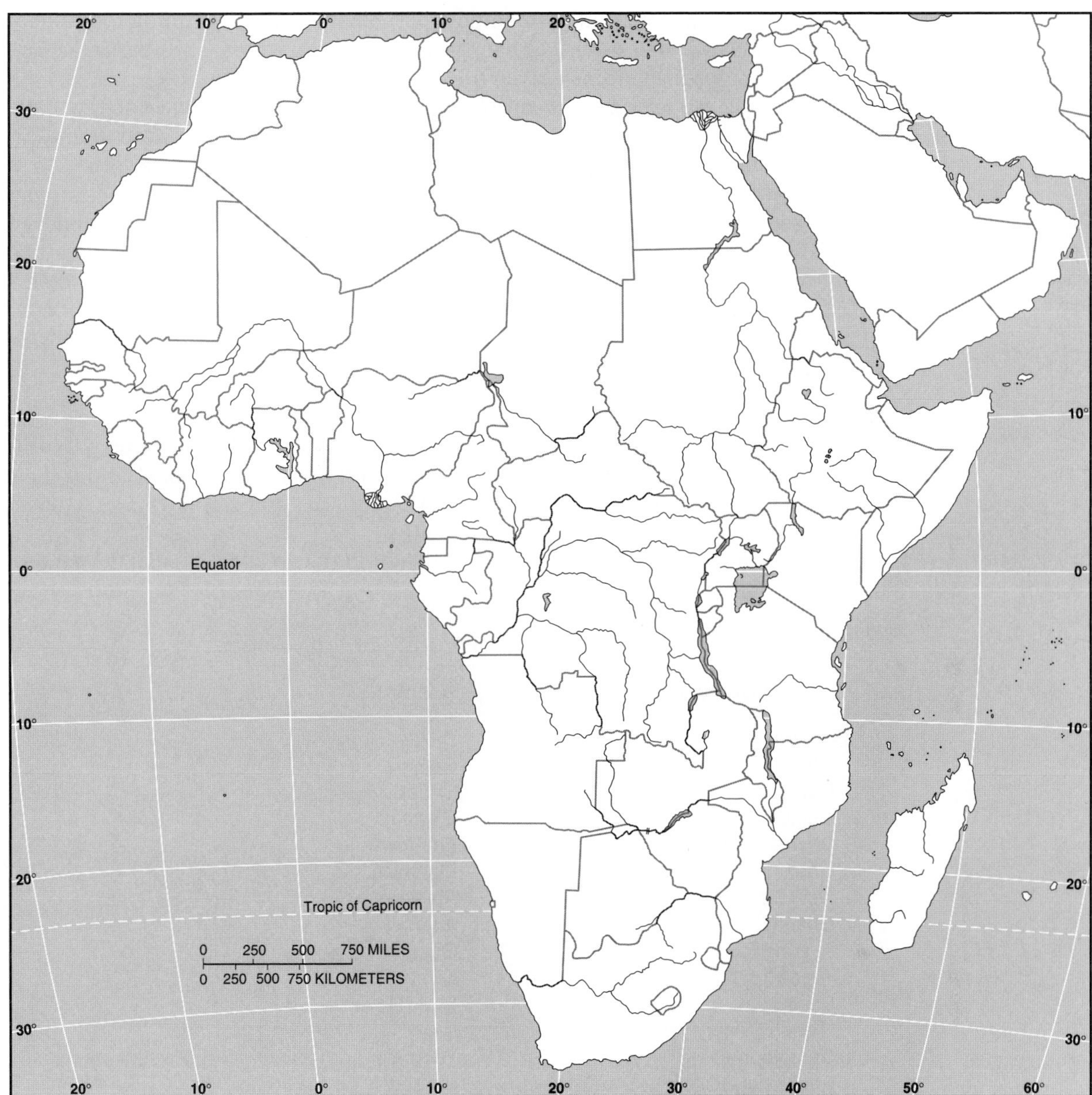


**FIGURE 11–1**
Africa—physical features.

Until the nineteenth century, Africa south of the Sahara was viewed by Europeans mostly as a large obstacle to sail around when traveling to India. (This obstacle to sea traffic was eliminated by the 1869 opening of the Suez Canal.) It was well into the nineteenth century before Europeans were sure which way the Niger River flowed! (The Niger, at 2590 miles [4168 km] in length, is the twelfth-longest in the world, longer than any river in Europe.) The Zaire (Congo) River was not mapped until 1876, and Europeans did not locate the source of the White Nile (the upper portion of the Nile) until 1862 (it's Lake Victoria, on the Equator). Much of Africa was literally unknown to non-Africans until late in the nineteenth century.

Although the physical geography of Africa now is well charted, the region south of the Sahara remains little known to outsiders in terms of its wealth of cultures, rich history, and contemporary problems. Africa is much closer geographically to North America than are China, Japan, Korea, and Australia. Yet, Africa remains mysterious and far off in our thinking.

This world region may require the most effort to become acquainted with its place geography and to recognize its geographic significance. These themes in Africa south of the Sahara all have important geographic aspects: As a major world region, it ranks among the most burdened by poverty and hunger. As a region overall, it is the fastest-expanding segment of humanity. Few states in this region have had time to build the traditions, loyalties, and strong identities of true nations. And finally, its political situation and many economic problems can be traced to the recency of colonialism and the fact of neocolonialism.

## POLITICAL GEOGRAPHY

The political geography of Africa south of the Sahara is bewildering to North Americans. Of its many independent countries, most did not exist under their present names and within their present boundaries until after the late 1960s; some achieved independence only recently.

On the outline map of the region in Figure 11–2, locate and label the nations and capitals listed in Table 11–1.

**TABLE 11–1** Countries and capitals of Africa south of the Sahara.

| Country | Capital |
|---|---|
| Senegal | Dakar |
| Gambia | Bathurst |
| Guinea-Bissau | Bissau |
| Guinea | Conakry |
| Sierra Leone | Freetown |
| Liberia | Monrovia |
| Ivory Coast (Côte d'Ivoire) | Abidjan |
| Ghana | Accra |
| Burkina Faso | Ouagadougou |
| Togo | Lome |
| Benin | Porto Novo |
| Nigeria | Lagos |
| Cameroon | Yaounde |
| Equatorial Guinea | Malabo |
| Gabon | Libreville |
| Congo | Brazzaville |
| Central African Republic | Bangui |
| Democratic Republic of the Congo (Zaire) | Kinshasa |
| Angola | Luanda |
| Namibia | Windhoek |
| Botswana | Gaborone |
| South Africa | Pretoria†,†† |
| Lesotho | Maseru |
| Swaziland | Mbabane |
| Zimbabwe | Harrare |
| Mozambique | Beira |
| Zambia | Lusaka |
| Malawi | L'longwe |
| Tanzania | Dar es Salaam‡ |
| Rwanda | Kigali |
| Burundi | Bujumbura |

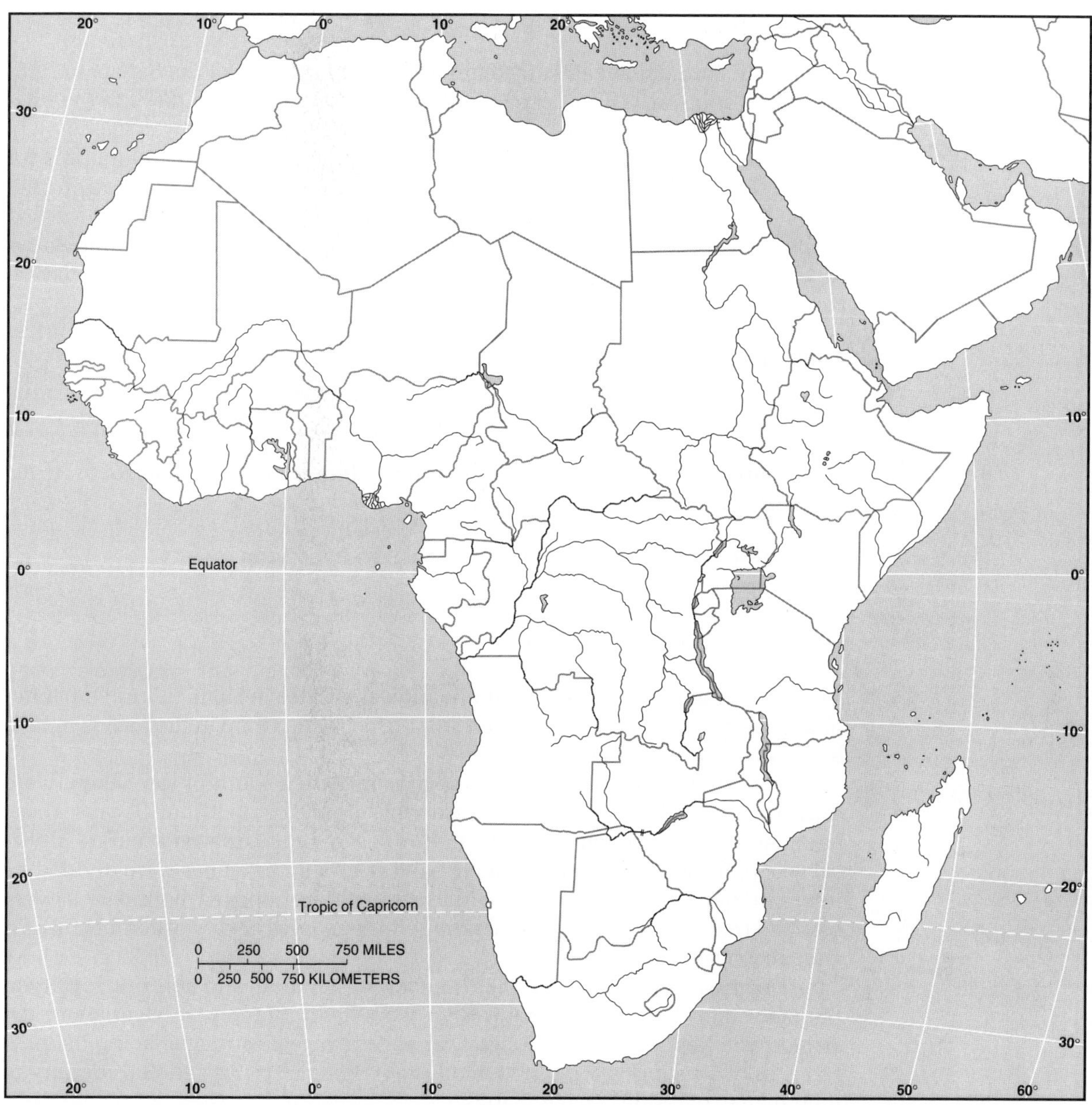


**FIGURE 11–2**
Countries and capitals of Africa south of the Sahara.

| | |
|---|---|
| Uganda | Kampala |
| Kenya | Nairobi |
| Somalia | Mogadishu (Muqdisho) |
| Ethiopia | Addis Ababa |
| Eritrea | Asmara |
| Djibouti | Arta (Djibouti) |
| Madagascar | Antananarivo |

†Pretoria is South Africa's administrative capital. Capetown is the seat of the legislature, and Bloemfontein is the judicial capital.

††In 2000, the South African government announced its intention to replace European placenames with names of African origin.

‡Tanzania intends to build a new capital at Dodoma.

### Colonialism

The tide of colonialism advanced quickly in Africa south of the Sahara and receded just as quickly. European colonial control advanced rapidly into the interior from fortified trade centers on the coast after 1870. In the late 1870s, a newly united Germany had entered the race for colonies, precipitating a scramble for territory in Africa. In 1884, lasting into 1885, the European colonial powers met in the Conference of Berlin to divide up Africa peaceably, without a European war erupting over colonial competition. Borders were drawn using inaccurate maps of African physical features; these borders ignored tribal and cultural boundaries. No Africans were invited to this conference.

**11–4** Refer to Figure 8–7, the map of colonialism in 1914, just before World War I. How many countries in all of Africa were independent, self-governing states? ____________________

Now, using Figure 11–3, make a similar map showing colonial control of Africa south of the Sahara in 1970:

Portuguese control: Guinea-Bissau; Angola; Mozambique
British control: Zimbabwe (Rhodesia)*
South African control: Namibia
Independent states: all the rest

Assign different colors to each colonial power and leave blank (white) the independent states. Be sure to label each color used on this map in the map legend, a box explaining what the colors mean.

As stated earlier, national boundaries in Africa south of the Sahara were drawn originally by Europeans for reasons of colonial competition. Because these boundaries paid no attention to the cultural, economic, or even accurate physical geography of Africa, traditional tribal territories sometimes were divided by international boundaries. Within these European-imposed boundaries were grouped traditional enemies, rival groups, and peoples of very different cultures and languages.

Creating a state is not the same as creating a nation, and building loyalty to the new states has not been easy. Civil wars nearly tore apart Nigeria and the Democratic Republic of the Congo (Zaire). The biggest struggle facing the new states in this region is the problem of establishing a strong national identity of their own to override tribal loyalties.

## ECONOMIC GEOGRAPHY

How can the poverty and malnutrition so unfortunately typical of Africa south of the Sahara be explained? A representative country of this region sells raw materials to the rest of the world—minerals, agricultural commodities, or both. The mineral resources are a gift of nature; "gold is where you find it." Agricultural commodities, on the other hand, are likely to reflect the opportunities provided by the natural environment (particularly climate) and the exploitation of these opportunities by people.

Any particular mineral resource, from iron ore to diamonds, *can* be a natural monopoly of a handful of countries. However, because many Latin

*In 1965, Rhodesia's white minority government declared itself independent of British control, a declaration not recognized by the British government. Majority rule as an independent state came in 1980.

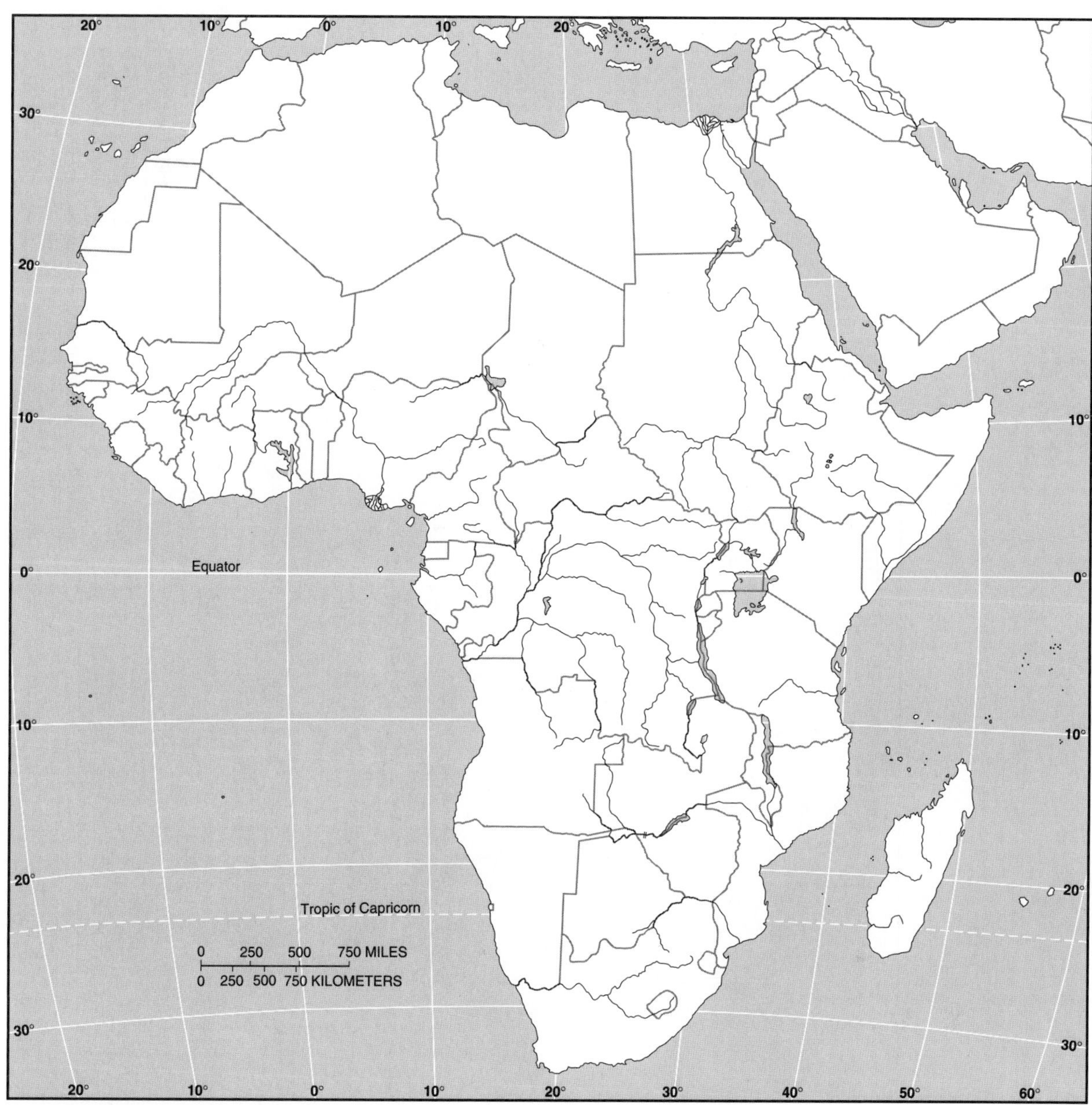


**FIGURE 11–3**
Colonial control in 1970.

American, South Asian, and Southeast Asian countries have tropical climates reasonably similar to those of Africa south of the Sahara, this region's countries do not have the potential advantages of monopolies in agricultural commodities markets. In most commodities, mineral or agricultural, competition has been so strong that prices have been drifting lower.

On the world outline map showing national boundaries in Figure 11–4, make a very general map of tropical crop production. Using different symbols for crops—a circle for cacao (chocolate), a square for coffee, and a triangle for palm oil—place the rank order of each country's share of world production within the symbol. For example, Nigeria leads the world in palm oil production, so Nigeria will have a triangle with the number 1 inside. Use the data provided in Table 11–2.

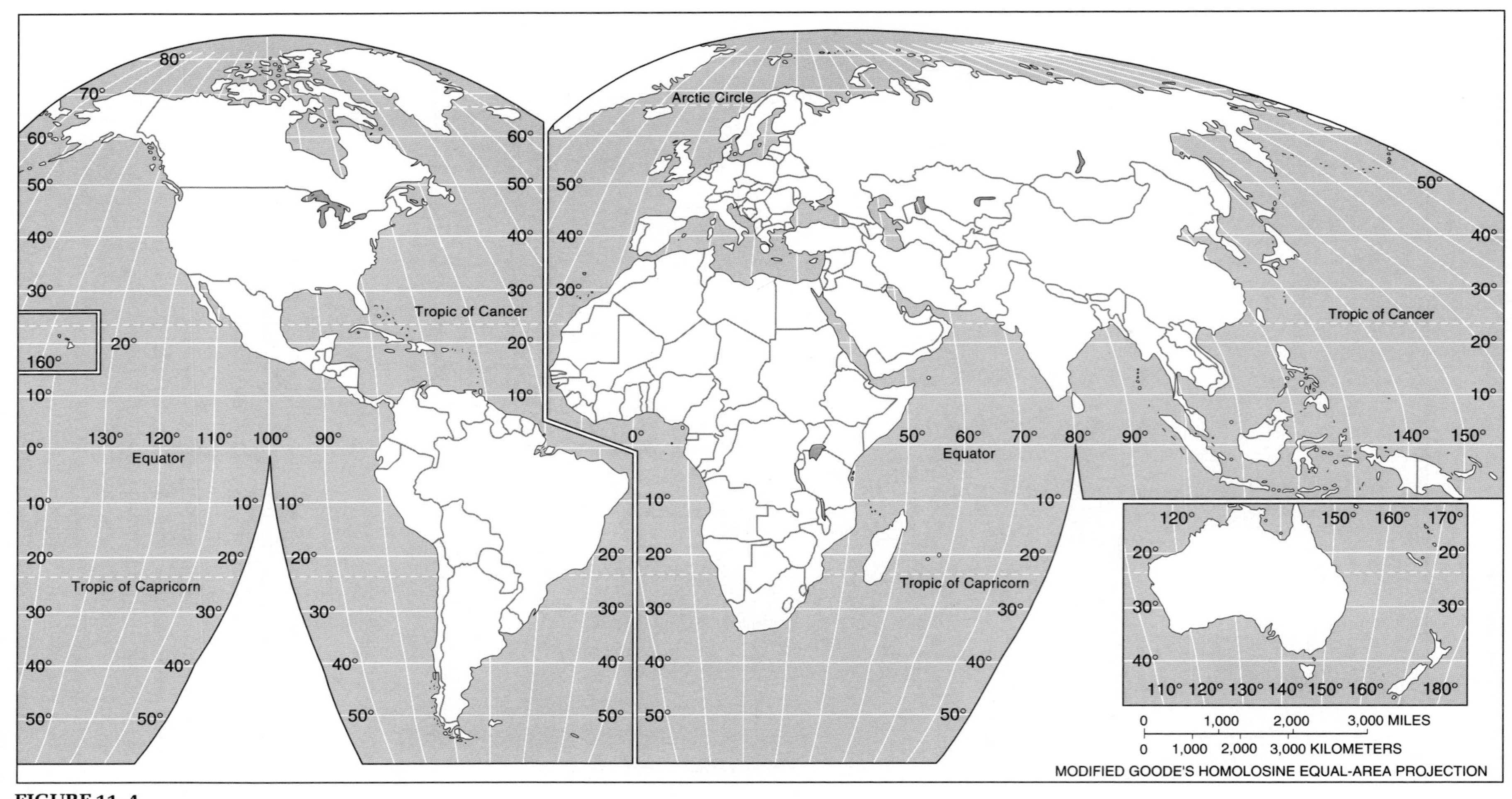


**FIGURE 11–4**
Tropical crop production.

**TABLE 11–2** Tropical crop producers and world rank.

| Cacao | Coffee | Palm Oil |
|---|---|---|
| 1. Ghana | 1. Brazil | 1. Nigeria |
| 2. Nigeria | 2. Colombia | 2. Malaysia |
| 3. Côte d'Ivoire | 3. Côte d'Ivoire | 3. Zaire |
| 4. Cameroon | 4. Angola | 4. Indonesia |
| 5. Brazil | 5. Uganda | 5. Sierra Leone |
| 6. Ecuador | 6. Ethiopia | 6. Benin |
| 7. Malaysia | 7. Mexico | 7. Cameroon |
| | 8. El Salvador | 8. Ghana |
| | 9. Guatemala | 9. Côte d'Ivoire |
| | 10. Indonesia | 10. Angola |
| | | 11. Liberia |

As noted, no country, region, or continent holds a monopoly on any of these tropical commodities. Similar physical environments make it possible for the same crops to be produced in Africa south of the Sahara, Latin America, and Southeast Asia. Coffee, for example, originally seems to be a gift to humanity from Ethiopia, but now Ethiopia is a relatively small producer, and the two largest producers are in South America (Brazil and Colombia).

On the outline map of Africa south of the Sahara in Figure 11–5, note the actual percentage of the total labor force engaged in agriculture shown in Table 11–3.

The countries of Africa south of the Sahara are outstanding examples of the agricultural employment paradox. High proportions of the total labor force engaged in agriculture are typical of poor, pre-industrial countries. These underdeveloped countries tend to have high rates of malnourishment, even starvation. On the other hand, countries with relatively few farmers in the labor force, such as the U.S. and Canada, turn out to be rich countries with, on average, well-fed citizens.

The fewer farmers, proportionately, the more food is available per person, and the reverse is true. Also, this seeming paradox can be explained, of course, by the fact that industrialized societies have modernized and industrialized agriculture as well.

**11–5** Which of the following countries does *not* have over 90% of its labor force in agriculture? (Circle the correct answer.): South Africa, Burundi, Burkina Faso, Rwanda.

**11–6** Referring to the statistical appendix, do any of the countries of this region with 90% or more of the labor force in agriculture have per capita GNP's over $300? (Circle the correct answer.): Yes, No.

**11–7** Which country is the exception to the rule that low proportions of farmers in the workforce generally are associated with higher average GNP's per capita? (The average GNP per capita for Africa south of the Sahara is $520.)—(Circle the correct answer.): South Africa, Madagascar, Botswana, Gabon.

**TABLE 11–3** African nations—percentage of labor force engaged in agriculture, 2000 (data not available for all countries).

| Country (in order of income per capita, lowest first) | Employment (in agriculture as a percentage of labor force) |
|---|---|
| Ethiopia | 89 |
| Djibouti | 60 |
| Democratic Republic of Congo | 65 |
| Burundi | 93 |
| Sierra Leone | 65 |
| Liberia | 70 |
| Guinea-Bissau | 77 |
| Eritrea | NA |
| Niger | 90 |
| Malawi | 86 |
| Mozambique | 80 |
| Tanzania | 90 |
| Burkina Faso | 92 |
| Rwanda | 93 |
| Madagascar | 31 |
| Central African Republic | 40 |
| Nigeria | 54 |
| Gambia | 75 |
| Uganda | 86 |
| Togo | 65 |
| Zambia | 85 |
| Kenya | 75 |
| Angola | 85 |
| Benin | 55 |
| Ghana | 61 |
| Senegal | 60 |
| Guinea | NA |
| Lesotho | 86 |
| Cameroon | 25 |
| Zimbabwe | 27 |
| Congo | 13 |
| Côte de'Ivoire | 51 |
| Equatorial Guinea | 72 |
| Swaziland | NA |
| Namibia | 49 |
| Botswana | 5 |
| South Africa | 30 |
| Gabon | 65 |
| **(For comparison purposes, not to be mapped)** | |
| Australia | 5 |
| Canada | 3 |
| Japan | 4 |
| UK | 2 |
| U.S. | 2 |

## CULTURAL GEOGRAPHY AND DEMOGRAPHICS

Although now one of the fastest-expanding regions in terms of population growth, Africa south of the Sahara was slow growing in the past. Table 11–4 shows estimated populations for continental regions from the year 1000 to 1960.

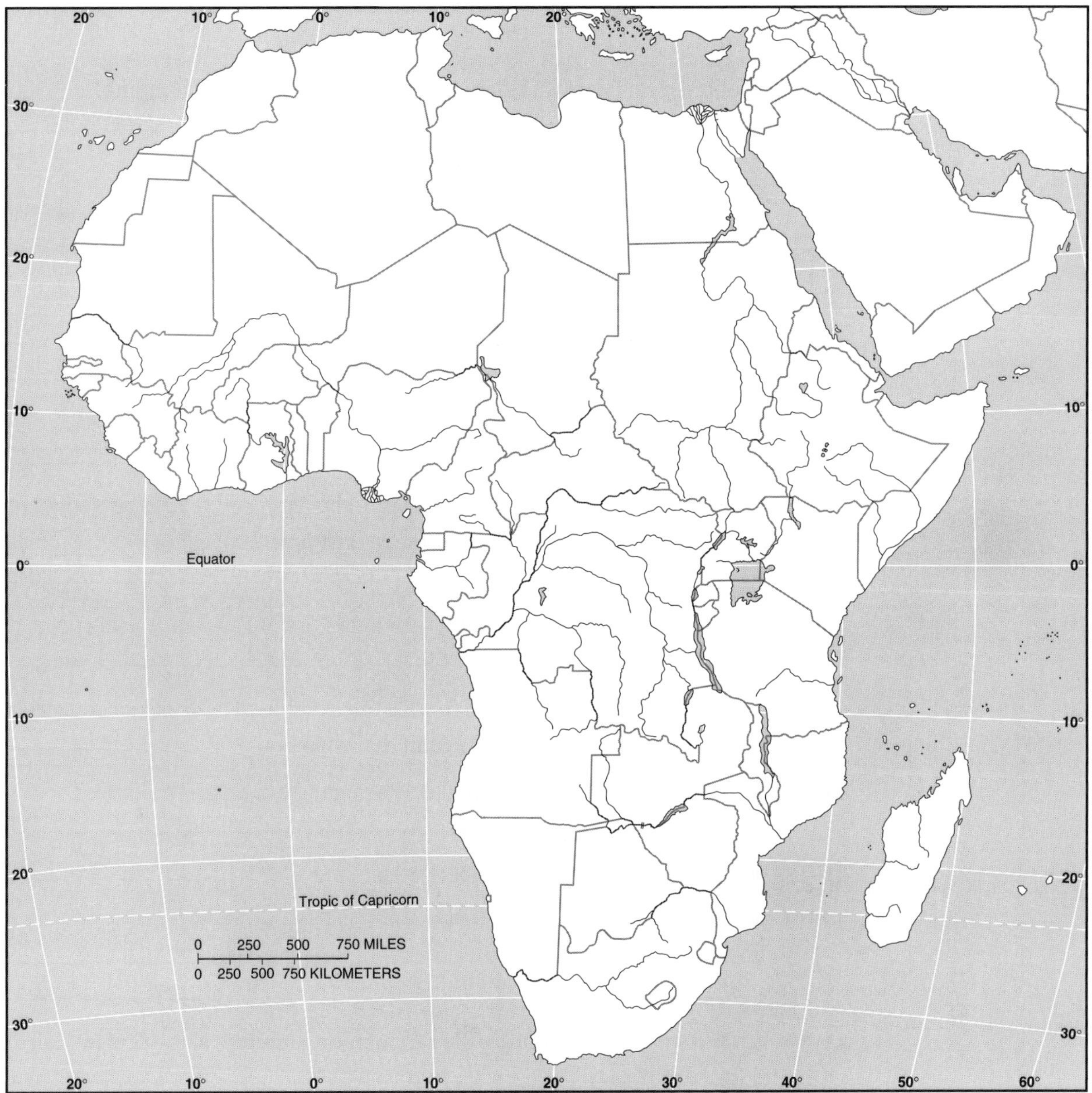


**FIGURE 11–5**
Percentage of labor force in agriculture.

**TABLE 11–4** Populations of continental regions, A.D. 1000–2000 (millions of people).

| Region | 1000 | 1600 | 1800 | 1900 | 1960 | 2000 |
|---|---|---|---|---|---|---|
| Asia and Oceania | 165 | 279 | 599 | 921 | 1700 | 3642 |
| Europe and newly independent states* | 47 | 102 | 192 | 423 | 641 | 801 |
| Africa | 50 | 90 | 90 | 120 | 244 | 800 |
| North and South America | 13 | 15 | 25 | 144 | 407 | 824 |
| Total | 275 | 486 | 906 | 1608 | 2992 | 6067 |

*Former USSR

## CHECK UP

**11–8** Which region has always had the highest proportion of humans?

**11–9** Which region grew at the fastest rate from 1000 to 1600?

**11–10** Why had Africa's proportion of the total declined by 1800?

**11–11** How was this related to the slave trade?

**11–12** What had happened to Africa's share by 1960, compared to 1800?

**11–13** Africa's share of total world population by the year 2010 is projected at 14%. How does that figure compare to past percentages?

**11–14** How many countries in Africa south of the Sahara are landlocked?

**11–15** How does this number compare with the number of landlocked countries in other world regions as defined in this workbook?

**11–16** In the colonial period, which were Portugal's two largest colonies?

**11–17** Which huge country once belonged to Belgium?

**11–18** Which countries were once controlled by Germany?

**11–19** Which two countries border the Democratic Republic of the Congo (Zaire) to the south?

**11–20** The Straits of Bab al Mandeb connect the Gulf of Aden with which sea?

**11–21** If a world power wished to acquire a naval base in a position to control shipping through the Bab al Mandeb (12° N, 43° E), which three states in Africa south of the Sahara would be likely possibilities?

**11–22** What African country is completely surrounded by South Africa?

**11–23** Which countries of Africa south of the Sahara front on the Indian Ocean?

**11–24** The Mozambique Channel separates which two countries?

**11–25** Surface transport through which of its neighbors could provide Zambia with an outlet to the sea? ____________________

**11–26** Circle the name of the country that would *not* have interested Europeans as a reprovisioning stopover on the way to and from India (pre-Suez Canal): South Africa, Zimbabwe, Mozambique.

**11–27** The West African subregion usually is defined as the area from Senegal to Nigeria. Circle the name of the state that is *not* located in West Africa: Sierra Leone, Benin, Ivory Coast, Liberia, Gabon.

**11–28** If you traveled due south from Burkina Faso, which countries could you visit? ____________________

____________________

____________________

**11–29** Which of these capital cities is *not* a seaport? (Circle the correct answer.): Luanda, Maputo, Nairobi, Mogadishu.

At present, the continent of Africa (Africa south of the Sahara plus North Africa) contains about 13% of the globe's population. Projections by the Population Reference Bureau show an increase in that continent's share of humans to 16% by 2025, and 20% by 2050.

Projections to 2050 indicate little change in their proportionate share of population on the parts of Asia, the Americas, or Australia, New Zealand and the Pacific Islands. Europe, however, may see a decline in its share of the human race from a current 12% to 7% by 2050.

Table 11–5 lists the "top 20" countries, as ranked by population size, for 2000 and as projected for 2050. On a world outline map, Figure 11–6, locate and color with a dark, solid color, the top 10 for 2000. Also using 2000 data, locate the countries ranked 11 through 20, labeling them with diagonal parallel lines. On another world map, Figure 11–7, do the same for the top 10 and 11–20 ranked states for 2050. Be sure to provide keys or legends on both maps explaining the meaning of the patterns.

**11–30** Which countries of Africa south of the Sahara appeared in the highest ten for 2000? ____________________

____________________

**11–31** Which countries of Africa south of the Sahara appeared in the second tier (11–20) list for 2000? ____________________

____________________

**11–32** Which countries of Africa south of the Sahara are projected to join the "top 10" by 2050? ____________________

____________________

**11–33** Which countries of Africa south of the Sahara are projected to join the "second 10" (11–20) by 2050? ____________________

____________________

____________________

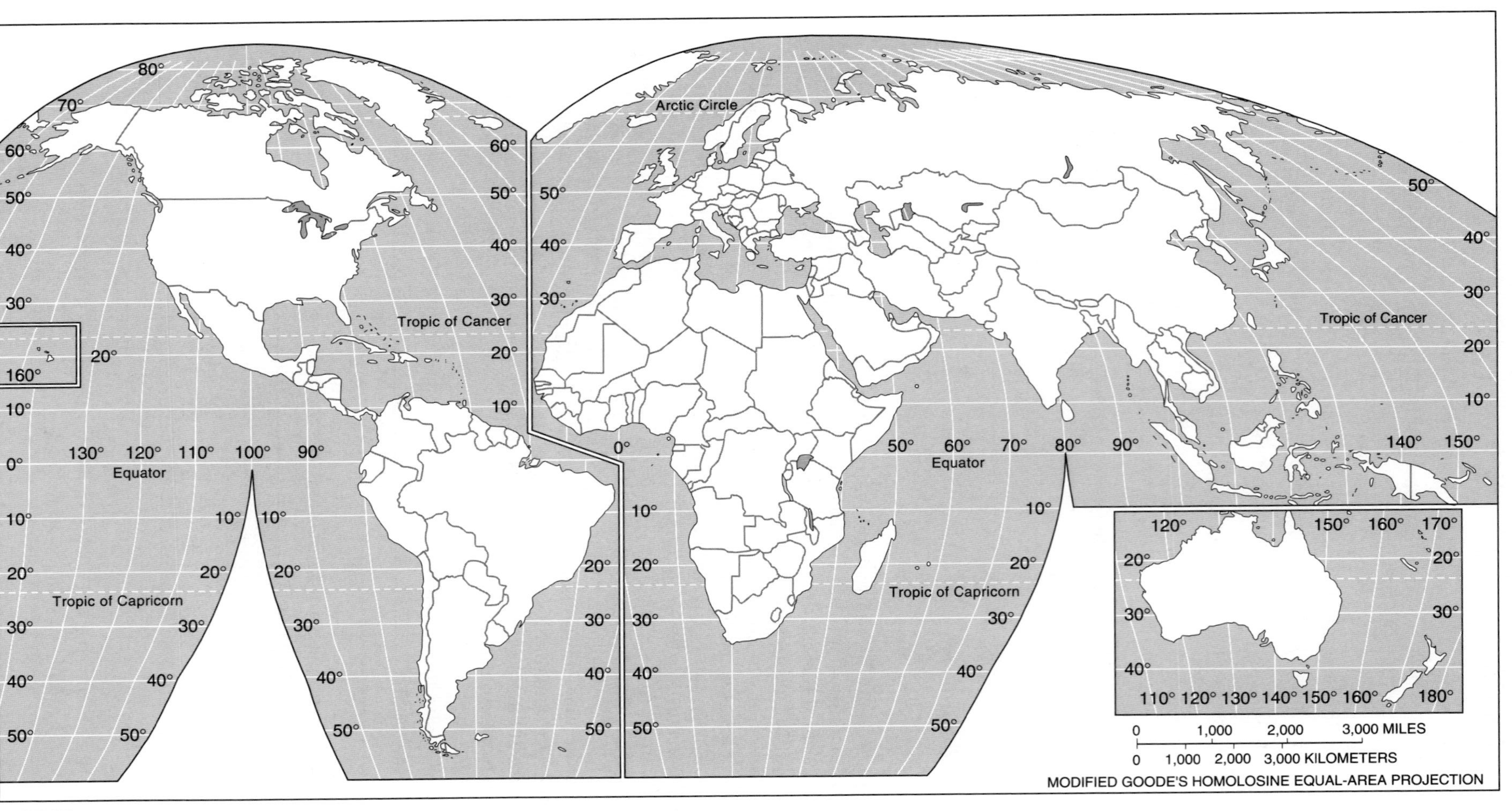


**FIGURE 11–6**
Top 20 nations in population, 2000.

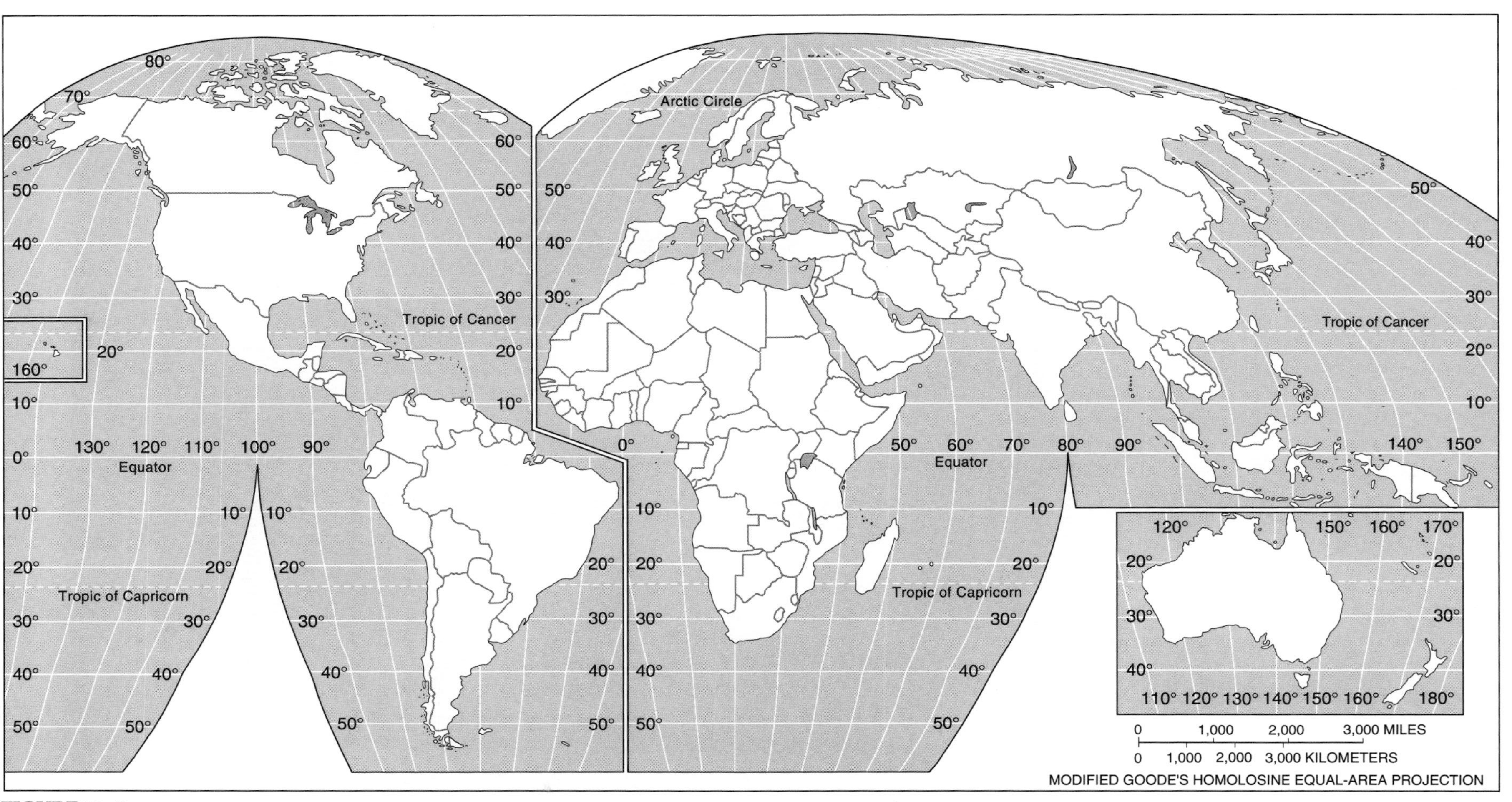


**FIGURE 11–7**
Top 20 nations in population, 2050.

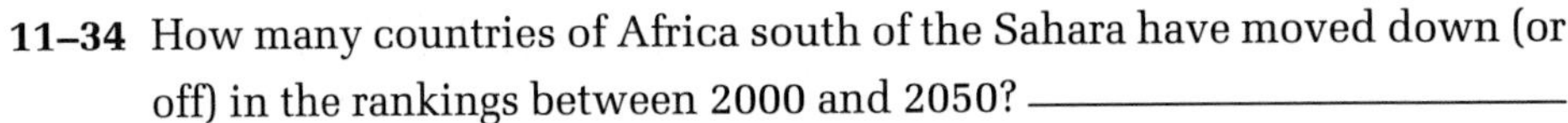

**11–34** How many countries of Africa south of the Sahara have moved down (or off) in the rankings between 2000 and 2050? ____________________

____________________

**11–35** How many European countries (do not count Russia) had moved down in the rankings, 2000 to 2050? ____________________

____________________

**TABLE 11–5** World's Largest Countries, by Population.

| 2000 | | | 2050 | | |
|---|---|---|---|---|---|
| **Rank** | **Country** | **Population (Millions)** | **Rank** | **Country** | **(Projected) Population (Millions)** |
| 1 | China | 1265 | 1 | India | 1628 |
| 2 | India | 1002 | 2 | China | 1369 |
| 3 | U.S. | 276 | 3 | U.S. | 404 |
| 4 | Indonesia | 212 | 4 | Indonesia | 312 |
| 5 | Brazil | 170 | 5 | Nigeria | 304 |
| 6 | Pakistan | 151 | 6 | Pakistan | 285 |
| 7 | Russia | 145 | 7 | Brazil | 244 |
| 8 | Bangladesh | 128 | 8 | Bangladesh | 211 |
| 9 | Japan | 127 | 9 | Ethiopia | 188 |
| 10 | Nigeria | 123 | 10 | Dem. Rep. Congo | 182 |
| 11 | Mexico | 100 | 11 | Mexico | 152 |
| 12 | Germany | 82 | 12 | Philippines | 140 |
| 13 | Philippines | 80 | 13 | Russia | 128 |
| 14 | Vietnam | 79 | 14 | Vietnam | 124 |
| 15 | Egypt | 68 | 15 | Egypt | 117 |
| 16 | Iran | 67 | 16 | Iran | 103 |
| 17 | Turkey | 65 | 17 | Turkey | 101 |
| 18 | Ethiopia | 64 | 18 | Japan | 100 |
| 19 | France | 59 | 19 | Tanzania | 88 |
| 20 | Italy | 58 | 20 | Myanmar | 88 |

Infant mortality rates—deaths of infants after birth and before the first birthday—are a classic measure of relative poverty or affluence within a society. Poor countries, where adequate nourishment and access to good medical care are less likely than in more prosperous societies, are characterized by high infant mortality, expressed per thousand live births. Table 11–6 lists the 20 countries of the world with the highest infant mortality rates, and the 20 with the lowest rates. On a world map, Figure 11–8, locate the nations with high infant mortality rates. Color them with a dark, solid color. Locate the 20 lowest rates, and use a diagonal line pattern to identify them on the map.

**11–36** Which countries with very high infant mortality rates (over 108) were *not* in the Africa south of the Sahara region? ____________________

____________________

**11–37** Which major region contains most of the "top 20" lowest infant mortality rates? ____________________

____________________

**11–38** Which Southeast Asian country appears on the list of lowest infant mortality rates? ______________________________

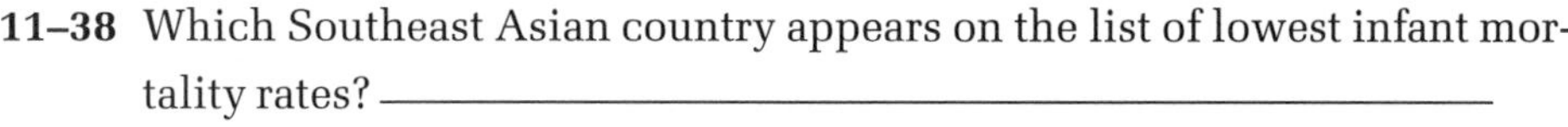

TABLE 11–6

| High Infant Mortality Rates—Top 20 | | Low Infant Mortality Rate—Top 20 | |
|---|---|---|---|
| World Average | 57/1000 live births | | |
| Sierra Leone | 157 | Iceland | 2.6 |
| Western Sahara | 150 | Singapore | 3.2 |
| Afghanistan | 150 | Hong Kong | 3.2 |
| Liberia | 139 | Japan | 3.5 |
| Gambia | 130 | Sweden | 3.5 |
| Guinea-Bissau | 130 | Norway | 4.0 |
| Iraq | 127 | Finland | 4.2 |
| Somalia | 126 | Czech Republic | 4.6 |
| Angola | 125 | Germany | 4.7 |
| Mali | 123 | Denmark | 4.7 |
| Niger | 123 | France | 4.8 |
| Rwanda | 121 | Switzerland | 4.8 |
| Ethiopia | 116 | Austria | 4.9 |
| Djibouti | 115 | Luxembourg | 5.0 |
| Côte d'Ivoire | 112 | Netherlands | 5.0 |
| Chad | 110 | Slovenia | 5.2 |
| Congo | 109 | Australia | 5.3 |
| Uganda | 109 | Malta | 5.3 |
| Dem. Rep. Congo | 109 | Portugal | 5.4 |
| Zambia | 109 | Italy | 5.5 |

# GEOCONCEPTS

## Colonialism

Almost the entire region—indeed, almost the entire continent of Africa—was controlled by European countries only 40 years ago. When Ethiopia temporarily was incorporated into the Italian empire (1936–1942), Liberia remained the only independent state north of minority-ruled South Africa. Since the collapse of the Spanish empire in mainland South America and Central America, no other continent was controlled so nearly completely by colonial powers—people from another continent.

What were the bad and good effects, if any, of colonialism? The negatives of life under colonial domination are well known. A typical list of complaints against a colonial power might include charges that that power represses attempts to establish a freely elected government; appoints judges who are not sympathetic to or accountable to the local population; sends swarms of administrators to harass the people; interferes with local trade and with overseas partners; imposes taxes without any local representation; encourages ethnic and racial conflict; forces local people to serve in the colonial power's military forces; and makes the occupying army legally superior to any locally chosen officials. All these wrongs, and more, were forced on colonies in Africa, though that particular

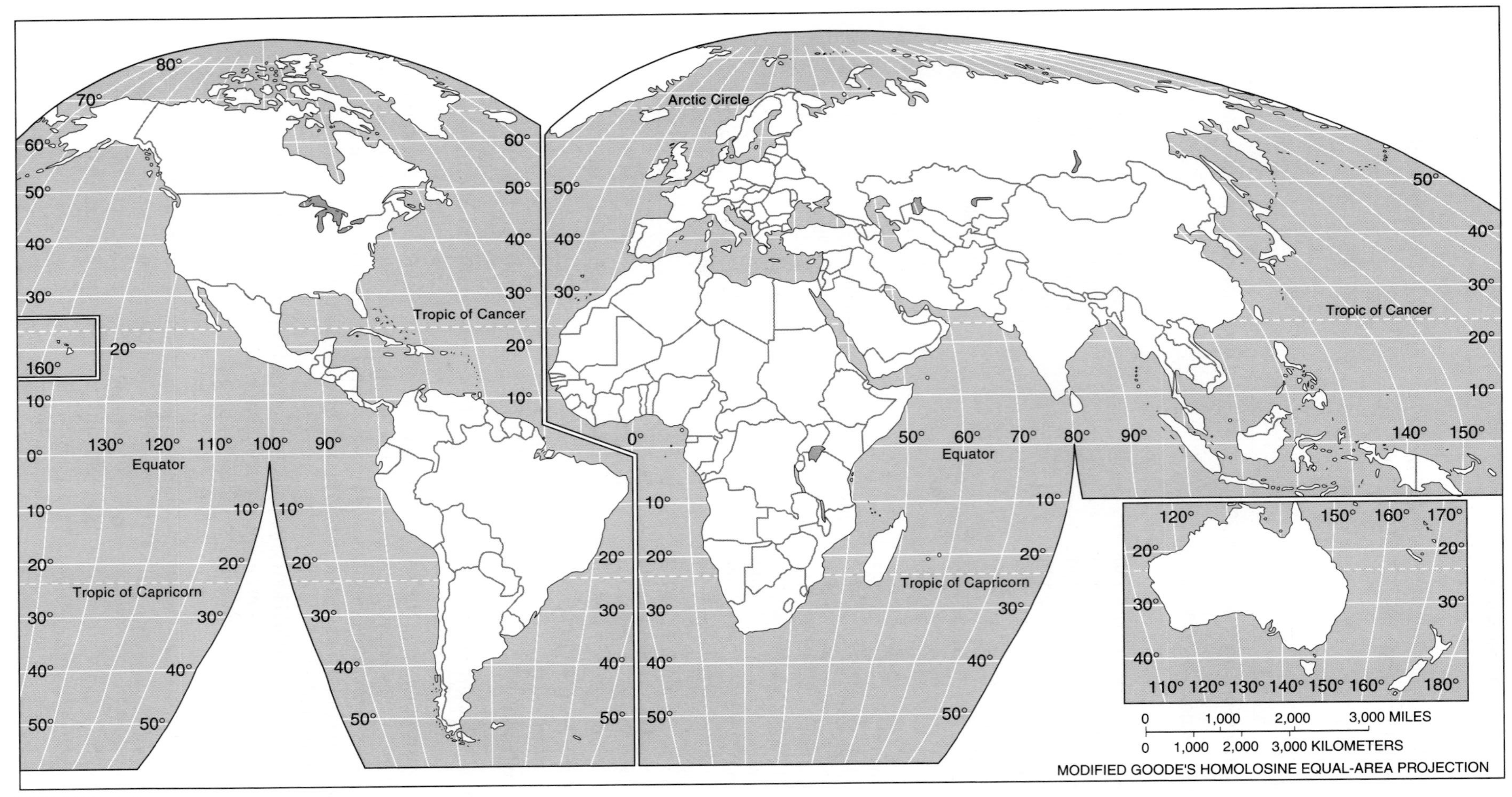


**FIGURE 11–8**
Infant mortality rates: 20 highest, 20 lowest.

list of complaints is paraphrased from the Declaration of Independence of the United States of America, July 4, 1776.

Were there any positive aspects, even if incidental, of having been a colony? A definite "yes" answers this question. Most African colonies were seen as suppliers of tropical commodities and minerals, where available. A few were important, at least at one time, for their strategic location, like South Africa. A handful of colonies even attracted Europeans as permanent settlers, including South Africa, Kenya, and Rhodesia (Zimbabwe). Whatever the relative importance of these motives, the colonizer had to make investments in its colony. Defending a colonial territory, as well as exploiting its resources and people, required *infrastructure*—port facilities like lighthouses, deep-water wharfs, refueling facilities, and warehouses. Developing mines or plantations in the interior usually meant building railroads, roads, bridges, tunnels, telephone and telegraph lines, and power lines. Airports, hydroelectric dams, even whole capital cities were other results of such developmental investments by the colonizing power. The value of these infrastructure investments is illustrated by capital cities. In virtually every case, the present capital of the now-independent state is the same city used as the administrative center for the colony.

**11–39** Except landlocked states, which countries in this region have capital cities that are *not* seaports? ______________________________

____________________________________________

To be sure, the local people of the colony paid for all these improvements, but the investments *were* made.

## Neocolonialism

Neocolonialism is the situation in which political independence does not seem to be matched by economic independence. What good does it do, critics ask, if legal independence is accompanied by continuing dependence on the former colonial power to provide capital and technology? If companies controlled by the former colonial power still handle a country's exports; supply critical imports; control banking; own local mines, plantations, and factories; and provide international transport and communications links, is the independent country really free to pursue it own policies and goals?

## Tribalism's Threat to Political Stability

"Tribalism" describes the political situation in which members of the tribe (an ethnic, linguistic, or cultural group with a strong identity) give their primary loyalty to their tribe rather than to the state of which they are citizens. Tribalism is both a threat to the internal peace and stability of a state and an international problem—not one limited to any one continent or region. The tragic civil wars between tribal or ethnic groups in the small central African states of Rwanda and Burundi [and influencing neighboring Democratic Republic of the Congo (Zaire)] are examples of the disastrous consequences of extreme tribalism. Long-standing intertribal rivalries and grievances, never supplanted by loyalty to the state, can lead to chaos and mass murder. Rwanda and Burundi territories were detached by Britain from the conquered German colony of Tanganyika (now Tanzania) and handed over to Belgian administration through the Belgian Congo [Democratic

Republic of the Congo (Zaire)]. Like most boundaries in Africa, Rwanda and Burundi are accidents of European politics rather than reflective of tribal boundaries.

## REGIONAL WATCHLIST

Africa south of the Sahara is a major world region of enormous potential. However, this bright potential may not emerge in the near future. Realizing the potential of this region largely depends on the states of this area triumphing over a powerful, if not necessarily related, group of political and economic handicaps. These include neocolonialism, low world-market prices for most tropical commodities, high probability of intertribal tensions leading to more civil wars and secessions, and the unresponsiveness of some undemocratic governments to their own citizens. But the long-range picture is bright with promise.

# 12

# The Middle East and North Africa

## INTRODUCTION

The sprawling region of the Middle East and North Africa extends from the heart of southwestern Asia to the Atlantic shores of northern Africa and includes the countries of Algeria, Bahrain, Chad, the island of Cyprus, Egypt, Iran, Iraq, Israel, Jordan, Kuwait, Lebanon, Libya, Mali, Mauritania, Morocco, Niger, Oman, Qatar, Saudi Arabia, Sudan, Syria, Tunisia, Turkey, United Arab Emirates, and Yemen. The region merits close attention from the rest of the world for several reasons. Historically, this region is the birthplace of the world's three great monotheistic religions—in historical order, Judaism, Christianity, and Islam. Culturally, as an ancient cradle of civilizations, this region created the basis for much of what is commonly thought of as "western" culture. This region lies at the crossroads of three continents—Asia, Africa, and Europe—a location that is the key to understanding this region's past, present, and future importance.

**12–1** In the system of regions used in this workbook, the Middle East–North Africa region is the only one sharing boundaries with *four* other regions, which are ________________________________________

____________________________________________________.

This crossroads position means that this is the meeting ground where people from three continents intermix and interchange ideas, technologies, and cultures. Oil provides another, more modern regional asset. This region holds 60% of the world's known oil reserves; Saudi Arabia alone controls one-quarter of the planet's reserves. At current output levels, this region supplies a third of all the oil consumed in the world. This makes the entire global economy extremely sensitive to any event in Middle East–North Africa that could affect the supply and price of petroleum products, such as the Iraqi invasion of Kuwait in 1990.

The term *Middle East* is immediately recognizable, though it is not free of a western, ethnocentric viewpoint: it is the Middle East only with reference to Europe. The term *Middle East* became widely used in Europe in the last century when *Near East* referred to Turkey and the eastern end of the Mediterranean, and the *Far East* was China, Korea, and Japan. *Middle East* then meant the Arabian Peninsula, Iraq, Iran, and perhaps Afghanistan. Now, the term *Near East* has almost disappeared from use, and *Middle East* has expanded to include the entire area from Asia's Mediterranean shores to Iran and often Afghanistan. European's Middle East, though, might as easily be called the "Middle West" by the Chinese, or the "Northeast" by central Africans! East–west terminology always depends on the location of the observer.

On the outline map of the region in Figure 12–1, locate and label the countries and cities in Table 12–1.

**TABLE 12–1** Middle Eastern and North African countries, capitals, and major cities.

| Country | Capital | Other Major Cities |
|---|---|---|
| Morocco | Rabat | Casablanca |
| Western Sahara (a territory of Morocco) | | |
| Mauritania | Nouakchott | |
| Mali | Bamako | |
| Algeria | Algiers (Alger) | |
| Niger | Niamey | |
| Chad | N'Djamena | |
| Tunisia | Tunis | |
| Libya | Tripoli (Tarabulus) | |
| Sudan | Khartoum | |
| Egypt | Cairo (El Qahira) | Alexandria (El Iskandariya) |
| Israel | Jerusalem | Tel Aviv |
| Jordan | Amman | |
| Saudi Arabia | Riyadh | Mecca (Mekka); Medina |
| Yemen | Sana | |
| Oman | Muscat | |
| United Arab Emirates | Abu Dhabi (Abu Zaby) | |
| Qatar | Doha (Ad Dawhah) | |
| Bahrain | Bahrain (Al Manamah) | |
| Kuwait | Kuwait City (Al Kuwayt) | |
| Iraq | Baghdad | |
| Iran | Tehran | |
| Lebanon | Beirut (Bayrut) | |
| Syria | Damascus (Dimashq) | |
| Turkey | Ankara (Angora) | Istanbul |
| Cyprus | Nicosia | |

## PHYSICAL GEOGRAPHY

All regions possess some degree of internal homogeneity, some characteristic that is typical of the region and less typical of other neighboring areas. The Middle East (southwest Asia) and North Africa region commonly is pictured as one big desert, awash in oil, and populated by Arabs. As it happens, none of these generalizations can be applied to all parts of this region. Arabs are an important, but certainly not the only, ethnic group in this region, and some countries here have no oil or very little of it.

Although desert and semidesert conditions prevail, there are important exceptions. Look at a world map with parallels of latitude labeled. On the west-facing coasts of North and South America, Europe, Africa, and Australia, find latitudes 32° or 33° to about 40° away from the Equator, both north and south. These are the general locations of the Mediterranean-type climates; sunny, dry summers are followed by mild, relatively warm winters in which the area receives most of its rainfall. Central and southern California are good examples of this very pleasant climate. Countries or areas with this climate are often major wine producers, because wine grapes flourish there. All of the countries bordering on the Mediterranean Sea, except Libya and Egypt, have some "Mediterranean" climate.

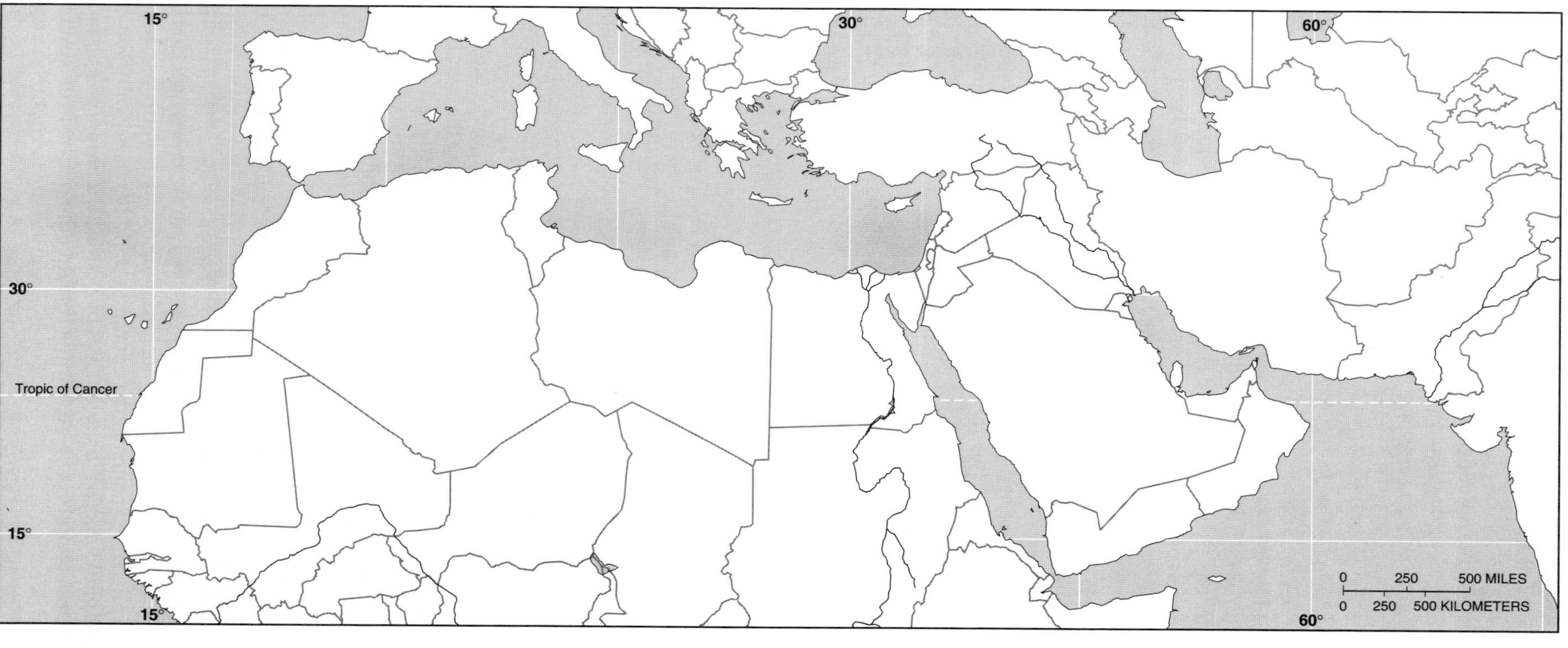


**FIGURE 12–1**
Middle Eastern and North African countries, capitals, and major cities.

**12–2** Keeping in mind the general association of wine production with this lower-midlatitude west coast climate, circle the names of the countries/areas that you would expect to be important wine or grape producers: Portugal, Spain, southern France, Greece, Indonesia, South Africa, Algeria, Brazil, Chile, Israel, Sudan, California, Australia.

Notice, in every case, the Mediterranean-type climates lie on the poleward side of deserts. If you leave a Mediterranean climate area and head directly toward the Equator, you will always come to a desert. On the west coasts of all continents at approximately 20° to 30° latitude, north or south, there is a desert. These deserts are flanked on both their poleward and equatorward sides by transitional, semidesert climates. Because the Middle East–North Africa region falls within these zones, the three most common climate types found there are desert, semidesert, and Mediterranean.

On the outline map of the region in Figure 12–2, locate and label the following:

| | | | |
|---|---|---|---|
| Mediterranean Sea: | 33° N, 15° E | Gulf of Gabes: | 34° N, 11° E |
| Black Sea: | 42° N, 35° E | Gulf of Aden: | 12° N, 46° E |
| Caspian Sea: | 40° N, 52° E | Persian Gulf: | 27° N, 52° E |
| Aegean Sea: | 38° N, 25° E | Strait of Hormuz: | 27° N, 56° E |
| Red Sea: | 22° N, 37° E | Arabian Sea: | 20° N, 65° E |
| Gulf of Suez: | 28° N, 33° E | Dardanelles Strait: | 40° N, 27° E |
| Gulf of Aqaba: | 27° N, 34° E | Atlantic Ocean | |
| Strait of Bab al Mandeb: | 12° N, 43° E | Nile River: | 28° N, 32° E |
| Strait of Gibraltar: | 36° N, 6° W | Arabian Peninsula: | 25° N, 45° E |
| Tigris River: | 34° N, 44° E | Sinai Peninsula: | 28° N, 34° E |
| Euphrates River: | 36° N, 40° E | Dead Sea: | 32° N, 36° E |
| Jordan River: | 32° N, 36° E | Isle of Cyprus: | 35° N, 33° E |
| Isle of Socotra: | 12° N, 54° E | Gulf of Oman: | 24° N, 58° E |
| Gulf of Sidra (Khalij Surt): | 31° N, 18° E | Indian Ocean | |
| Sea of Marmara: (Marmara Denizi) | 41° N, 27° E | Bosporus Strait: | 41° N, 28° E |

## OBJECTIVES AND STUDY HINTS

Geography is concerned with locations, the characteristics of locations, and the significance of relative location. As noted before, geographers describe locations at two levels, site and situation. *Site* describes the specifics of the immediate surroundings. For example, the site of Istanbul (its more ancient name is Constantinople) is at the eastern end of a narrowly tapering peninsula between the Black Sea to the north, the Sea of Marmara to the south, and the Bosporus Strait to the east at about 41° N, 29° E.

*Situation* deals with *relative location,* location relative to a country, a larger region, or to the entire world. The situation of Istanbul underlines the importance of that city—it lies on the route between the Ukraine's Black Sea ports and the Mediterranean. Istanbul and the city opposite it on the other side of the narrow Bosporus Strait, Iskandar (or Uskudar), are a short ferry ride apart or a quick trip across a bridge; they also are on two different continents. Many Americans commute daily across state lines; many Turks commute daily between continents! The *situation* of Istanbul helps us to understand its historic importance as a port and exchange center. Istanbul was where ships carrying the products of Russia and Ukraine met ships carrying the products of the Mediterranean countries and the world beyond.

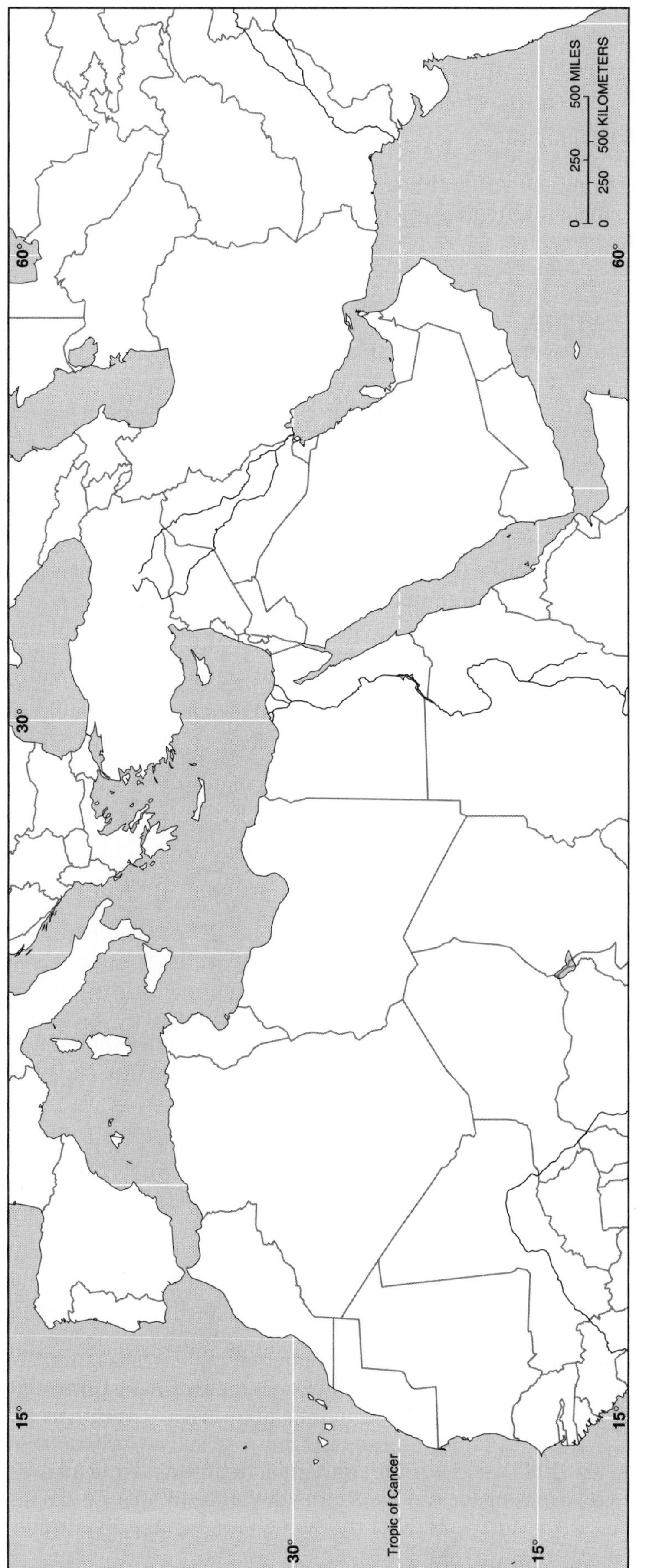


**FIGURE 12–2**
Major physical features of the Middle East–North Africa region.

To the Russians, the relative location of Istanbul always has been very important. In wartime, Istanbul was the lock on the gate between the Black Sea and the rest of the world-ocean.

An interesting thing about situation is that the relative importance of situation can, and does, change. Understanding changes in the significance of situation is a key to understanding the changing importance of the Middle East and North Africa, especially the changing importance of specific locations like the Bosporus, the Isthmus of Suez, the Strait of Hormuz, the Persian Gulf, and the Mediterranean and Red Seas.

The relative importance of the Mediterranean Sea has changed repeatedly in history, and with it, the significance of North African and Middle Eastern locations. The name *Mediterranean* in Latin is wonderfully descriptive; it means "the sea in the middle, or in the midst, of the land." Like the Black Sea, the Mediterranean has only one narrow, natural connection (the Strait of Gibraltar) to larger seas and oceans. For much of the ancient world, the Mediterranean was *the* sea, the center of their world. The Romans made it a "Roman lake" and included all the shores of the Mediterranean in their empire. Although the Romans certainly knew about the North Atlantic and navigated its European coastal waters, the Mediterranean was more important to them by far.

For western Europeans, a major reorientation in their thinking about the world took place from the fifteenth century forward. Prince Henry the Navigator and Christopher Columbus initiated a revolution in European perceptions of the world. The Atlantic Ocean suddenly was promoted from a somewhat threatening mystery into a busy east–west connection to the riches and potentials of a whole new world, the Indian Ocean and India. The Atlantic and Indian oceans became vitally important to European trade, prosperity, and colonial ambitions. At the same time, the Mediterranean Sea was demoted to a small backwater—a sea that dead-ended to the east where hostile shores were controlled by Muslims. To the Europeans of the time, the way to wealth and power was the Atlantic, and no longer the Mediterranean.

If Columbus's dramatic voyages of discovery reduced the relative importance of the Middle East–North Africa region, the builders of the Suez Canal reversed that situation again. The Suez Canal (opened in 1869) made a tremendous change in the importance of the Mediterranean because it converted two single-outlet seas, the Mediterranean and Red seas, into connected, through routes. It is as if one had been living on a quiet cul-de-sac or dead-end street that suddenly became a busy interstate highway.

Keep in mind throughout the chapter that the relative importance of situation can, and does, change over time. The dynamics of this change sometimes are the initiative of trade strategies, military goals, and colonial ambitions of countries from outside the region, whichever region it is.

## POLITICAL GEOGRAPHY

The Middle East–North Africa region contains some of the world's last monarchies in which the monarch holds real power. Whereas the surviving European monarchs have only ceremonial, symbolic functions, the kings of Saudi Arabia, Morocco, and Jordan rule with almost unlimited authority. On the world outline map in Figure 12–3, use different colors to represent constitutional monarchies (in which the monarch reigns but does not rule) and ruling monarchies. Countries left white on this map are organized as republics (whether or not the people participate in genuine elections).

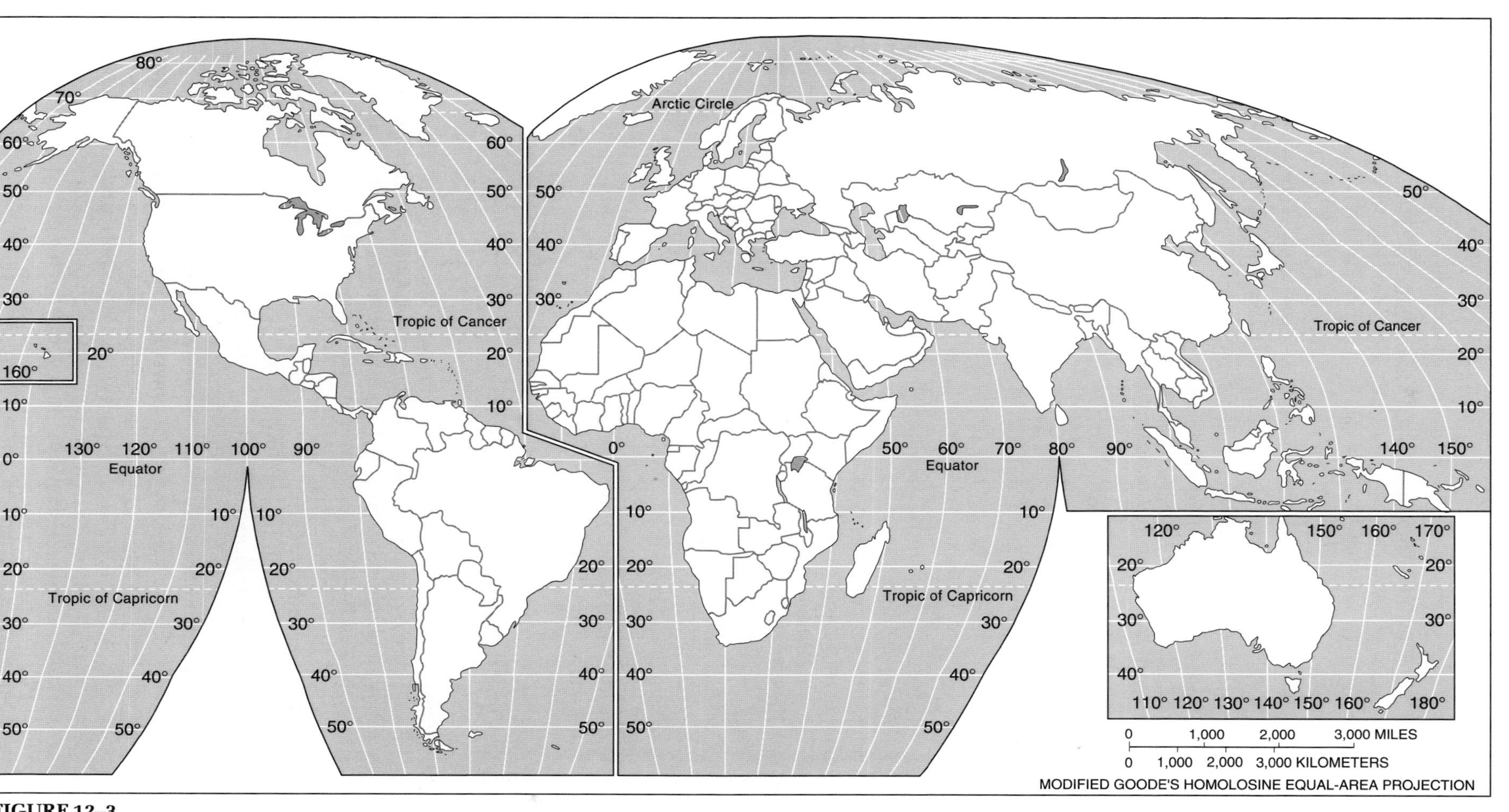


**FIGURE 12–3**
Constitutional and ruling monarchies and Islamic republics.

**Constitutional Monarchies**

| | |
|---|---|
| United Kingdom | Netherlands |
| Sweden | Luxembourg |
| Norway | Spain |
| Denmark | Malaysia |
| Belgium | Japan |

The microstates of Liechtenstein and Monaco are constitutional monarchies but are too small to appear on the world map.

**Ruling Monarchies, Sultanates, Emirates (monarch has real power)**

| | |
|---|---|
| Saudi Arabia | Bahrain |
| Morocco | Qatar |
| Jordan | Kuwait |
| United Arab Emirates | Lesotho |
| Thailand | Swaziland |
| Nepal | Brunei |
| Bhutan | Oman |

The holy Koran and the Hadith (reports of what the Prophet Muhammad said and did) provide the basis for Islam and outline its religious teachings. Furthermore, the holy writings establish a legal system and laws for administering secular society as well, including a listing of appropriate punishments for criminals. Several states in this region have incorporated the adjective *Islamic* in their official names to emphasize the country's adherence to the legal philosophies of Islam.

Using the same outline map of the region (Figure 12–3), choose a color to indicate the official Islamic status of these governments:

Islamic Republic of Iran
Islamic Republic of Pakistan
Federal and Islamic Republic of the Comoros (islands off east coast of Africa)
Islamic Republic of Mauritania

## ECONOMIC GEOGRAPHY

Petroleum and natural gas are the most famous resources of this region, though not the only mineral resources present. The Persian Gulf oil fields are a particularly happy combination of quantity, quality, and accessibility. "Arabian light" crude oil, the standard by which all other crude oils are judged, is low in polluting sulfur and flows freely in the wells. In a world ranking of countries by size of oil reserves, countries bordering the Persian Gulf account for five of the top six. In decreasing order, they are Saudi Arabia, Kuwait, Russia and the newly independent states, Iran, Iraq, and the United Arab Emirates.

Another way to look at the significance of oil to a country is to express known reserves in proportion to that country's population. Using the outline map of the region in Figure 12–4, construct a map of oil reserves (barrels) per capita. Choose different colors to represent the following categories: above 10,000 barrels per capita; 1,000 to 9,999 barrels per capita; 100 to 999 barrels per capita; and less than 99 barrels per capita. Use the data in Table 12–2. As these data indicate, oil is distributed extremely unevenly throughout the region, especially in relation to population size. For each Kuwaiti, for example, there is over 10,000 times as much oil as there is for each Turk.

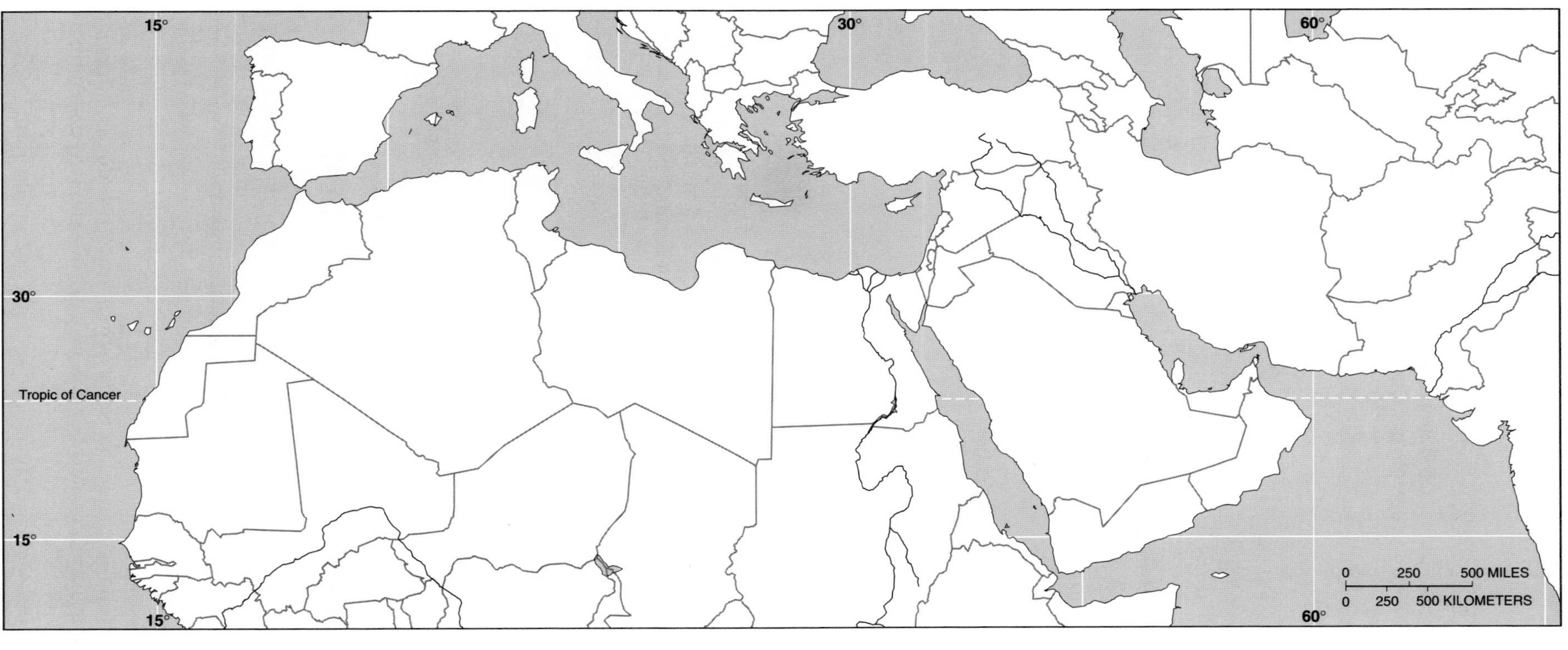


**FIGURE 12–4**
Oil reserves per capita in the Middle East–North Africa region.

TABLE 12–2 Oil reserves by country.

| Country | Oil Reserves (barrels per Capita) |
|---|---|
| Algeria | 438 |
| Bahrain | 285 |
| Egypt | 54 |
| Iran | 1330 |
| Iraq | 4316 |
| Kuwait | 42,500 |
| Libya | 5274 |
| Oman | 1583 |
| Qatar | 7000 |
| Saudi Arabia | 12,212 |
| Syria | 139 |
| Tunisia | 31 |
| Turkey | 5 |
| United Arab Emirates | 22,928 |
| **(For Comparison Purposes Only)** | |
| United States | 82 |
| Canada | 178 |
| Mexico | 410 |

What these facts mean is that Kuwait can sustain high oil output per capita, and thus high income per capita, for a long time. Turkey, by contrast, could not expect to do more than reduce its imports of foreign oil for a relatively short time. Turkey and Egypt, the two largest countries in the region in population, have low ratios of oil reserves. Tiny sheikdoms like the United Arab Emirates (less than 3 million people) could maintain lavish incomes for their small populations for generations.

**12–3** Which Middle East–North Africa states listed have *lower* ratios of oil reserves per capita than the U.S.? ______

**12–4** Which North African state has the highest ratio of oil reserves per capita?

______

**12–5** Which of the oil-rich (over 10,000 barrels per person) countries has the largest population? ______

## CULTURAL GEOGRAPHY AND DEMOGRAPHICS

Some of this region's countries are experiencing very rapid rates of population increase. Population growth rates are the product of two forms of addition (births and in-migration) and two of subtraction (deaths and out-migration). Two reasons for the relatively high rates of population increase for the oil-rich states is that their people seldom have a strong economic incentive to move out, and many foreigners see good reasons to try to move in.

Trends in birth and death rates, however, provide most of the explanation for growth rates. If we could eliminate migration in or out of a country, then the balance between the two *vital rates,* birth and death, would be the growth rate. Under these controlled circumstances of no migration, a net gain results from a surplus of births over deaths, or a net loss reflects a death rate higher than the birth rate.

Imagine a society in which no migration occurs. On January first, the population is 1 million. During the year, 10,000 births and 10,000 deaths occur. On December 31st, the population would still be 1 million. The addition of 10,000 births would have exactly balanced the 10,000 deaths, or subtractions. The natural increase rate would be zero. Notice that a change in either, or both, of these rates could produce a net gain—a rise in birth rates and/or a decrease in death rates. On the other hand, a net loss could result from a dropping birth rate and/or a rising death rate, again not considering the effects of migration.

It is common for developing countries to have high rates of increase, and just as common for wealthy, industrialized countries to have low rates of increase. Typically, the surge in population of developing countries is a result of death rates that are falling much faster than birth rates are declining (if, indeed, birth rates drop at all).

Table 12–3 shows the *rates of change* in both birth and death rates over an eight-year period for states in Middle East–North Africa. On the outline map of the region in Figure 12–5, map the generalized changes in the crude death rates over a recent eight-year period. Because Morocco has absorbed Western Sahara (Spanish Sahara), use the same color for Western Sahara as for Morocco. As no reliable data exist for Lebanon, label that country ND (no data).

**TABLE 12–3** Average annual natural increase rates.

| Country | Average Annual Natural Increase Rate* | Percentage Change in Birth Rates Over 8 Years | Percentage Change in Death Rates Over 8 Years |
|---|---|---|---|
| Chad | 2.3 | –2% | –29% |
| Mali | 2.4 | +2% | –26% |
| Niger | 3.0 | +6% | –31% |
| Sudan | 3.1 | –6% | –33% |
| Yemen | 2.8 | –4% | –41% |
| Mauritania | 2.7 | +2% | –30% |
| Morocco | 2.7 | –12% | –44% |
| Egypt | 2.7 | –18% | –47% |
| Tunisia | 2.6 | –32% | –59% |
| Turkey | 2.3 | –27% | –40% |
| Jordan | 3.9 | –19% | –66% |
| Syria | 3.6 | –6% | –56% |
| Algeria | 3.1 | –22% | –50% |
| Libya | 4.3 | –10% | –50% |
| Oman | 4.6 | –8% | –50% |
| Iran | 3.0 | –11% | –50% |
| Iraq | 3.6 | –12% | –55% |
| Saudi Arabia | 4.3 | –12% | –60% |
| Israel | 1.7 | –15% | +16% |
| Kuwait | 4.5 | –30% | –62% |
| United Arab Emirates | 5.2 | –44% | –73% |
| (For Comparison Purposes) | | | |
| Germany | –0.1 | –44% | 0% |
| Japan | 0.6 | –42% | 0% |
| Canada | 1.0 | –29% | 0% |
| United States | 1.0 | –16% | 0% |

*Average annual natural increase rate differs from the *current* natural increase rate listed in the Appendix.

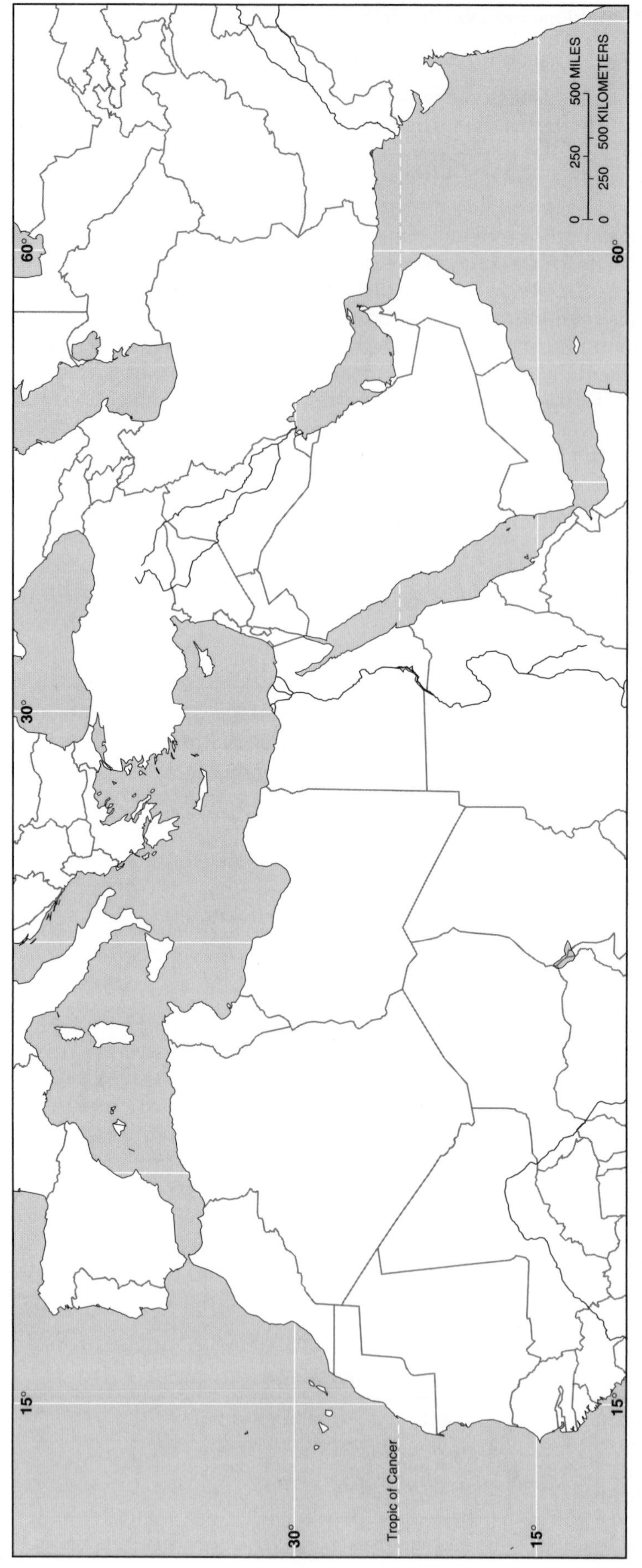


**FIGURE 12–5**
Changes in death rates in the Middle East–North Africa region.

Note on the table that changes in the death rates are listed for 21 countries in this region. Divide these 21 countries into three groups—highest, middle, and lowest—in rates of decline. To produce a map with three categories of comparative data, use a dark color to show the group with the sharpest percentage decline (over –55%); use a lighter color to show the middle group of percentage declines (–41% to –50% decline), and leave blank (white) those countries with a smaller than 41% decline in death rates.

As a generalization, the poorest countries have the smaller rates of decline in death rates.

**12–6** Which relatively rich state does not fit the trend for wealthy states to have large drops in the death rate? ____________________

**12–7** The highest average annual natural increase rate occurs in ____________________.

**12–8** As a region, Middle East–North Africa's percentage change in birth rates could be characterized as (Circle the correct answer.): mostly negative; mostly positive; no change.

**12–9** The steepest decline in death rates within the region has occurred in ____________________.

## CHECK UP

**12–10** Which states in this region have a Mediterranean coastline? ____________________

**12–11** Which have an Atlantic coastline? ____________________

**12–12** Which three countries in this region lack direct access to the sea? ____________________

**12–13** Control of which strait is vital to the movement of oil tankers out of the Persian Gulf? ____________________

**12–14** Which two countries in this region share borders with the newly independent states (former USSR)? ____________________

**12–15** Which are the two "population giants" of this region? ____________________

**12–16** Which countries in this region still are ruled by kings or sheiks? ____________________

**12–17** Which five countries in this region concentrate most of the oil reserves per capita? ____________________

**12–18** In general, the region's poorest countries are those located in/on: Persian Gulf, eastern Mediterranean, northernmost in region, southernmost in region.

**12–19** Cities considered holy by at least one of the three major monotheistic religions are found in what *two* countries in this region? ______________

______________________________

**12–20** Which state in the region is *least* typical in regard to religion and degree of industrialization? ______________________________

# GEOCONCEPTS

## Desertification

There is an ongoing argument among scientists as to why the world's deserts seem to be expanding. Some see this desert expansion as proof that the world's climate is changing; gradually increasing temperatures around the world would increase evaporation of moisture from deserts and their fringes, leading to the drying of both deserts and their semidesert margins. Others believe that most of the desertification going on around the world is the result of human error in managing the environment. There is little doubt that people can contribute to the outward spread of deserts through poor conservation practices.

Because moisture is scarce in desert and semidesert environments, any change that increases the evaporation of moisture or decreases the supply of moisture to plant roots will contribute to desertification. Picture a mountain slope on the fringes of the desert. There is barely enough moisture to support trees and shrubby plants, but while these trees and shrubs are in place, they provide shade, they reduce wind speed near the ground, and their intertwined roots help keep soil in place.

If people cut the trees for lumber, hack down even little trees for firewood, and allow their domestic animals to *overgraze* (put too many animals in an area at the same time, resulting in destruction of most remaining edible plants), this slope may be stripped almost bare of plant cover. Without shade, the sun heats the slope more than before, and evaporation increases. Without trees to slow surface winds, windspeed increases and so does evaporation. When rain does fall on this now-bare slope, soil erosion speeds up, perhaps removing the possibility of the forest ever growing back.

Such *degradation* of the semiarid margins of deserts has been observed often enough to indicate that human mismanagement is *a* cause, if not *the* cause, of desertification.

**12–21** The thematic map of desertification in Figure 12–6 shows which major world region is without some major desertification?

______________________________

**12–22** The Middle East–North Africa region is most completely affected. Are *any* countries in this region *not* affected by desertification?

______________________________

Those areas of semiarid climates on the southern edges of the Sahara, the so-called *Sahel,* have suffered serious droughts leading to crop failures and the loss of many domestic animals. Unfortunately, poverty and environmental mismanagement seem to collaborate in a downward spiral. Poor people in developing countries do not slip dinner into the microwave—they go scavenging for firewood, and so put heavy pressure on nearby trees and woody shrubs. Then, too, as human numbers climb, more and more cattle, sheep, and goats must be grazed in an effort to feed everyone. In their effort to feed themselves, the domestic animals search out and destroy what is left of sparse, overgrazed vegetation. Far richer societies permit overgrazing, too, without as understandable an excuse as poverty.

**12–23** Highly industrialized, relatively rich countries with large areas of desertification include ______________________________________________

______________________________________________________________.

## Old Empires, New Empires? Politics and Islamic Fundamentalism

Islam is a world religion; large numbers of its adherents live in all major regions except Australia, New Zealand, and the Pacific Islands. Islam is the dominant religion not just in the Middle East and North Africa, but also in the former Soviet Union's Azerbaijan and Central Asia republics (see Chapter 6). Islam also is the major religion in Indonesia, Malaysia, and Bangladesh. The world of Islam presently is undergoing both political and religious change. Within a century, the "Heart of Islam," from the Caucasus and Central Asia to the Middle East and North Africa (Figure 12–7), was dominated by now-defunct empires—Turkish, Russian/Soviet, British, and French. Only the Turkish empire was ruled by fellow Muslims. When imperial power collapses, a "power vacuum" may tempt states to expand and create their own empires from fragments of old empires. Both Iraq and Iran have expansionist ambitions toward their neighbors. On Figure 12–7, place a capital letter "B" (one time part of or "protectorate" of the British Empire) in Egypt, Sudan, Aden, Kuwait, United Arab Emirates, Iraq, Jordan, Cyprus, and Israel. Place an "F" (former French colony, protectorate, or mandate) in Syria, Djibouti, and Lebanon. Place a "T" (onetime Turkish Empire) in Egypt, Eastern Saudi Arabia (Red Sea coasts), Jordan, Israel, Syria, Kuwait, and Iraq. Azerbaijan and the Central Asian republics should be marked with "R/S," former Russian/Soviet empires.

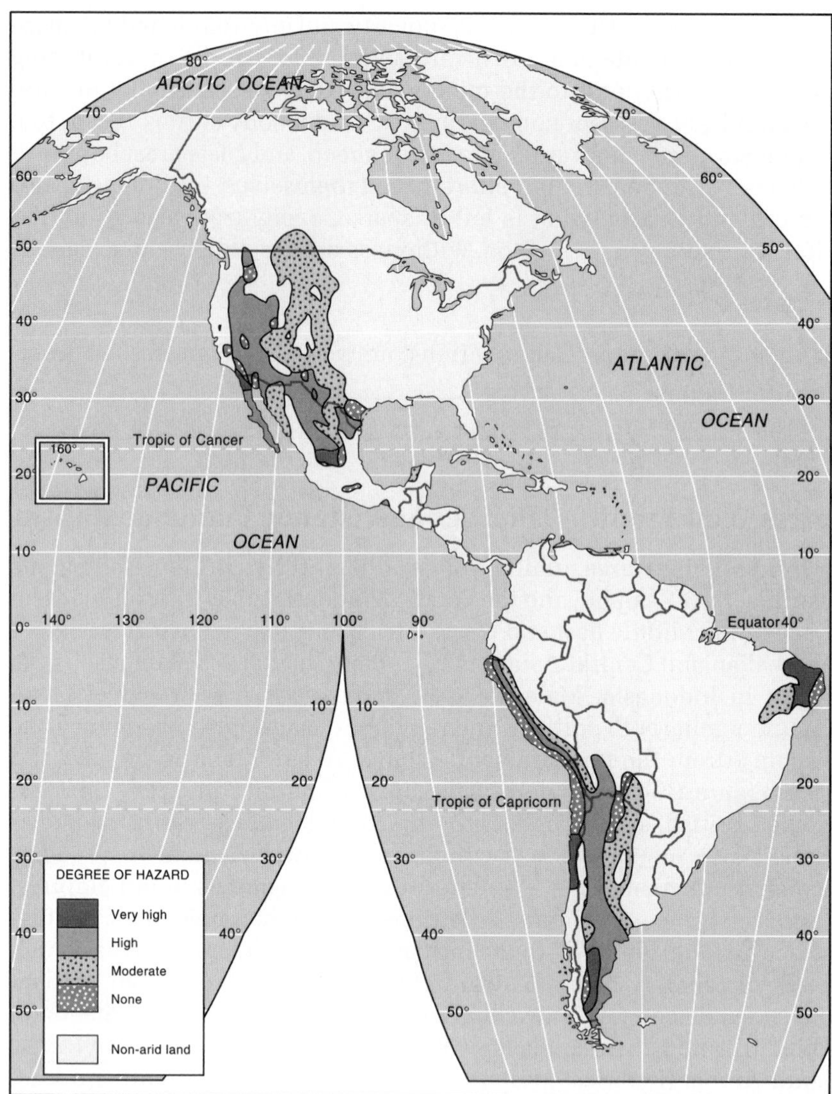


**FIGURE 12–6**
Rapid population growth, increasingly large livestock herds, deforestation, and bad agricultural practices have contributed to an expansion of deserts. Many scientists suggest that these secondary causal factors are less important than ongoing climatic change on a global scale.

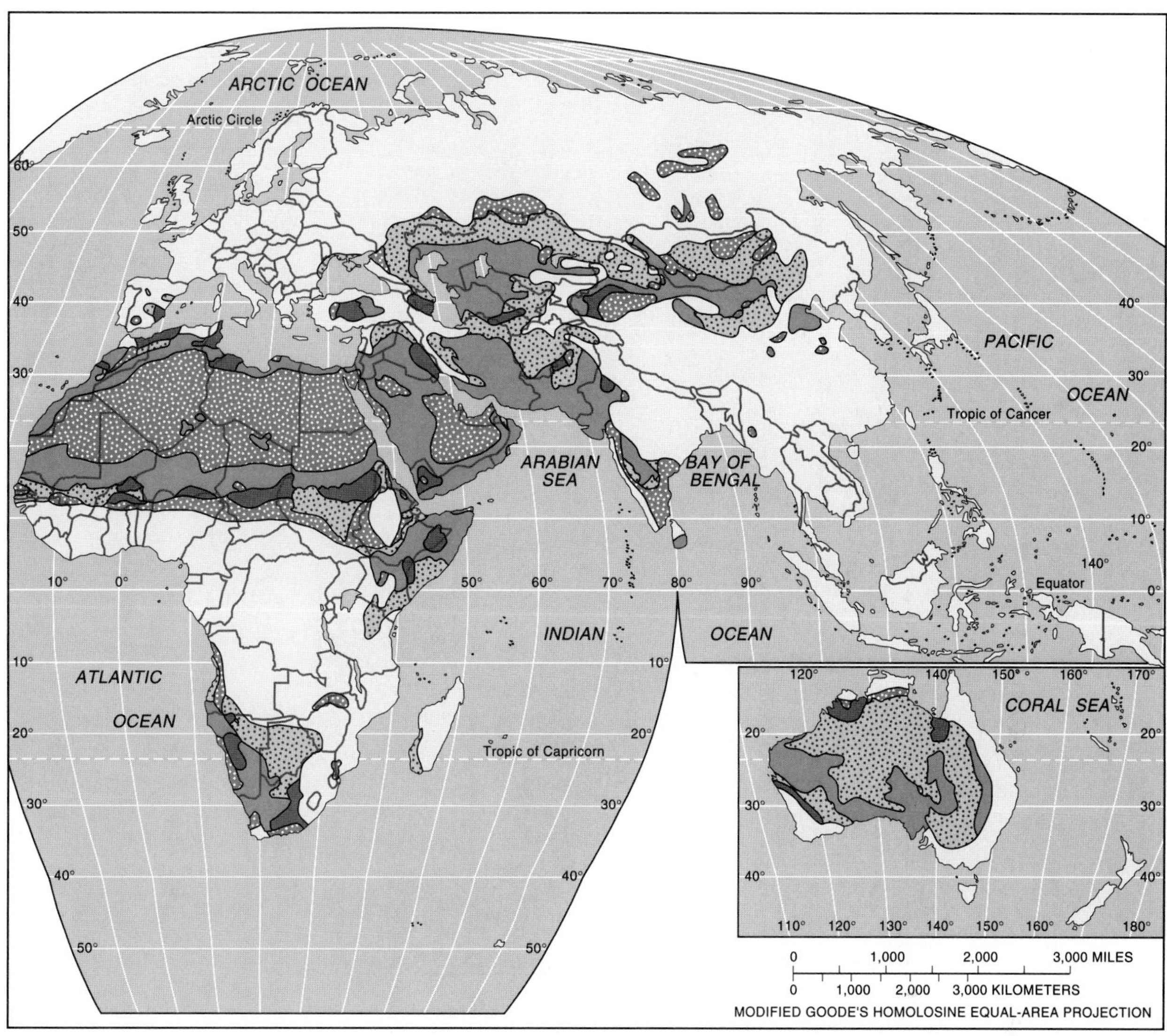
ARCTIC OCEAN
Arctic Circle
PACIFIC
OCEAN
Tropic of Cancer
ARABIAN
SEA
BAY OF
BENGAL
Equator
INDIAN
OCEAN
ATLANTIC
OCEAN
Tropic of Capricorn
CORAL SEA
0 1,000 2,000 3,000 MILES
0 1,000 2,000 3,000 KILOMETERS
MODIFIED GOODE'S HOMOLOSINE EQUAL-AREA PROJECTION

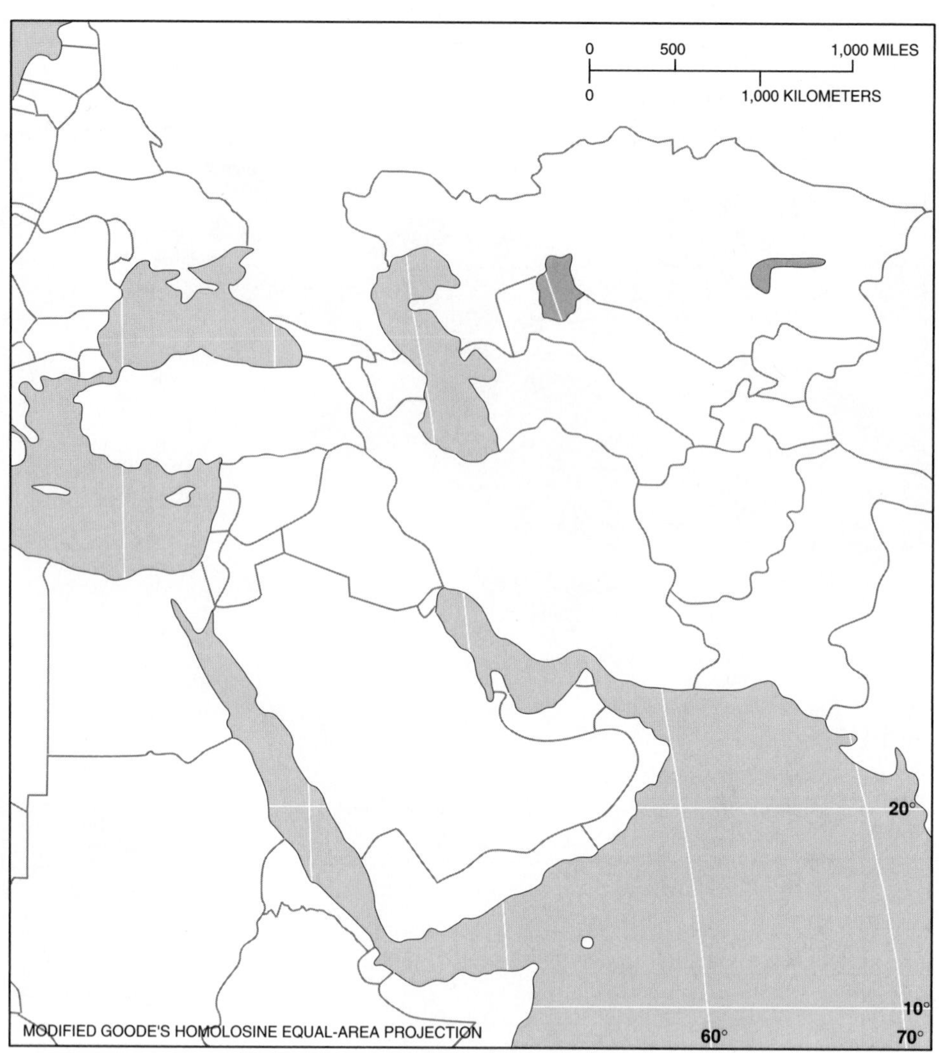


**FIGURE 12–7**
Historic colonial control.

## REGIONAL WATCHLIST

The Middle East–North Africa region may be the least stable major world region, though there is close competition for this title from several others, notably Russia and the newly independent states. Progress toward a peaceful resolution of the Palestinian–Israeli dispute is likely to be uneven and halting. Iraq may continue to threaten its neighbors despite the lessons to be learned in its long war with Iran and its unsuccessful invasion of Kuwait. Events in Saudi Arabia and the Persian Gulf sheikdoms may determine if absolute monarchies can persist in the modern world. And, of course, there remains the problem of nationalist ethnic minorities like the Kurds, whose envisioned "Kurdistan" (Figure 12–8) would disrupt Iraq, Iran, and Turkey.

**12–24** Which four Middle Eastern countries would lose territory to a new state envisioned by the Kurdish people? ____________________________

________________________________________________

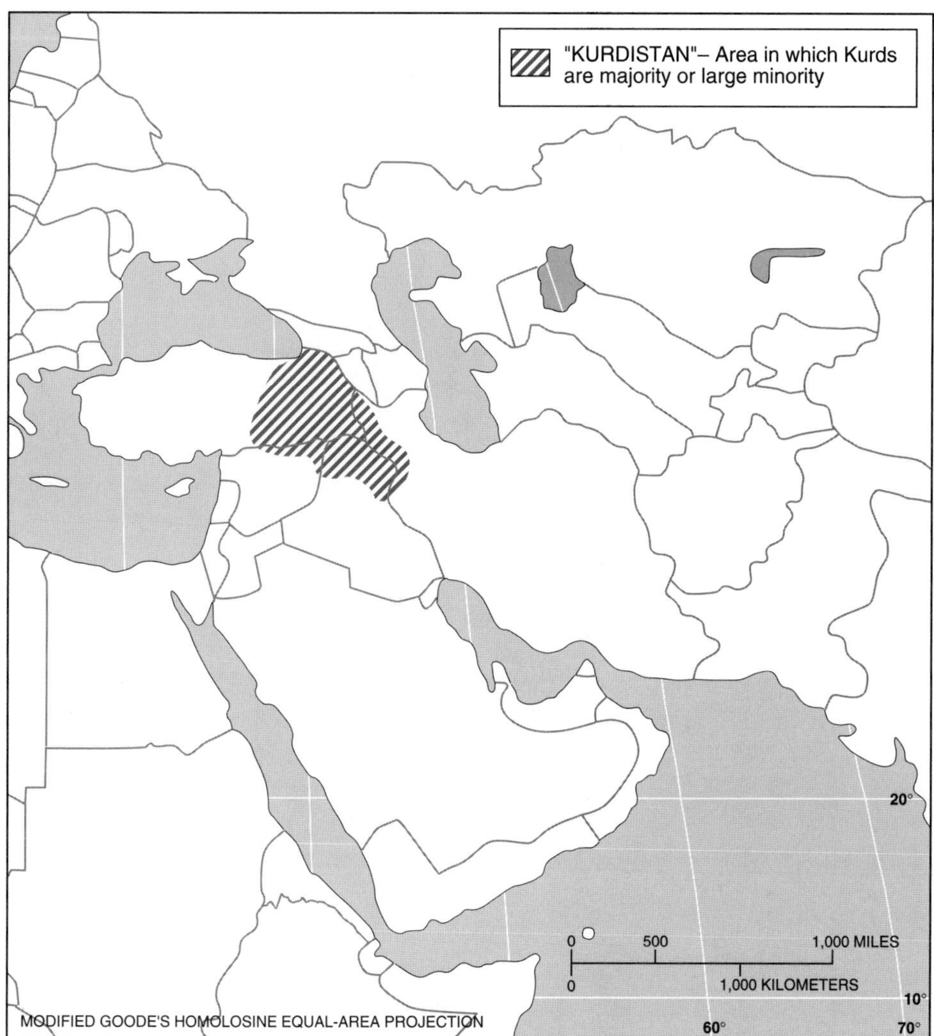


**FIGURE 12–8**
Kurdistan: A nation without a state.

# 13

# Latin America

## INTRODUCTION

Latin America extends through the greatest latitudinal span of any world region, from the northernmost part of Mexico at about 32° N to the southern coast of Tierra del Fuego at 55° S. The very name *Latin America* identifies this as a cultural rather than a physical region. Two Latin-based languages, Spanish and Portuguese, help give this huge region a strong cultural unity. Much smaller groups of Latin Americans speak Amerindian languages, mostly in rural areas of Brazil, Peru, Bolivia, Paraguay, and Colombia. French, another Latin-based language, is spoken in some of the Caribbean islands, Haiti, and French Guiana, whereas English is official only in Guyana, Belize, Jamaica, and former British colonies among the smaller Caribbean islands.

As in the United States, Canada, and Australia, European colonial control frequently was accompanied by large-scale migrations from Europe, creating a distinctive cultural blend with Amerindian peoples. In proportion to total population, the European-origin people are most prominent in Argentina, Chile, Uruguay, Costa Rica, and Brazil. The slave trade brought many Africans to those parts of Latin America that produced cane sugar for European markets—the Caribbean islands, British, Dutch, and French Guiana (the three longest-lived European colonies in South America), and Brazil.

Geographers commonly divide Latin America into Middle America and South America; the outline maps used in this chapter reflect the basic subdivision. South America is composed of twelve independent states—Argentina, Bolivia, Brazil, Chile, Colombia, Ecuador, Guyana, Paraguay, Peru, Suriname, Uruguay, and Venezuela, plus French Guiana, an overseas department of France, and the British colony of the Falkland Islands, known to the Argentines, who also claim them as the Islas Malvinas. Brazil clearly is the giant of South America. Larger than the entire continent of Australia, Brazil ranks fifth in area among the world's nations, after Russia, Canada, China, and the United States. Brazil alone accounts for 50% of South America's land area and 50% of the continent's population.

Middle America is that subdivision of Latin America that consists of Mexico, Central America, and the Caribbean islands. Middle America, like South America, is dominated by one large political unit, Mexico. Mexico accounts for an astonishing 58% of Middle America's population. Like Brazil, Mexico is well on its way to becoming a modern industrial state with a large, industrious population and a huge and varied natural resource base.

Sugar and spices led to strong European competition for the Caribbean islands, making that area a complex mosaic of colonial possessions whose peoples generally were slower to achieve independence than mainland countries. Among the tiniest Caribbean islands are the last remnants of colonialism, including territories still controlled by the UK, France, and the United States. On the outline maps of Middle America and South America, Figures 13–1 and 13–2, locate and label the countries and cities listed in Table 13–1.

**TABLE 13–1** Countries, capitals, and major cities of Latin America.

| Country | Capital | Other Major Cities |
|---|---|---|
| *Middle America* | | |
| Bahamas | Nassau | |
| Barbados | Bridgetown | |
| Belize | Belmopan | Belize City |
| Costa Rica | San Jose | |
| Cuba | Havana | |
| Dominica | Roseau | |
| Dominican Republic | Santo Domingo | |
| El Salvador | San Salvador | |
| Grenada | St.George's | |
| Guadeloupe | Basse-Terre | |
| Guatemala | Guatemala City | |
| Haiti | Port-au-Prince | |
| Honduras | Tegucigalpa | |
| Jamaica | Kingston | |
| Martinique | Fort-de-France | |
| Mexico | Mexico City | Monterrey; Vera Cruz |
| Netherlands Antilles | Willemstad | |
| Nicaragua | Managua | |
| Panama | Panama City | |
| Puerto Rico | San Juan | |
| St. Lucia | Castries | |
| St. Vincent and the Grenadines | Kingston | |
| Trinidad and Tobago | Port of Spain | |
| U.S. Virgin Islands | St. Thomas | |
| *South America* | | |
| Argentina | Buenos Aires | Rosario |
| Bolivia | La Paz | |
| | (seat of government) | Sucre (legal capital) |
| Brazil | Brasilia | Rio de Janiero; Sao Paulo |
| Chile | Santiago | Valparaiso |
| Colombia | Bogota | Medellin |
| Ecuador | Quito | |
| Falkland Islands (Islas Malvinas) | Stanley | |
| French Guiana | Cayenne | |
| Guyana | Georgetown | |
| Paraguay | Asuncion | |
| Peru | Lima | |
| Suriname | Paramaribo | |
| Uruguay | Montevideo | |
| Venezuela | Caracas | |

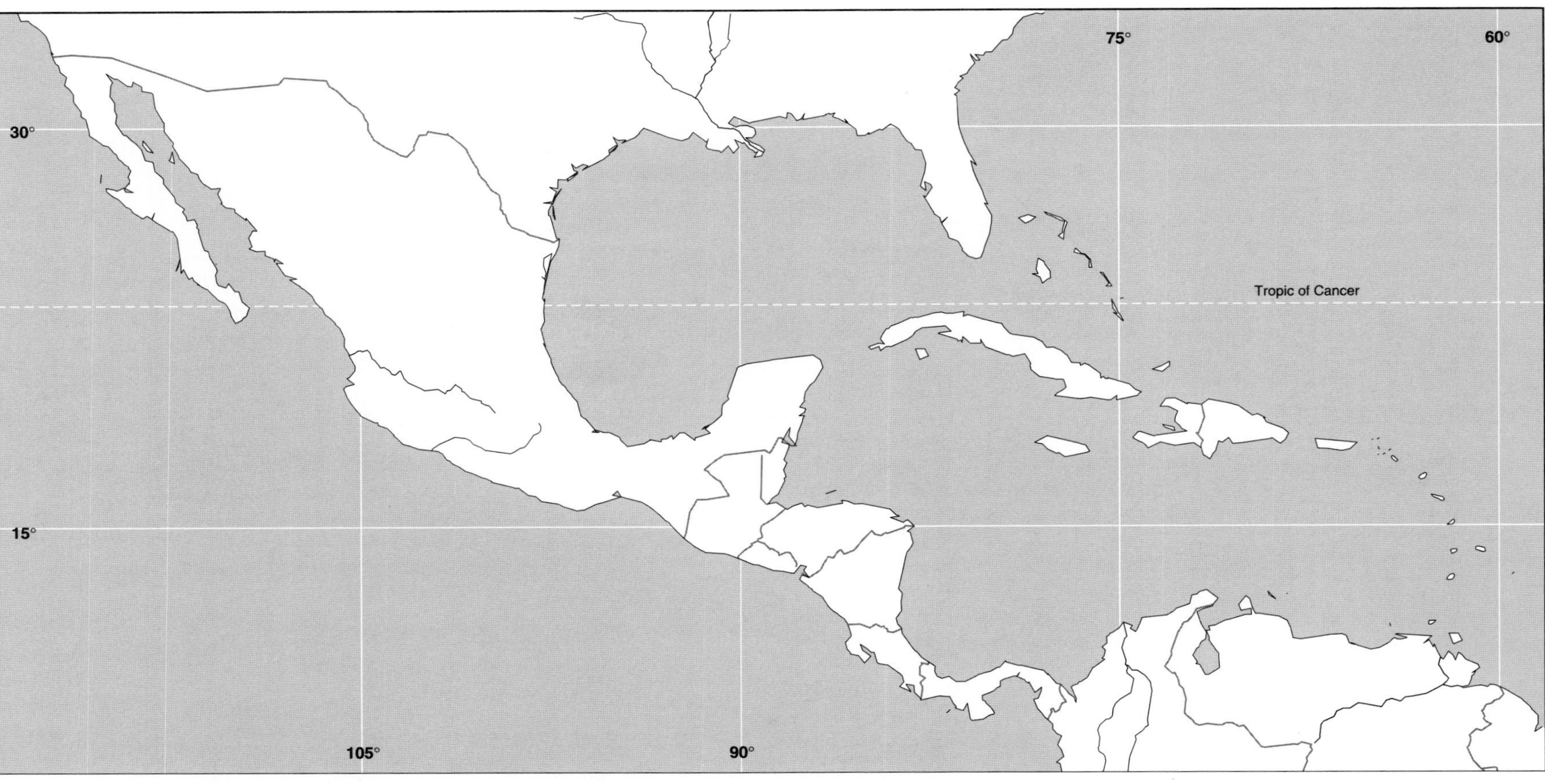


**FIGURE 13–1**
Middle American countries, capitals and other important cities.

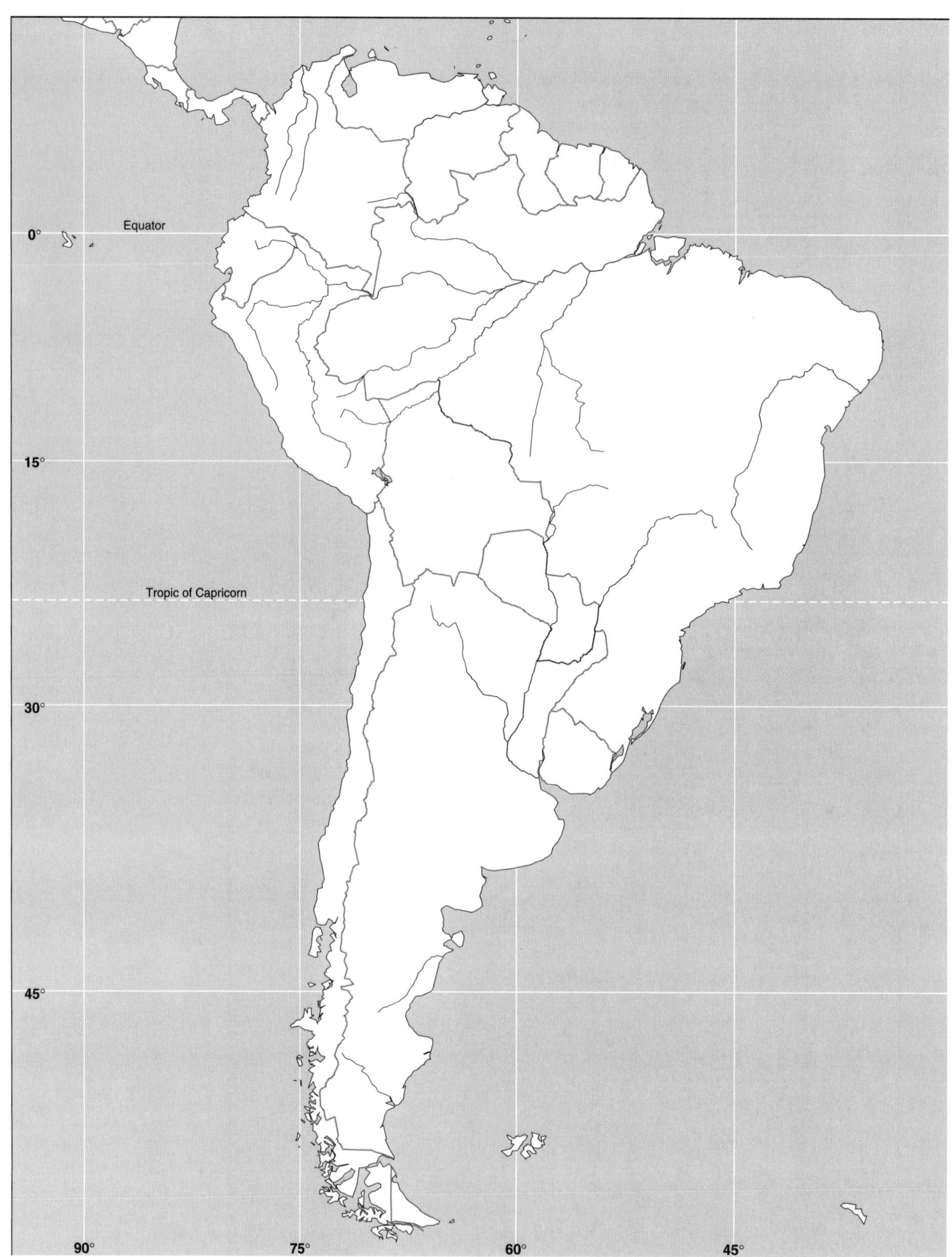


**FIGURE 13–2**
South American countries, capitals and other important cities.

## PHYSICAL GEOGRAPHY

The west coasts of South America and Middle America are part of the *Ring of Fire* that circles the Pacific basin. It involves great rigid plates of solid crust about 60 miles (96.5 km) thick that move over Earth's molten mantle. Sometimes, these plates collide head on, sometimes split apart, and sometimes slowly grind past one another. Earthquakes and volcanoes outline these crustal plate splits, collisions, and "sideswipe" contacts. A smaller line of frequent earthquakes and many volcanoes branches eastward from the Pacific Ring of Fire through Colombia and Venezuela to Trinidad, arching northward through the Lesser Antilles to the larger islands of Puerto Rico, Hispaniola, and Jamaica.

South America's Andes Mountains are extremely rugged, with the second-highest average heights in the world, after the Himalayas of the South Asia–East Asia borderlands. There is a relatively narrow continental shelf off South and Middle America's west coast, with broader shelves north of Mexico's Yucatan Peninsula and off Argentina's coast.

Northern Mexico has Latin America's largest desert (Great American Desert) and semidesert area, and lying along the northern coast of Chile is a narrow strip of desert (Atacama Desert) known for its extraordinary dryness. Aside from the high mountains and the deserts of the region, the only other sizeable non-tropical areas of Latin America are found in the extreme south of Chile and Argentina, as South America extends much closer to the Antarctic Circle than Africa or Australia.

Four thousand miles (6437 km) long, the Amazon River ranks second only to the Nile in length, but the Amazon carries the greatest volume of water of any river. Oceangoing ships can sail to Manaus in the heart of Amazonas, Brazil's fast-developing rainforest frontier. South America has both the world's largest area of tropical rainforest and the world's most controversial destruction of rainforest, a topic more fully explored as a geoconcept.

On the outline map of Middle America, Figure 13–3, locate and label the following features:

|  |  |
|---|---|
| Pacific Ocean | |
| Gulf of California: | 30° N, 114° W |
| Gulf of Tehuantepec: | 16° N, 95° W |
| Baja California: | 27° N, 113° W |
| Yucatan Peninsula: | 20° N, 90° W |
| Isthmus of Tehuantepec: | 17° N, 95° W |
| Isthmus of Panama: | 10° N, 80° W |
| Lesser Antilles: | 15° N, 62° W |
| Bahamas Islands: | 24° N, 76° W |
| Cayman Islands: | 19° N, 81° W |
| Cuba: | 23° N, 80° W |
| Hispaniola: | 19° N, 72° W |
| Puerto Rico: | 18° N, 67° W |
| Virgin Islands: | 18° N, 65° W |
| "ABC" Islands (Aruba, Bonaire, Curacao): | 18° N, 70° W |
| Strait of Florida: | 24° N, 81° W |
| Mona Passage: | 18° N, 68° W |
| Atlantic Ocean | |
| Jamaica: | 18° N, 78° W |
| Guadeloupe: | 16° N, 62° W |
| Antigua: | 18° N, 62° W |
| Montserrat: | 17° N, 62° W |
| St. Christopher: (St. Kitts): | 17° N, 67° W |
| Barbuda: | 18° N, 62° W |
| Turks and Caicos Islands: | 22° N, 72° W |
| Yucatan Channel: | 22° N, 84° W |
| Windward Passage: | 20° N, 74° W |
| Caribbean Sea: | 16° N, 80° W |
| Gulf of Mexico: | 24° N, 90° W |
| Bay of Campeche: | 18° N, 94° W |
| Gulf of Honduras: | 16° N, 88° W |
| Grenada: | 12° N, 62° W |
| St. Vincent: | 13° N, 62° W |
| Barbados: | 14° N, 59° W |
| St. Lucia: | 14° N, 61° W |
| Martinique: | 15° N, 61° W |
| Dominica: | 16° N, 61° W |

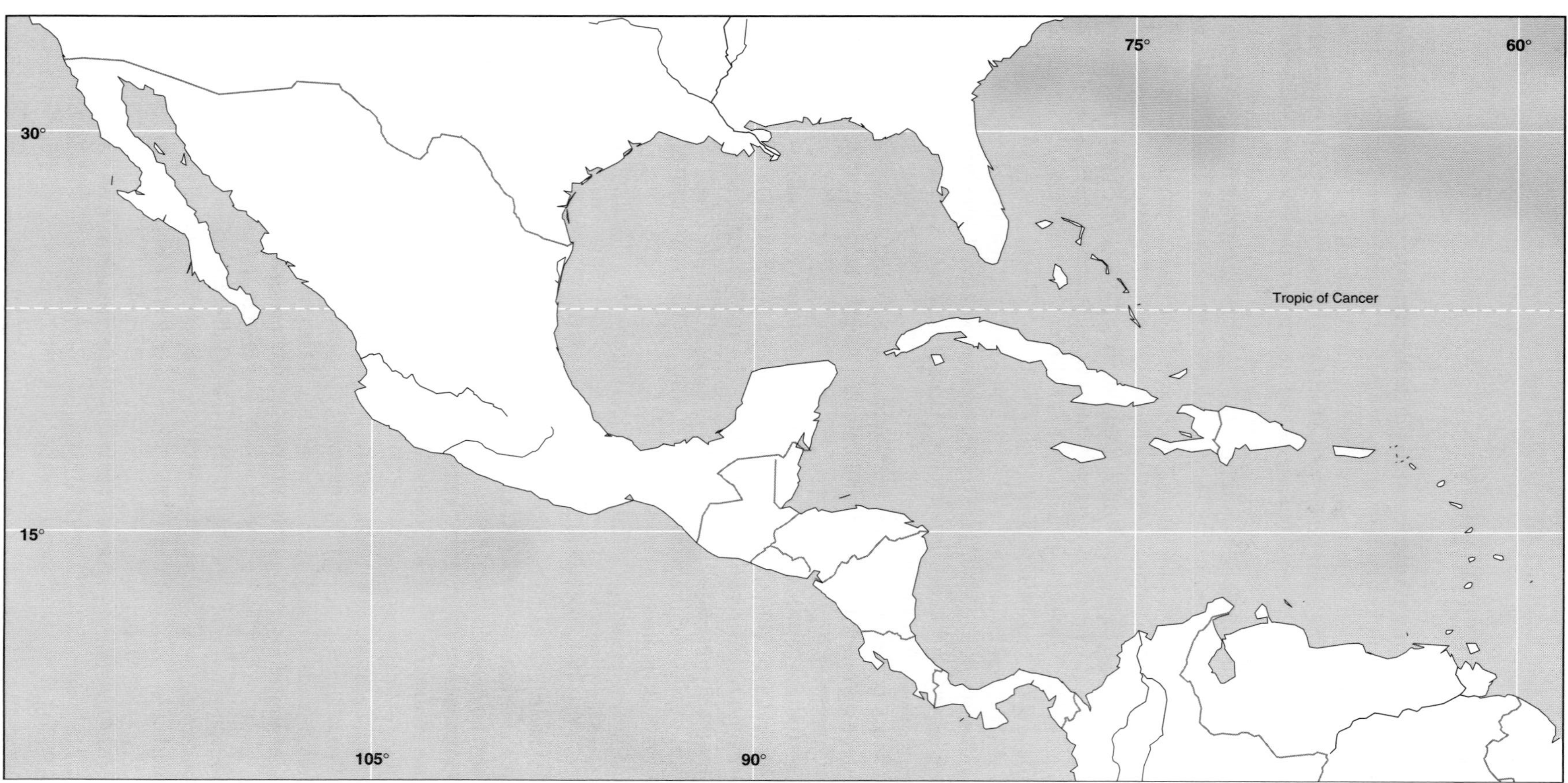


**FIGURE 13–3**
Middle America: Physical features.

On the outline map of South America, Figure 13–4, locate and label the following features:

| | | | |
|---|---|---|---|
| Pacific Ocean | | Caribbean Sea: | 16° N, 80° W |
| Atlantic Ocean | | Falkland Islands | |
| Amazon River: | 3° S, 59° W | (Islas Malvinas): | 52° S, 60° W |
| Rio de la Plata: | 35° S, 57° W | Cape Horn: | 54° S, 68° W |
| Andes Mountains: | 10° S, 77° W | Drake Passage: | 56° S, 67° W |
| Strait of Magellan: | 53° S, 70° W | Galapagos Islands: | 0° , 91° W |
| Tierra del Fuego: | 54° S, 70° W | Scotia Sea: | 55° S, 50° W |

## OBJECTIVES AND STUDY HINTS

An understanding of Latin America and its geographic problems and potentials will be enhanced by keeping five general themes in mind: (1) the blending of native American peoples and their cultures with those of both Europe and Africa; (2) the early independence of most of mainland Latin America from European imperialism; (3) the assertion of a special relationship and protective role in the region by the United States; (4) the relatively moderate population density of most mainland Latin American states; and (5) the middle-income status of most Latin American countries and the implications of this in present and future development.

Non-native population movements, cultural impacts, and political interests in Latin America can be summarized in three words—gold, sugar, and liberty. Gold and silver in fabulous amounts motivated early Spanish imperial efforts. Present patterns of languages spoken in Latin America reflect the early demarcation of Spanish and Portuguese colonial spheres. In the fifteenth century, Spanish and Portuguese navigators were the most active Europeans in exploring the world beyond Europe. Basically, Spanish interests lay westward across the Atlantic following the world-shaking discoveries of Columbus, whereas Portuguese ambitions lay more to the east. The Portuguese were more interested in developing trade routes to India than in colonizing the New World.

On the outline map of the world in Figure 13–5, draw a large arrow from Spain to the Caribbean and on to Mexico. Draw another arrow from Spain to Peru and Bolivia. Mexico, Peru, and Bolivia were the main sources of gold and silver for Spain. Many of Spain's other colonial territories were important mostly as supply and military bases to control shipping routes from these areas back to Spain. On that same map, draw an arrow from Portugal to India. What directions must this arrow take, as this was centuries before the Suez Canal? Shade in coastal areas in Guinea-Bissau, Angola, and Mozambique as Portuguese bases on the way to India.

To avoid a lengthy war over colonial spheres of influence, the Spanish and Portuguese asked the Pope to arbitrate their conflicting colonial claims. It was agreed in the Treaty of Tordesillas that a line would be drawn at approximately 50° W longitude, separating Spanish claims to the west from Portuguese interests to the east. 50° W is the longitude of the northern channel mouth of the Amazon River. This treaty from the year 1494 explains why the citizens of Brazil speak Portuguese, whereas most of the people in the rest of Latin America speak Spanish. Notice how control over the mouth of the easily navigable Amazon River enabled the Portuguese to expand rapidly into the heart of the Amazon Basin, far west of the original treaty boundary.

Sugar and, to a lesser extent, spices were second only to precious metals as an attraction to European colonization in Latin America. Tiny "sugar islands" were highly valued by European powers, who transported large numbers of African slaves to do the work on their new plantations. On the map (Figure 13–5), draw arrows from Senegal to Cameroon on the African coast to the Caribbean

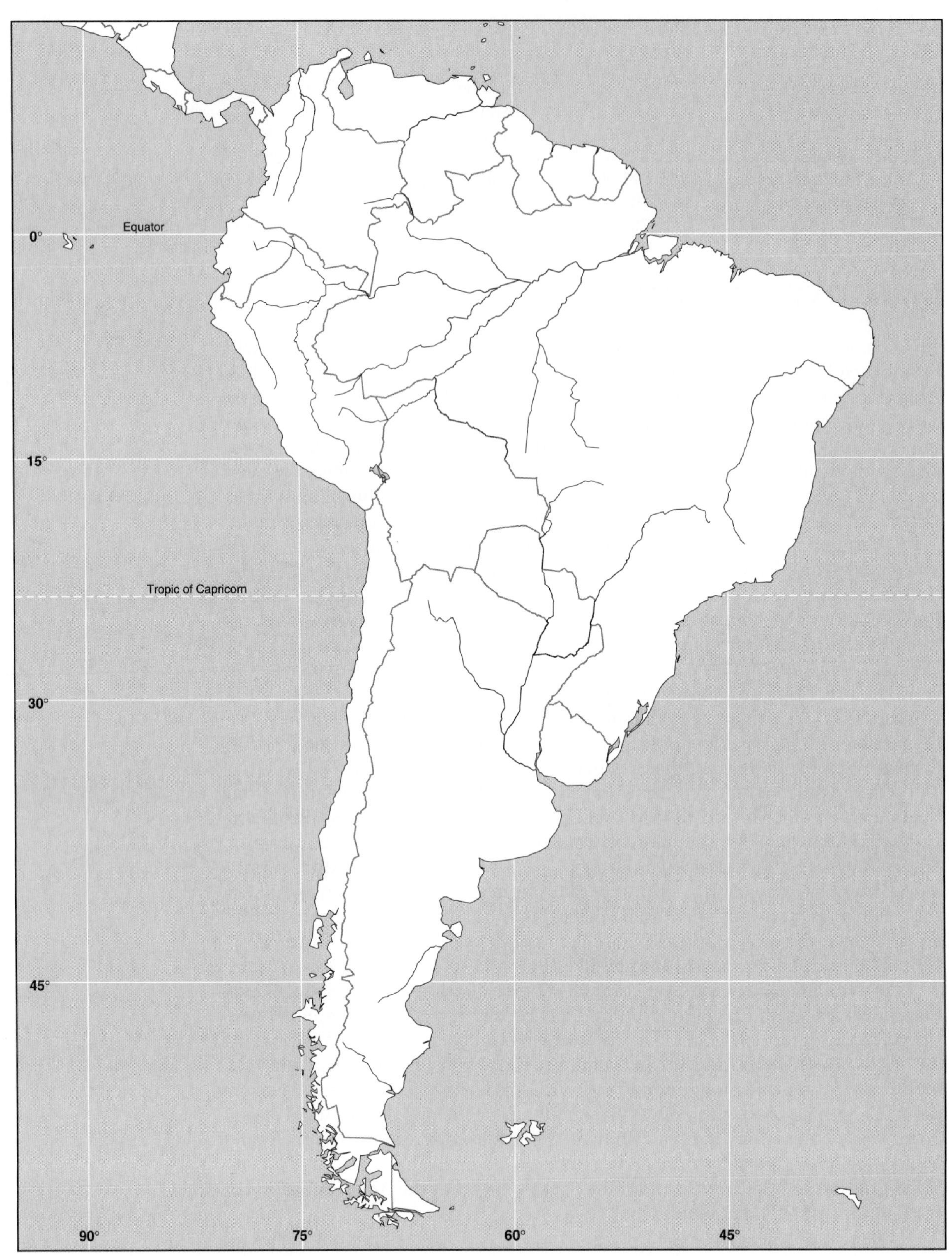


**FIGURE 13–4**
South America: Physical features.

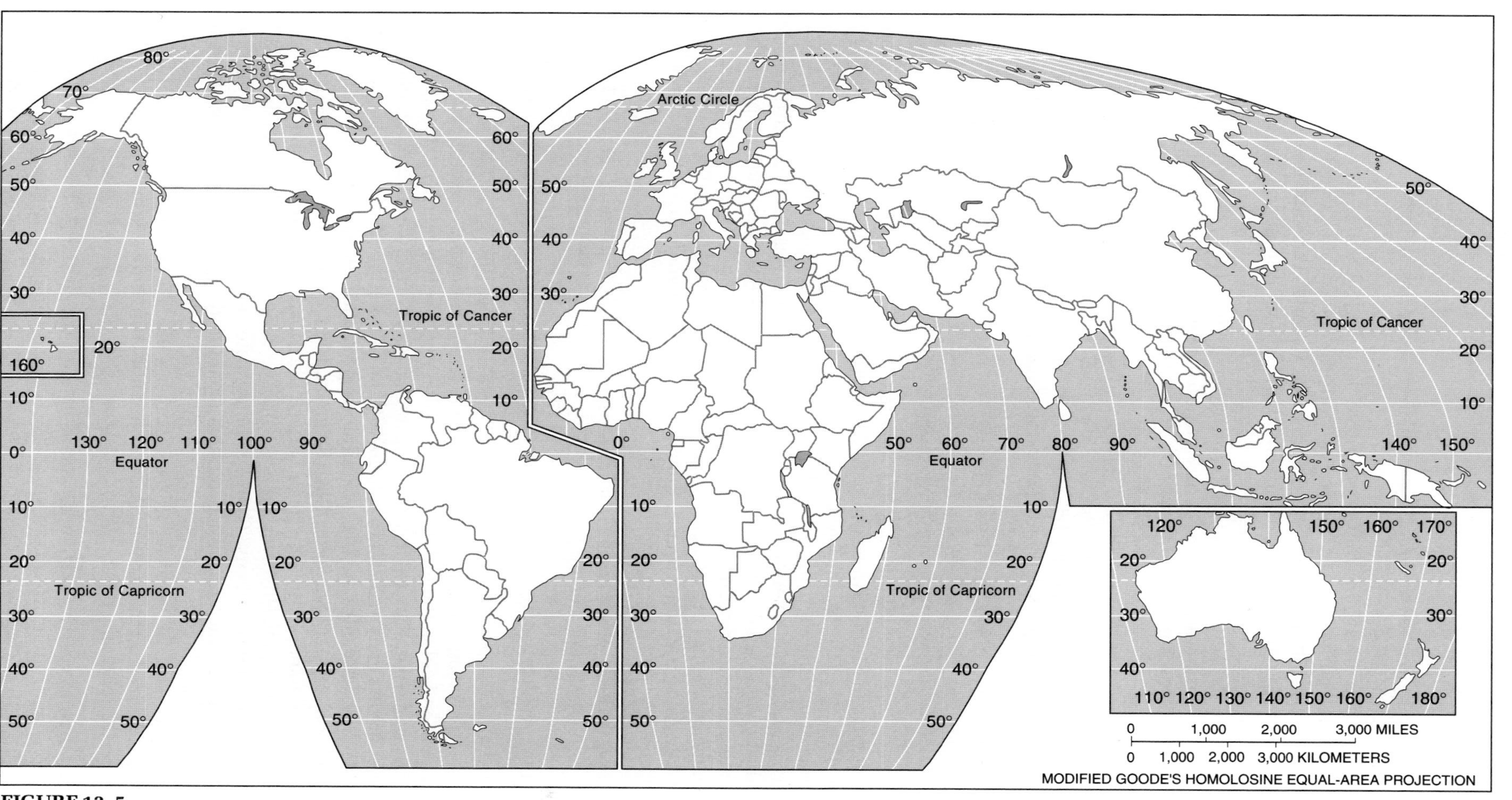


**FIGURE 13–5**
Colonial objectives.

islands, Belize, the three Guianas, and the coast of Brazil from Fortaleza to Rio de Janiero.

A third wave of in-migrants arrived in the late nineteenth and twentieth centuries, attracted by economic and political liberty much like the waves of European migration to the United States and Canada. You can show this liberty-motivated migration with a very broad arrow from Europe to Latin America, in particular to southern Brazil, Argentina, Chile, and Uruguay.

On the outline map of Latin America in Figure 13–6, show the following as the only European colonies to survive into the twentieth century (or almost to the twentieth century in the case of Cuba and Puerto Rico, which Spain lost in the Spanish-American War of 1898).

*British colonies:* Jamacia, Belize, Guyana, Bahamas, Turks and Caicos Islands, Dominica, Grenada, Trinidad and Tobago, Barbados, St. Lucia, St. Vincent, Barbuda, St. Kitts and Nevis, the Cayman Islands, and the British Virgin Islands.
*Spanish colonies:* Cuba, Puerto Rico.
*French colonies:* French Guiana, Martinique, Guadeloupe.
*Dutch colonies:* Suriname, "ABC" islands (Aruba, Bonaire, Curacao).

The United States long has regarded Middle America as its own backyard. Citizens of Middle American countries believe that the United States treats the Gulf of Mexico and the Caribbean as "American lakes" and expects to play the role of Big Brother throughout the region. At least one major motive for repeated U.S. military intervention in Middle American countries has been the defense of the Panama Canal. The canal is still important to the United States in permitting cheap ocean transport between the U.S. east and west coasts, in addition to speeding movements between U.S. Atlantic and Pacific naval fleets. The U.S. Virgin Islands were acquired from Denmark to aid in safeguarding approaches to the Canal.

**13–1** Which wide passage from the Atlantic to the Caribbean is guarded by U.S. bases in Puerto Rico? ______________________________

**13–2** Which passage is defended by the U.S. base at Guantanamo Bay, Cuba?

________________________________________

On an outline map of Middle America in Figure 13–7, locate and label, with a date and color, the following U.S. military occupations: Haiti, 1915–1934; Dominican Republic, 1916–1924; Nicaragua, 1911–1932; and Cuba, 1900–1905. Use arrows to show military actions without occupation: Veracruz, Mexico, 1914; Mexican border areas, 1917; Dominican Republic, 1965; Cuba (U.S.-supported Cubans), 1961; El Salvador, 1981; Grenada, 1983; Panama, 1989; and Haiti, 1994.

As noted, Latin American nations are mostly middle income rather than very poor or rich. However, there is usually a wide disparity between the wealthiest and poorest citizens. The World Bank categorizes countries as low-income, lower-middle income, upper-middle income, and high-income. Using the world outline map in Figure 13–8, construct a map showing these four groups of countries. Use a dark color for upper-income, a medium color for upper-middle, a lighter color for lower-middle, and leave low-income countries blank (white). Use the following data:

*Upper-income:* United States, Canada, Japan, Finland, United Arab Emirates, Iceland, Switzerland, Austria, UK, France, Belgium, Netherlands, Luxembourg, Germany, Sweden, Denmark, Norway, Australia, New Zealand, Israel, Spain, Ireland, Singapore, Hong Kong, Italy.

**FIGURE 13–6**
Last surviving colonies.

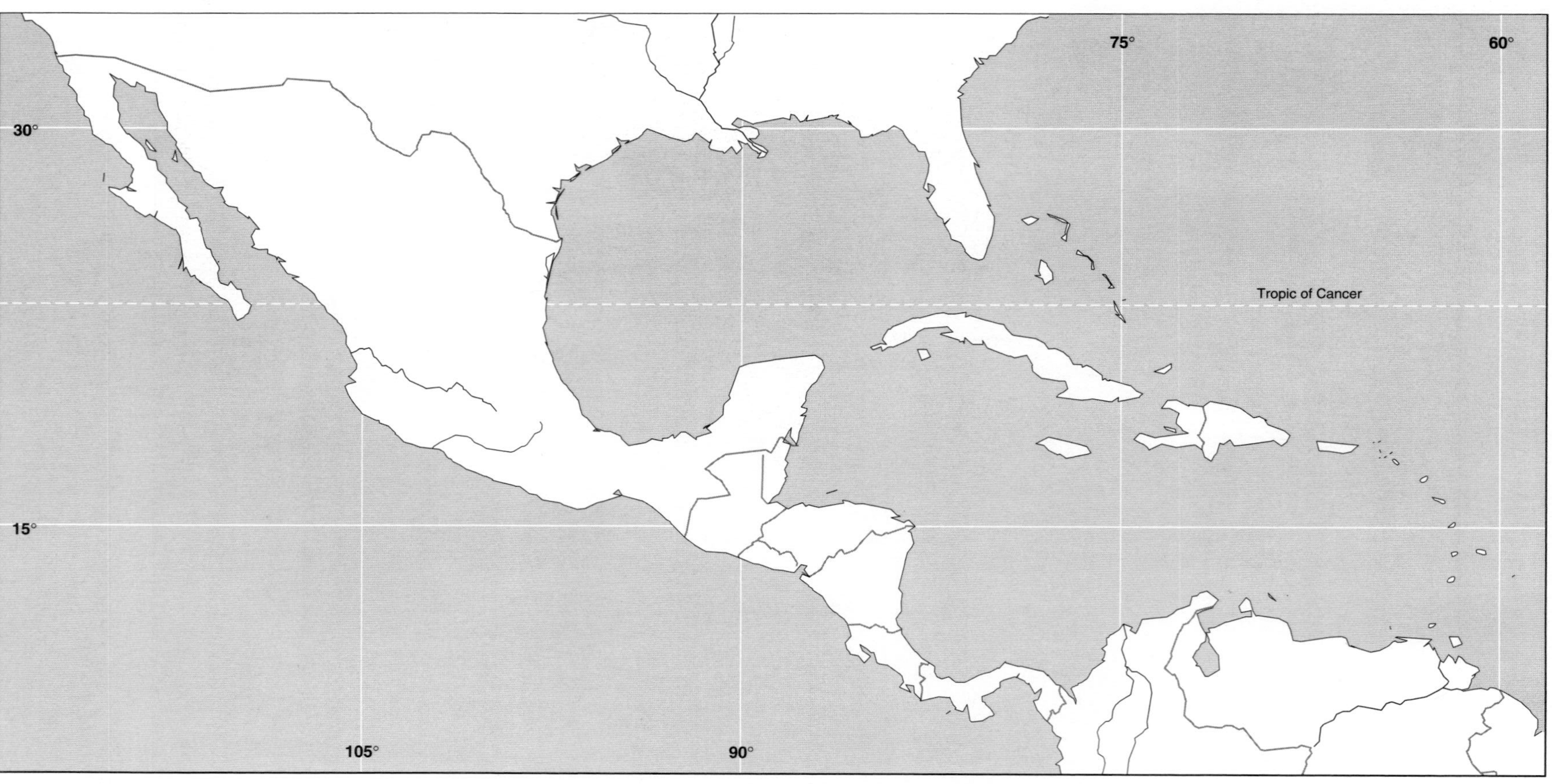


**FIGURE 13–7**
U.S. military occupations.

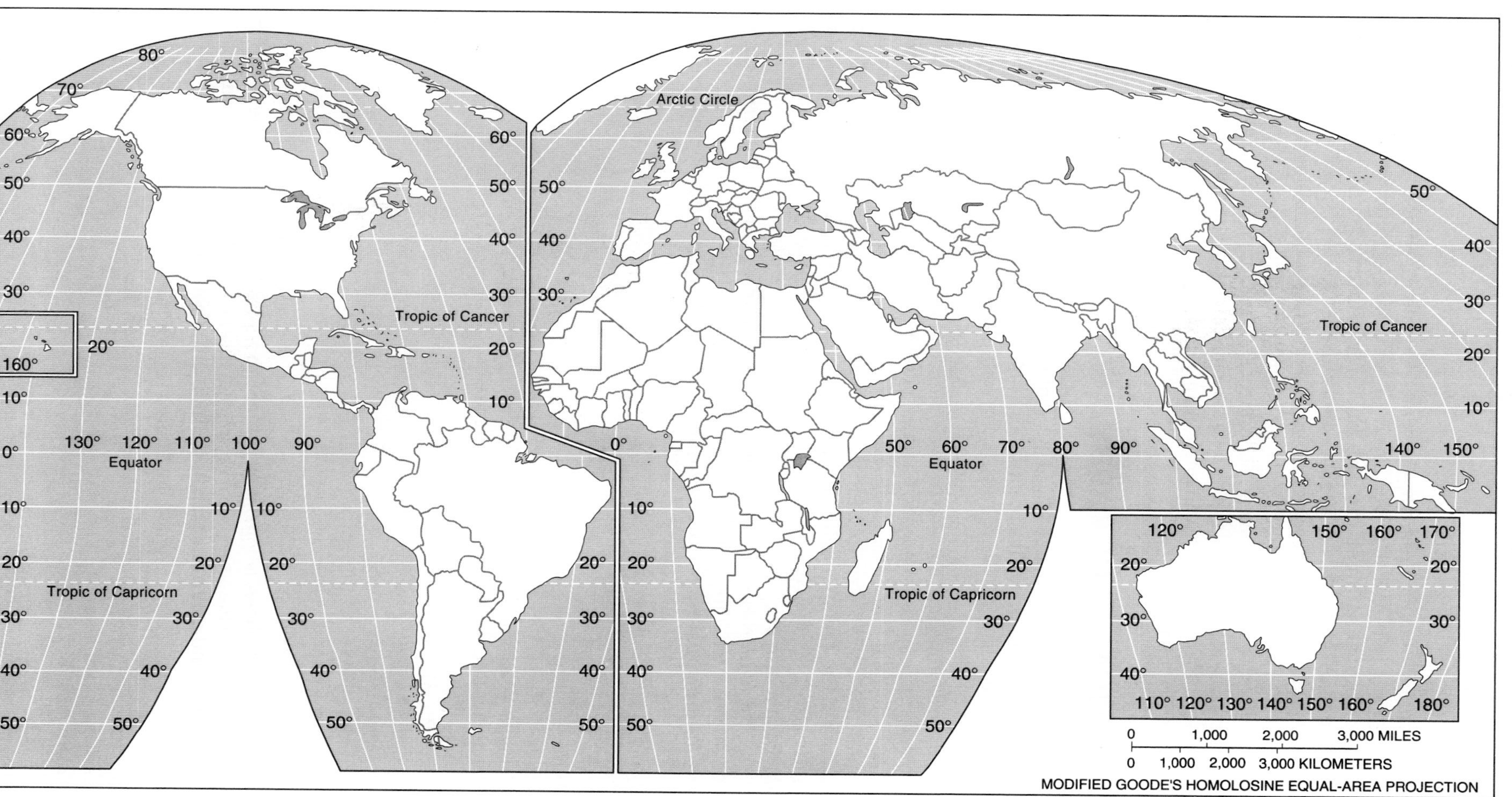


**FIGURE 13–8**
Income levels.

*Upper-middle income:* Brazil, Uruguay, Argentina, Hungary, South Korea, Gabon, Portugal, Greece, Trinidad and Tobago, Puerto Rico, Bosnia-Herzegovna, Antigua and Barbuda, Chile, Mexico, Cyprus, Bahrain, Saudi Arabia, Kuwait, Czech Republic, Poland, Malta.
*Lower-middle income:* Russia, Colombia, Peru, Costa Rica, Thailand, Tunisia, Turkey, Belize, Panama, Malaysia, South Africa, Lebanon, Venezuela, Dominica, Grenada, St. Lucia, St. Vincent and the Grenadines, Estonia, Latvia, Lithuania, Belarus, Slovakia, Namibia, El Salvador, Guatemala, Dominican Republic, Paraguay, Surinam, Iran, Jamaica.
*Low-income:* all the rest.

## POLITICAL GEOGRAPHY

In mainland Latin America, many present-day international boundaries reflect the colonial administrative boundaries established centuries ago by Spain. The colonial viceroyalty of New Grenada, for example, was originally subdivided into three units that closely correspond with modern Colombia, Venezuela, and Ecuador. The colonial administrative centers tended to become the capitals of independent countries—Bogota, Caracas, and Quito. Although briefly united in Gran Colombia under the great leader Simon Bolivar, Ecuador and Venezuela eventually separated again. Note that Panama was part of the territory ruled from Bogota under Spanish colonial administration. It remained part of Colombia until its independence in 1903, encouraged by the United States in order to build the Panama Canal.

The countries of Latin America have frequently disputed territory and fought wars with one another over the years. On the regional outline map in Figure 13–9, use two-way arrows across international borders to show past conflicts:

Honduras–El Salvador
Honduras–Nicaragua
Panama–Colombia
Haiti–Dominican Republic
Chile–Bolivia
Paraguay–Bolivia
Paraguay–Brazil, Argentina, and Uruguay
Chile–Argentina
Peru–Brazil
Uruguay–Brazil
Bolivia–Brazil
Ecuador–Peru
Colombia–Brazil
Venezuela–Brazil

Brazil has been a consistent winner in its territorial disputes with neighbors. Bolivia and Paraguay have been major losers of territory. Presently disputed boundaries and conflicting territorial claims are between:

Honduras–Nicaragua
Venezuela–Guyana
Guyana–Suriname
Guatemala–Belize
Paraguay–Bolivia
Suriname–French Guiana

Shade these border areas as the sites of possible future wars.

**FIGURE 13–9**
International border conflicts in Latin America.

## ECONOMIC GEOGRAPHY

As a region, Latin America has achieved the greatest percentage increase in GNP per capita, 1991 to 2000. The "middle class" region of the world has increased its per capita output of goods and services at an astonishing three times the world average.

Table 13–2 lists GNP per capita for world regions, in both 1991 and 2000, and indicates the percentage of change over that decade. As can be observed on Table 13–2, world GNP per capita has increased by 30%. However, there is an enormous difference when contrasting the more developed world with the less developed. The more developed countries saw positive change but at less than one-quarter the rate of change in less developed countries. This suggests that, on the basis of developmental level, less developed countries typically are moving on up the productivity/income ladder at a faster rate than are the already developed countries.

**TABLE 13–2** GNP per capita, world regions*

| | 1991 | 2000 | % Change |
|---|---|---|---|
| World | $ 3,760 | $ 4,890 | 30% |
| More Developed World | 16,990 | 19,480 | 14.6 |
| Less Developed World | 750 | 1,260 | 68 |
| Africa | 610 | 670 | 9.8 |
| U.S. & Canada | 20,900 | 28,230 | 35 |
| Asia | 1,580 | 2,910 | 84 |
| Latin America | 1,990 | 3,880 | 95 |
| Europe (including Russia**) | 11,990 | 13,420 | 12 |
| Australia, New Zealand, Pacific Islands | 14,440 | 15,400 | 6.6 |

***Note**: World regions as organized on this table do *not* all coincide with this workbook's definitions *except* Latin America, U.S. and Canada, and Australia, New Zealand and Pacific Islands

**1991 data not available for Russia; 1994 data for Russia used instead.

**13–3** Which of the primarily developing regions has the lowest percentage increase in GNP per capita? ____________

**13–4** Which of the listed regions has the smallest percentage increase?

____________

**13–5** Is that region characteristically developed or less developed?

____________

**13–6** Did the U.S. and Canada exceed the "more developed world" average?

____________

**13–7** Which region saw the smallest absolute (not percentage) gain in GNP per capita? ____________

## CULTURAL GEOGRAPHY AND DEMOGRAPHICS

Most South American countries have a peripheral distribution of population, that is, most people live on or close to the seacoast. Brazil is the classic example, with a densely populated coastal region from Fortaleza south to the border with

Uruguay, and a much lower density in the interior. Brazil's construction of its new capital, Brasília, was motivated by a need to psychologically turn Brazilians' eyes toward the interior and its possibilities. Some Brazilians are convinced that their national population is poorly distributed in terms of economic potential. They would especially like to move unemployed people from the cities of the northeast "bulge" (Fortaleza, Natal, Recife, and Salvador) into the job opportunities of the great inland frontier.

All of the countries that share the immense Amazon Basin with Brazil—Venezuela, Colombia, Ecuador, Peru, Bolivia, and the three Guianas—also have relatively few people in their portion of Amazonia; they all share the general lack of enthusiasm for this great tropical lowland rainforest. One factor that helps explain population distribution in the area from Venezuela to Ecuador is a preference for living in the cooler highlands when near the Equator. In Ecuador (the name means Equator), the seaport of Guayaquil has daily high temperatures that vary only from 84°F to 88°F (28.9°C to 31°C), with very little seasonal variation that close to the Equator (about 2° S latitude). The capital, Quito, is almost exactly on the Equator but, at an elevation of 9350 feet (2850 m), enjoys springlike weather with daily high temperatures between 70°F and 73°F (21°C and 23°C). Before air conditioning, Quito was a lot less sweaty than Guayaquil; but with air conditioning, think of the electric bill in Guayaquil compared to that in Quito!

## CHECK UP

**13–8** Which are the only Central American countries that lack "two-ocean frontage"? ____________________

**13–9** Which two countries share the island of Hispaniola?

____________________

**13–10** The giant state of Middle America is __________; the giant of South America is __________.

**13–11** Which two Latin American countries front on both the Gulf of Mexico and the Caribbean? ____________________

**13–12** The largest English-speaking Caribbean island is __________.

**13–13** The first black republic in the world, Haiti, once was a colony of what European country? (Haiti's official language is a clue.) __________

**13–14** The only two South American countries that do not share a border with Brazil are ____________________.

**13–15** The only nonself-governing state on mainland South America is

____________________.

**13–16** Which two South American nations would be most likely to claim parts of Antarctica? ____________________

**13–17** The only two landlocked Latin American countries are

____________________.

**13–18** Which Latin American country is farthest east? ______________________
Farthest north? ______________________

**13–19** Which Latin American country extends farthest south?
______________________

**13–20** Which seven South American capitals are *not* also seaports?
______________________

**13–21** The South American country with the highest ratio of seacoast to total area is ______________________.

**13–22** The richest Latin American country in per capita GNP is ____________.

**13–23** The only two South American countries that have common land borders with only two other countries are ______________________.

**13–24** Which is the only South American country with frontage on both the Caribbean and Pacific? ______________________

**13–25** If you traveled directly east from Sydney, Australia, which Latin American country would you come to? ______________________
______________________

**13–26** If you wanted to travel the shortest distance over water between Liberia in west Africa and an airport in Latin America, which Latin American country would you plan to land in? ______________________

**13–27** Most of the continent of South America lies (circle all that apply) south of the United States, north of Europe, east of the United States, west of Africa south of the Sahara.

## GEOCONCEPTS

### Primate Cities

In some countries, a single very large city, the *primate city,* dominates the economic, cultural, and usually the political life of the entire country. In true primate city situations, the largest city is far larger than the second-largest; there is no real rivalry for first place. Primate cities almost always are national capitals (though national capitals are not always primate cities). Primate cities are always the leading centers of transport and communications within their countries, and because they are the largest and richest markets in their countries, they also are centers of industrial and service activities.

In general, the poorer, less-industrialized countries are less urbanized, whereas the richer, industrial nations are highly urbanized. For example, the proportion of total population living in cities in low-income countries averages 30%; in lower-middle-income countries, 51%; in upper-middle-income countries, 66%; and in high-income economies, 77% of the populations live in cities. Latin America, the most "middle income" of the developing world regions, also is the most urbanized of those developing regions. At 74% urban, Latin America has more than twice as high an urban proportion than Africa south of the Sahara, South Asia, or Southeast Asia.

One indicator, though not the only one, of primate city situations is the proportion of a country's urban population that is concentrated in the largest city. When 56% of Haiti's city dwellers live in Port-au-Prince, this is definitely an example of a primate city. On the other hand, while only a third of Mexican urbanites live in Mexico City, that city is recognized as a primate city also, just as Paris, at 23% of France's urban population, also dominates France. In the United States, which does not have a primate city, the largest city concentrates only 12% of the urbanites.

On the world outline map in Figure 13–10, use four different colors to indicate the four economic levels for the countries in which 40% or more of the urban people live in the largest city. (If 40% of the U.S. urban population lived in the largest city—the New York/northeastern New Jersey metropolitan area—there would be 78 million people there!) Countries with 40% or more of their urban populations living in the largest city are as follows:

| Low-Income Economies | Lower-Middle-Income | Upper-Middle-Income | High Income |
|---|---|---|---|
| Laos | Dominican Republic | Uruguay | Ireland |
| Mozambique | Thailand | Argentina | |
| Tanzania | Jamaica | South Korea | |
| Burkina Faso | Paraguay | Portugal | |
| Uganda | Panama | Greece | |
| Togo | Lebanon | Chile | |
| Sierra Leone | | | |
| Benin | | | |
| Kenya | | | |
| Zimbabwe | | | |
| Nicaragua | | | |
| Congo | | | |
| Libya | | | |
| Iraq | | | |
| Senegal | | | |
| Bolivia | | | |
| Haiti | | | |
| Yemen | | | |
| Guinea | | | |

**13–28** Are countries with 40% or more of the urban population living in the largest city also likely to be relatively small countries in territory? Are there notable exceptions to this generalization? ____________________

____________________________________________

## Destruction of Tropical Rainforests

The thematic map showing the world's rainforests in Figure 13–11 indicates that the largest area of rainforest lies in Latin America, mostly in Brazil.

**13–29** What high-income industrialized country also has some rainforest areas?

____________________________________________

**13–30** Which islands of Southeast Asia still have large areas of rainforest?

____________________________________________

**13–31** Which South American countries have *no* areas of rainforest, past or present? ____________________________________________

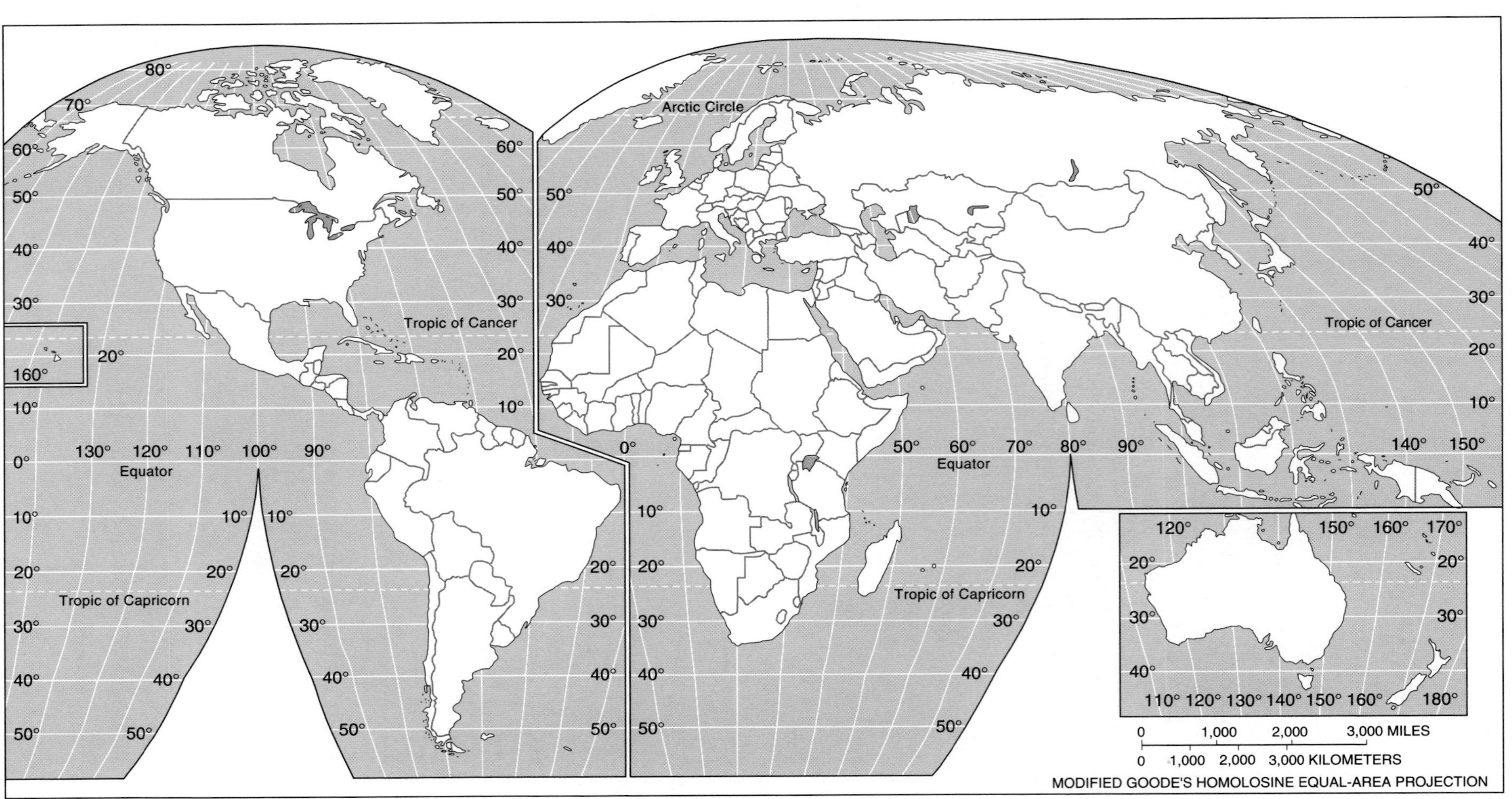


**FIGURE 13–10**
Percentage of total population in largest metropolitan area.

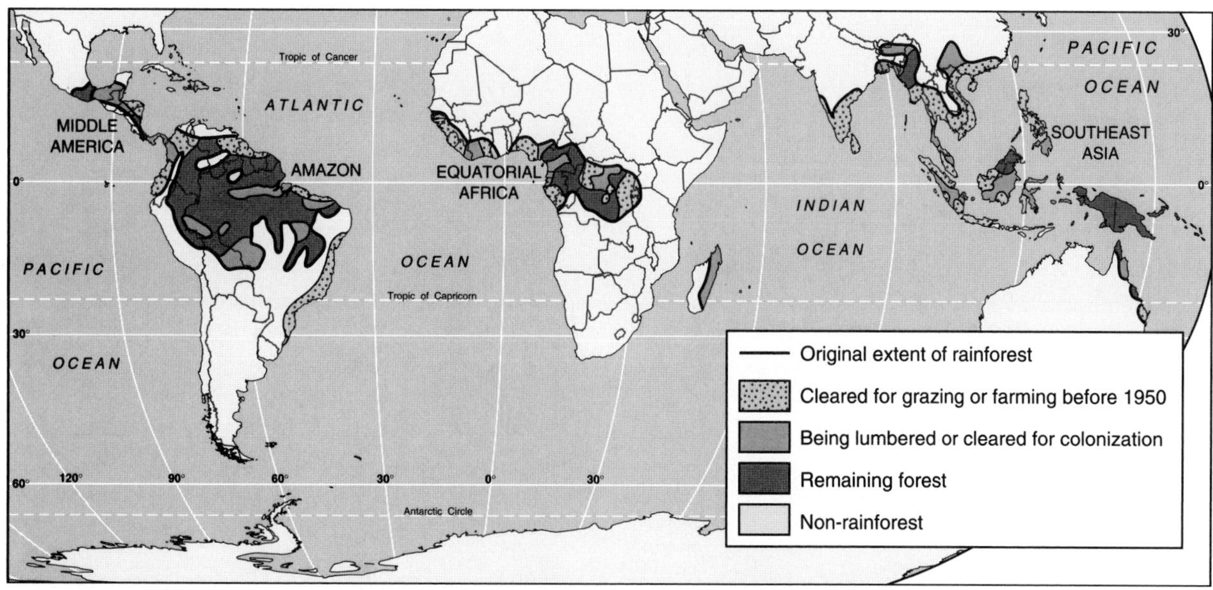


**FIGURE 13–11**
The destruction of the world's rainforests through lumbering and the extension of agriculture underscores the changeable nature of physical region boundaries as well as the larger environmental problem associated with that change.

Only about 6% of Earth's surface is covered by rainforest, but this 6% is extraordinarily important. The vast majority of all the 8 million plant and animal species on Earth are found in the rainforest. Wiping out the rainforest habitat could destroy forever many plants and animals that carry potentially useful genetic traits. There is no doubt that the climates of the tropics and even mid-latitude areas will be negatively affected (less precipitation) by rainforest destruction.

Brazil has become the focus of criticism in world public opinion for the amount of rainforest destruction occurring there. Ironically, this may be because Brazil has only recently begun to clear large areas of rainforest; forest clearance progressed faster and earlier in much of Africa south of the Sahara and Southeast Asia. It is Brazil's relatively recent determination to redistribute its population from the nonrainforest coastal areas to the interior that has placed it on a collision course with conservationists.

## REGIONAL WATCHLIST

Latin America is a region that is more likely to see internal political turmoil within countries rather than international wars or boundary changes. Native Americans, or Indios as they are known in the region, have participated in riots and insurrections of various scales against the central governments of Latin American states from Mexico to Peru. International boundary disputes focus on Guyana and its neighbors, and Guatemala occasionally has asserted a claim to Belize. Haiti and the Dominican Republic are uneasy neighbors on the large island of Hispaniola.

# Final Check Up

You can refer to the maps in this book, but count how many times you check them! Fill in the blanks or circle the correct answer(s).

1 When you fly from New York to London, which ocean do you cross?

______________________________

2 List all of the (U. S.) states in which you could visit an Atlantic Ocean or Gulf of Mexico beach. ______________________________

______________________________

3 If you were to drive as directly as possible from Chicago to San Francisco, which state would you *not* cross? Nevada, California, Illinois, Texas, Colorado, Utah.

4 If you were to fly from Denver to Anchorage, Alaska, the plane would fly in which general direction? South, southeast, northwest, northeast.

5 You could dip your toes in the Pacific Ocean from which five U.S. states?

______________________________

6 Which foreign country would you encounter first if you sailed due south from Miami, Florida? ______________________________

7 Which country has the largest number of land boundaries with other countries? Canada, Mexico, United States, Brazil, South Africa.

8 If you were to drive as directly as possible from Washington, D.C., to Orlando, Florida, which states would you pass through? ______________________________

______________________________

9 The four easternmost Canadian provinces are ______________________________

______________________________.

10 If you take a Caribbean cruise from Miami, which countries could you *not* visit on that cruise? Venezuela, Panama, Haiti, Bolivia, Nigeria, Guatemala, Jamaica, Grenada

11 If Muslims in Texas wished to face toward Mecca, Saudi Arabia, as they prayed, in which general direction would they face? North, northwest, east, southwest, west.

12 If you wished to visit an Indian Ocean beach, which country could you *not* visit to do so? India, Australia, Somalia, South Africa, Sudan, Mozambique.

13 The four oceans are ______.

14 The seven traditional continents are ______ ______.

15 The continent with the largest number of people is ______.

16 The continent with the fewest people is ______.

17 List the continents on which you would find the following cities:

Mexico City ______

Sydney ______

Beijing (Peking) ______

Paris ______

Brasilia ______

Montreal ______

Berlin ______

New Delhi ______

Buenos Aires ______

Hanoi ______

Rome ______

Los Angeles ______

18 If you were convening a conference of African states, which countries would *not* be invited? Zambia, Zaire, Iraq, Malawi, Laos, Indonesia, Nigeria, Mauritania, Tunisia, Egypt, Angola, Paraguay, Chad, Tanzania, Zimbabwe.

19 If you wanted to organize a Central American Common Market, which countries would you *not* invite? Nicaragua, Belize, Panama, Costa Rica, Senegal, Uruguay, Guatemala, Malaysia, Honduras, El Salvador.

20 The five northernmost countries on the African continent are ______

21 Which countries are *not* considered part of Southeast Asia? Thailand, Myanmar, Bhutan, Indonesia, Korea, Malaysia, Libya, Laos, Vietnam, Brunei, the Philippines, Sri Lanka.

22 Name at least two neighboring countries sharing land borders with the following:

Russia ______

United States ______

Venezuela ______

Israel ______

India ______

China ______

Brazil ______

Egypt ______

Bolivia ______

Zaire ______

Thailand ______

Honduras ______

France ______

Poland ______

South Africa ______

Iraq ______

Argentina ______

Germany ______

**23** You've won a big prize in a state lottery and decided on a luxury, around-the-world cruise. There are two mistakes in the following cruise itinerary—two ports of call that are not located on a reasonable direct route between the preceding and succeeding ports. Circle the two ports that are obvious mistakes: The cruise begins at Miami and ends in New York. Ports of call are Miami; Nassau, Bahamas; Kingston, Jamaica; Panama Canal; Tahiti, French Polynesia; Trinidad; Sydney, Australia; Bali, Indonesia; Singapore; Madras, India; Hong Kong; Suez Canal; Alexandria, Egypt; Athens, Greece; Nice, France; Gibraltar; New York.

**24** As the vice-president in charge of international trade of a major U.S. corporation, you must categorize your trading partners into the four groups listed. Unfortunately, your assistant has made a mistake in each category. Circle the country that does *not* fit the category definition.

*Rich oil-exporting, but otherwise preindustrial countries:* United Arab Emirates, Kuwait, South Africa, Saudi Arabia.

*Fast-developing Pacific Rim economies:* South Korea, Singapore, Hong Kong, Taiwan, Bangladesh.

*Newly industrializing countries whose major asset is cheap labor:* Indonesia, Sweden, India, Malaysia, Thailand.

*Wealthy, highly industrialized countries:* Germany, Japan, Bolivia, Canada, Italy.

**25** Russia has commissioned your petroleum geology consultants to survey their offshore waters for possible undersea oil fields. For this purpose, you will *not* need good maps of which sea? Black Sea, White Sea, Caspian Sea, Sea of Okhotsk, Tasmanian Sea.

**26** Which three of the following are *not* landlocked countries? Paraguay, Uganda, Nepal, Iraq, Afghanistan, Bolivia, Zambia, Namibia, Laos, Zimbabwe, Austria, Bhutan, Slovakia, Cameroon, Mongolia.

27 Which of the following are *not* naturally and directly connected to the world ocean? Baltic Sea, Philippine Sea, Persian Gulf, Aral Sea, Coral Sea, Mediterranean Sea, Caspian Sea, North Sea, Arabian Sea.

28 Which of the following is/are *not* really part(s) of the Pacific Ocean? Bering Sea, South China Sea, Bay of Bengal, Sea of Japan, Red Sea, Sea of Okhotsk.

29 List *all* of the countries you would visit if you flew around the world along the Equator: ______________________________

______________________________

30 Many countries consist of *all* of one main island. Which countries do *not* fit this description? Iceland, Malagasy (Madagascar), Indonesia, Cuba, Japan, Sri Lanka, Jamaica, Denmark, Taiwan, Liberia.

31 How many questions could you answer *without* looking at the map? ________

# Appendix A

## 2000 World Population Data

| | Population mid-2000 (millions) | Births per 1,000 pop. | Deaths per 1,000 pop. | Natural Increase (annual, %) | "Doubling Time" in Years at *Current* Rate | Projected Population (millions) 2025 | 2050 | Infant Mortality Rate[a] | Percent Urban | GNP Per Capita, 1998 (US$) |
|---|---|---|---|---|---|---|---|---|---|---|
| **WORLD** | **6,067** | **22** | **9** | **1.4** | **51** | **7,810** | **9,039** | **57** | **45** | **4,890** |
| **More Developed** | **1,184** | **11** | **10** | **0.1** | **809** | **1,236** | **1,232** | **8** | **75** | **19,480** |
| **Less Developed** | **4,883** | **25** | **9** | **1.7** | **42** | **6,575** | **7,808** | **63** | **38** | **1,260** |
| **Less Developed (Excl. China)** | **3,619** | **29** | **9** | **1.9** | **36** | **5,144** | **6,439** | **69** | **40** | **1,450** |
| **AFRICA** | **800** | **38** | **14** | **2.4** | **29** | **1,258** | **1,804** | **88** | **33** | **670** |
| **Sub-Saharan Africa** | **657** | **41** | **16** | **2.5** | **27** | **1,053** | **1,556** | **94** | **29** | **520** |
| Northern Africa | 173 | 27 | 7 | 2.0 | 34 | 251 | 306 | 51 | 46 | 1,200 |
| Algeria | 31.5 | 29 | 6 | 2.4 | 29 | 46.6 | 57.7 | 44 | 49 | 1,550 |
| Egypt | 68.3 | 26 | 6 | 2.0 | 35 | 97.4 | 117.1 | 52 | 44 | 1,290 |
| Libya | 5.1 | 28 | 3 | 2.5 | 28 | 8.3 | 10.7 | 33 | 86 | — |
| Morocco | 28.8 | 23 | 6 | 1.7 | 41 | 39.3 | 46.1 | 37 | 54 | 1,240 |
| Sudan | 29.5 | 33 | 12 | 2.2 | 32 | 46.3 | 59.2 | 70 | 27 | 290 |
| Tunisia | 9.6 | 22 | 7 | 1.6 | 44 | 12.9 | 15.0 | 35 | 61 | 2,060 |
| Western Sahara | 0.3 | 46 | 18 | 2.9 | 24 | 0.4 | 0.6 | 150 | — | — |
| Western Africa | 234 | 42 | 14 | 2.8 | 25 | 390 | 567 | 89 | 35 | 340 |
| Benin | 6.4 | 45 | 17 | 2.8 | 24 | 11.7 | 18.1 | 94 | 38 | 380 |
| Burkina Faso | 11.9 | 47 | 18 | 2.9 | 24 | 21.6 | 34.3 | 105 | 15 | 240 |
| Cape Verde | 0.4 | 37 | 9 | 2.8 | 25 | 0.5 | 0.4 | 77 | 44 | 1,200 |
| Côte d'Ivoire | 16.0 | 38 | 16 | 2.2 | 32 | 23.3 | 30.5 | 112 | 46 | 700 |
| Gambia | 1.3 | 43 | 19 | 2.4 | 29 | 2.2 | 2.8 | 130 | 37 | 340 |
| Ghana | 19.5 | 34 | 10 | 2.4 | 29 | 26.5 | 32.0 | 56 | 37 | 390 |
| Guinea | 7.5 | 42 | 18 | 2.4 | 29 | 12.6 | 18.1 | 98 | 26 | 530 |
| Guinea-Bissau | 1.2 | 42 | 20 | 2.2 | 31 | 1.9 | 2.7 | 130 | 22 | 160 |
| Liberia | 3.2 | 50 | 17 | 3.2 | 21 | 6.0 | 10.0 | 139 | 45 | — |
| Mali | 11.2 | 47 | 16 | 3.1 | 22 | 21.3 | 31.4 | 123 | 26 | 250 |

| | Population mid-2000 (millions) | Births per 1,000 pop. | Deaths per 1,000 pop. | Natural Increase (annual, %) | "Doubling Time" in Years at *Current* Rate | Projected Population (millions) 2025 | 2050 | Infant Mortality Rate[a] | Percent Urban | GNP Per Capita, 1998 (US$) |
|---|---|---|---|---|---|---|---|---|---|---|
| Mauritania | 2.7 | 41 | 13 | 2.7 | 25 | 4.8 | 6.6 | 92 | 54 | 410 |
| Niger | 10.1 | 54 | 24 | 3.0 | 23 | 18.8 | 28.5 | 123 | 17 | 200 |
| Nigeria | 123.3 | 42 | 13 | 2.8 | 24 | 204.5 | 303.6 | 77 | 36 | 300 |
| Senegal | 9.5 | 41 | 13 | 2.8 | 25 | 16.7 | 23.1 | 68 | 41 | 520 |
| Sierra Leone | 5.2 | 47 | 21 | 2.6 | 26 | 9.9 | 15.7 | 157 | 37 | 140 |
| Togo | 5.0 | 42 | 11 | 3.1 | 23 | 7.6 | 9.7 | 80 | 31 | 330 |
| Eastern Africa | 246 | 42 | 18 | 2.4 | 29 | 390 | 584 | 102 | 20 | 260 |
| Burundi | 6.1 | 42 | 17 | 2.5 | 28 | 10.5 | 16.1 | 75 | 8 | 140 |
| Comoros | 0.6 | 38 | 10 | 2.8 | 25 | 1.1 | 1.8 | 77 | 29 | 370 |
| Djibouti | 0.6 | 39 | 16 | 2.3 | 30 | 1.0 | 1.3 | 115 | 83 | — |
| Eritrea | 4.1 | 43 | 13 | 3.0 | 23 | 8.4 | 13.7 | 82 | 16 | 200 |
| Ethiopia | 64.1 | 45 | 21 | 2.4 | 29 | 115.0 | 187.9 | 116 | 15 | 100 |
| Kenya | 30.3 | 35 | 14 | 2.1 | 33 | 34.4 | 38.7 | 74 | 20 | 350 |
| Madagascar | 14.9 | 44 | 14 | 2.9 | 24 | 28.5 | 46.9 | 96 | 22 | 260 |
| Malawi | 10.4 | 41 | 22 | 1.9 | 36 | 12.6 | 14.7 | 127 | 20 | 210 |
| Mauritius | 1.2 | 17 | 7 | 1.1 | 66 | 1.4 | 1.5 | 19.4 | 43 | 3,730 |
| Mozambique | 19.1 | 41 | 19 | 2.2 | 32 | 20.6 | 22.9 | 134 | 28 | 210 |
| Reunion | 0.7 | 20 | 5 | 1.4 | 49 | 1.0 | 1.2 | 9 | 73 | — |
| Rwanda | 7.2 | 43 | 20 | 2.3 | 30 | 8.0 | 8.9 | 121 | 5 | 230 |
| Seychelles | 0.1 | 18 | 7 | 1.1 | 65 | 0.1 | 0.1 | 9 | 59 | 6,420 |
| Somalia | 7.3 | 47 | 18 | 2.9 | 24 | 14.9 | 25.5 | 126 | 24 | — |
| Tanzania | 35.3 | 42 | 13 | 2.9 | 24 | 59.8 | 88.3 | 99 | 20 | 220 |
| Uganda | 23.3 | 48 | 20 | 2.9 | 24 | 48.0 | 84.1 | 81 | 15 | 310 |
| Zambia | 9.6 | 42 | 23 | 2.0 | 35 | 14.3 | 20.3 | 109 | 38 | 330 |
| Zimbabwe | 11.3 | 30 | 20 | 1.0 | 69 | 9.5 | 9.3 | 80 | 32 | 620 |
| Middle Africa | 96 | 46 | 16 | 3.0 | 23 | 185 | 303 | 106 | 32 | 320 |
| Angola | 12.9 | 48 | 19 | 3.0 | 23 | 25.1 | 36.9 | 125 | 32 | 380 |
| Cameroon | 15.4 | 37 | 12 | 2.6 | 27 | 24.7 | 34.7 | 77 | 44 | 610 |
| Central African Republic | 3.5 | 38 | 18 | 2.0 | 34 | 4.9 | 6.4 | 97 | 39 | 300 |
| Chad | 8.0 | 50 | 17 | 3.3 | 21 | 17.3 | 31.5 | 110 | 22 | 230 |
| Congo | 2.8 | 40 | 16 | 2.4 | 29 | 4.6 | 6.9 | 109 | 41 | 680 |
| Congo, Dem. Rep. of (Zaire) | 52.0 | 48 | 16 | 3.2 | 22 | 105.3 | 181.9 | 109 | 29 | 110 |
| Equatorial Guinea | 0.5 | 41 | 16 | 2.5 | 28 | 0.8 | 1.1 | 108 | 37 | 1,110 |
| Gabon | 1.2 | 38 | 16 | 2.2 | 32 | 2.0 | 2.7 | 87 | 73 | 4,170 |
| Sao Tome and Principe | 0.2 | 43 | 9 | 3.4 | 20 | 0.3 | 0.5 | 51 | 44 | 270 |
| Southern Africa | 50 | 26 | 13 | 1.3 | 52 | 43 | 43 | 51 | 42 | 3,100 |
| Botswana | 1.6 | 32 | 17 | 1.6 | 45 | 1.2 | 1.2 | 57 | 49 | 3,070 |
| Lesotho | 2.1 | 33 | 13 | 2.1 | 33 | 2.4 | 2.8 | 85 | 16 | 570 |
| Namibia | 1.8 | 36 | 20 | 1.7 | 42 | 2.3 | 3.8 | 68 | 27 | 1,940 |
| South Africa | 43.4 | 25 | 12 | 1.3 | 55 | 35.1 | 32.5 | 45 | 45 | 3,310 |
| Swaziland | 1.0 | 41 | 22 | 1.9 | 37 | 1.6 | 3.1 | 108 | 22 | 1,400 |
| **NORTH AMERICA** | **306** | **14** | **9** | **0.6** | **124** | **374** | **444** | **7** | **75** | **28,230** |
| Canada | 30.8 | 11 | 7 | 0.4 | 178 | 36.0 | 40.2 | 5.5 | 78 | 19,170 |

| | Population mid-2000 (millions) | Births per 1,000 pop. | Deaths per 1,000 pop. | Natural Increase (annual, %) | "Doubling Time" in Years at *Current* Rate | Projected Population (millions) 2025 | 2050 | Infant Mortality Rate[a] | Percent Urban | GNP Per Capita, 1998 (US$) |
|---|---|---|---|---|---|---|---|---|---|---|
| United States | 275.6 | 15 | 9 | 0.6 | 120 | 337.8 | 403.7 | 7.0 | 75 | 29,240 |
| **LATIN AMERICA & THE CARIBBEAN** | **518** | **24** | **6** | **1.8** | **39** | **703** | **823** | **35** | **74** | **3,880** |
| Central America | 136 | 26 | 5 | 2.1 | 33 | 192 | 232 | 34 | 67 | 3,230 |
| Belize | 0.3 | 32 | 5 | 2.7 | 26 | 0.4 | 0.5 | 34 | 50 | 2,660 |
| Costa Rica | 3.6 | 22 | 4 | 1.8 | 39 | 5.8 | 7.0 | 13 | 45 | 2,770 |
| El Salvador | 6.3 | 30 | 7 | 2.4 | 29 | 9.8 | 13.6 | 35 | 58 | 1,850 |
| Guatemala | 12.7 | 37 | 7 | 2.9 | 24 | 22.3 | 32.2 | 45 | 39 | 1,640 |
| Honduras | 6.1 | 33 | 6 | 2.8 | 25 | 8.6 | 11.0 | 42 | 45 | 740 |
| Mexico | 99.6 | 24 | 4 | 2.0 | 36 | 132.5 | 152.1 | 32 | 74 | 3,840 |
| Nicaragua | 5.1 | 36 | 6 | 3.0 | 23 | 8.7 | 11.6 | 40 | 63 | 370 |
| Panama | 2.9 | 22 | 5 | 1.7 | 41 | 3.8 | 4.3 | 21 | 56 | 2,990 |
| Caribbean | 36 | 22 | 8 | 1.3 | 52 | 46 | 51 | 47 | 61 | — |
| Antigua and Barbuda | 0.1 | 22 | 6 | 1.6 | 45 | 0.1 | 0.1 | 17 | 37 | 8,450 |
| Bahamas | 0.3 | 21 | 5 | 1.5 | 45 | 0.4 | 0.5 | 18.4 | 84 | — |
| Barbados | 0.3 | 14 | 9 | 0.5 | 130 | 0.3 | 0.3 | 14.2 | 38 | — |
| Cuba | 11.1 | 14 | 7 | 0.7 | 103 | 11.7 | 10.6 | 7 | 75 | — |
| Dominica | 0.1 | 16 | 8 | 0.8 | 83 | 0.1 | 0.1 | 14.6 | — | 3,150 |
| Dominican Republic | 8.4 | 28 | 6 | 2.2 | 32 | 12.1 | 14.9 | 47 | 62 | 1,770 |
| Grenada | 0.1 | 29 | 6 | 2.3 | 30 | 0.2 | 0.2 | 14 | 34 | 3,250 |
| Guadeloupe | 0.4 | 17 | 6 | 1.1 | 61 | 0.5 | 0.5 | 10.0 | 48 | — |
| Haiti | 6.4 | 33 | 16 | 1.7 | 40 | 9.6 | 11.9 | 103 | 34 | 410 |
| Jamaica | 2.6 | 22 | 7 | 1.6 | 45 | 3.3 | 3.8 | 24 | 50 | 1,740 |
| Martinique | 0.4 | 15 | 6 | 0.9 | 81 | 0.5 | 0.5 | 9 | 81 | — |
| Netherlands Antilles | 0.2 | 17 | 6 | 1.1 | 62 | 0.3 | 0.3 | 14 | — | — |
| Puerto Rico | 3.9 | 17 | 8 | 0.9 | 75 | 4.2 | 4.2 | 11.3 | 71 | — |
| St. Kitts-Nevis | 0.04 | 20 | 11 | 0.9 | 82 | 0.1 | 0.1 | 24 | 43 | 6,190 |
| Saint Lucia | 0.2 | 19 | 6 | 1.2 | 56 | 0.2 | 0.2 | 16.8 | 48 | 3,660 |
| St. Vincent & the Grenadines | 0.1 | 19 | 7 | 1.2 | 59 | 0.1 | 0.2 | 20.4 | 44 | 2,560 |
| Trinidad and Tobago | 1.3 | 14 | 7 | 0.7 | 103 | 1.5 | 1.5 | 16.2 | 72 | 4,520 |
| South America | 345 | 23 | 6 | 1.7 | 42 | 465 | 540 | 34 | 78 | 4,270 |
| Argentina | 37.0 | 19 | 8 | 1.1 | 62 | 47.2 | 54.5 | 19.1 | 90 | 8,030 |
| Bolivia | 8.3 | 30 | 10 | 2.0 | 34 | 12.2 | 15.5 | 67 | 62 | 1,010 |
| Brazil | 170.1 | 21 | 6 | 1.5 | 45 | 221.2 | 244.2 | 38 | 78 | 4,630 |
| Chile | 15.2 | 18 | 5 | 1.3 | 54 | 19.5 | 22.2 | 10.5 | 85 | 4,990 |
| Colombia | 40.0 | 26 | 6 | 2.0 | 34 | 58.3 | 73.3 | 28 | 71 | 2,470 |
| Ecuador | 12.6 | 27 | 6 | 2.1 | 33 | 17.8 | 21.2 | 40 | 63 | 1,520 |
| French Guiana | 0.2 | 27 | 3 | 2.4 | 29 | 0.4 | 0.6 | 18 | 79 | — |
| Guyana | 0.7 | 24 | 7 | 1.7 | 40 | 0.8 | 0.8 | 63 | 36 | 780 |
| Paraguay | 5.5 | 32 | 6 | 2.7 | 26 | 9.4 | 12.6 | 27 | 52 | 1,760 |
| Peru | 27.1 | 27 | 6 | 2.1 | 32 | 39.2 | 47.9 | 43 | 72 | 2,440 |
| Suriname | 0.4 | 26 | 7 | 1.9 | 37 | 0.5 | 0.4 | 29 | 69 | 1,660 |
| Uruguay | 3.3 | 16 | 10 | 0.7 | 107 | 3.9 | 4.2 | 14.5 | 92 | 6,070 |

| | Population mid-2000 (millions) | Births per 1,000 pop. | Deaths per 1,000 pop. | Natural Increase (annual, %) | "Doubling Time" in Years at *Current* Rate | Projected Population (millions) 2025 | 2050 | Infant Mortality Rate[a] | Percent Urban | GNP Per Capita, 1998 (US$) |
|---|---|---|---|---|---|---|---|---|---|---|
| Venezuela | 24.2 | 25 | 5 | 2.0 | 34 | 34.8 | 42.2 | 21.0 | 86 | 3,530 |
| **OCEANIA** | **31** | **18** | **7** | **1.1** | **65** | **39** | **44** | **29** | **70** | **15,400** |
| Australia | 19.2 | 13 | 7 | 0.6 | 110 | 22.8 | 24.9 | 5.3 | 85 | 20,640 |
| Fed. States of Micronesia | 0.1 | 33 | 7 | 2.6 | 27 | 0.2 | 0.3 | 46 | 27 | 1,800 |
| Fiji | 0.8 | 22 | 7 | 1.5 | 46 | 1.1 | 1.3 | 13 | 46 | 2,210 |
| French Polynesia | 0.2 | 21 | 5 | 1.6 | 44 | 0.3 | 0.4 | 10 | 54 | — |
| Guam | 0.2 | 28 | 4 | 2.4 | 29 | 0.2 | 0.3 | 9.1 | 38 | — |
| Kiribati | 0.1 | 33 | 8 | 2.5 | 28 | 0.2 | 0.2 | 62 | 37 | 1,170 |
| Marshall Islands | 0.1 | 26 | 4 | 2.2 | 31 | 0.2 | 0.3 | 31 | 65 | 1,540 |
| Nauru | 0.01 | 19 | 5 | 1.4 | 48 | 0.02 | 0.02 | 25 | 100 | — |
| New Caledonia | 0.2 | 21 | 5 | 1.7 | 42 | 0.3 | 0.3 | 7 | 59 | — |
| New Zealand | 3.8 | 15 | 7 | 0.8 | 89 | 4.4 | 4.5 | 5.5 | 85 | 14,600 |
| Palau | 0.02 | 18 | 8 | 1.0 | 68 | 0.03 | 0.03 | 19 | 71 | — |
| Papua-New Guinea | 4.8 | 34 | 10 | 2.4 | 29 | 7.7 | 9.5 | 77 | 15 | 890 |
| Solomon Islands | 0.4 | 37 | 6 | 3.1 | 23 | 0.8 | 1.1 | 25 | 13 | 760 |
| Tonga | 0.1 | 27 | 6 | 2.1 | 33 | 0.2 | 0.2 | 19 | 32 | 1,750 |
| Vanuatu | 0.2 | 35 | 7 | 2.8 | 25 | 0.3 | 0.3 | 39 | 18 | 1,260 |
| Western Samoa | 0.2 | 31 | 6 | 2.5 | 28 | 0.2 | 0.2 | 25 | 21 | 1,070 |
| **ASIA** | **3,684** | **22** | **8** | **1.4** | **48** | **4,723** | **5,267** | **56** | **35** | **2,130** |
| **Asia (Excl. China)** | **2,420** | **26** | **8** | **1.7** | **40** | **3,292** | **3,898** | **64** | **38** | **2,910** |
| Western Asia | 189 | 28 | 7 | 2.1 | 33 | 300 | 396 | 55 | 65 | 3,620 |
| Armenia | 3.8 | 10 | 6 | 0.4 | 161 | 4.1 | 3.8 | 15 | 67 | 460 |
| Azerbaijan | 7.7 | 15 | 6 | 0.9 | 77 | 9.8 | 11.5 | 17 | 52 | 480 |
| Bahrain | 0.7 | 22 | 3 | 1.9 | 37 | 1.7 | 2.9 | 8 | 88 | 7,640 |
| Cyprus | 0.9 | 14 | 8 | 0.6 | 124 | 1.0 | 1.1 | 8 | 64 | 11,920 |
| Georgia | 5.5 | 9 | 8 | 0.2 | 462 | 4.8 | 4.2 | 15 | 56 | 970 |
| Iraq | 23.1 | 38 | 10 | 2.8 | 25 | 41.0 | 54.9 | 127 | 68 | — |
| Israel | 6.2 | 22 | 6 | 1.5 | 45 | 8.3 | 9.4 | 6.0 | 90 | 16,180 |
| Jordan | 5.1 | 33 | 5 | 2.9 | 24 | 8.8 | 12.0 | 34 | 78 | 1,150 |
| Kuwait | 2.2 | 24 | 2 | 2.2 | 32 | 3.8 | 4.4 | 13 | 100 | — |
| Lebanon | 4.2 | 23 | 7 | 1.6 | 43 | 5.6 | 6.5 | 35 | 88 | 3,560 |
| Oman | 2.4 | 43 | 5 | 3.9 | 18 | 5.2 | 9.0 | 25 | 72 | — |
| Palestinian Territory | 3.1 | 41 | 5 | 3.7 | 19 | 7.4 | 11.2 | 27 | — | 1,560 |
| Qatar | 0.6 | 20 | 2 | 1.8 | 38 | 0.8 | 0.8 | 20 | 91 | — |
| Saudi Arabia | 21.6 | 35 | 5 | 3.0 | 23 | 40.0 | 54.5 | 46 | 83 | 6,910 |
| Syria | 16.5 | 33 | 6 | 2.8 | 25 | 26.9 | 35.3 | 35 | 51 | 1,020 |
| Turkey | 65.3 | 22 | 7 | 1.5 | 46 | 88.0 | 100.7 | 38 | 66 | 3,160 |
| United Arab Emirates | 2.8 | 24 | 2 | 2.2 | 32 | 3.8 | 4.2 | 16 | 84 | 17,870 |
| Yemen | 17.0 | 39 | 11 | 2.8 | 25 | 38.6 | 69.3 | 75 | 26 | 280 |
| South Central Asia | 1,475 | 28 | 9 | 1.9 | 37 | 2,037 | 2,451 | 75 | 29 | 510 |
| Afghanistan | 26.7 | 43 | 18 | 2.5 | 28 | 48.0 | 76.2 | 150 | 20 | — |

| | Population mid-2000 (millions) | Births per 1,000 pop. | Deaths per 1,000 pop. | Natural Increase (annual, %) | "Doubling Time" in Years at *Current* Rate | Projected Population (millions) 2025 | 2050 | Infant Mortality Rate[a] | Percent Urban | GNP Per Capita, 1998 (US$) |
|---|---|---|---|---|---|---|---|---|---|---|
| Bangladesh | 128.1 | 27 | 8 | 1.8 | 38 | 177.3 | 210.8 | 82 | 20 | 350 |
| Bhutan | 0.9 | 40 | 9 | 3.1 | 22 | 1.4 | 2.0 | 71 | 15 | 470 |
| India | 1,002.1 | 27 | 9 | 1.8 | 39 | 1,363.0 | 1,628.0 | 72 | 28 | 440 |
| Iran | 67.4 | 21 | 6 | 1.4 | 48 | 90.8 | 102.9 | 31 | 63 | 1,650 |
| Kazakhstan | 14.9 | 14 | 10 | 0.4 | 161 | 14.6 | 13.0 | 21 | 56 | 1,340 |
| Kyrgyzstan | 4.9 | 22 | 7 | 1.5 | 47 | 5.8 | 6.1 | 26 | 34 | 380 |
| Maldives | 0.3 | 35 | 5 | 3.0 | 23 | 0.5 | 0.7 | 27 | 25 | 1,130 |
| Nepal | 23.9 | 36 | 11 | 2.5 | 28 | 38.0 | 49.3 | 79 | 11 | 210 |
| Pakistan | 150.6 | 39 | 11 | 2.8 | 25 | 227.0 | 285.0 | 91 | 33 | 470 |
| Sri Lanka | 19.2 | 18 | 6 | 1.2 | 60 | 23.9 | 25.9 | 17 | 22 | 810 |
| Tajikistan | 6.4 | 21 | 5 | 1.6 | 43 | 8.4 | 9.5 | 28 | 27 | 370 |
| Turkmenistan | 5.2 | 21 | 6 | 1.5 | 48 | 6.8 | 7.5 | 33 | 44 | — |
| Uzbekistan | 24.8 | 23 | 6 | 1.7 | 40 | 31.5 | 33.8 | 22 | 38 | 950 |
| Southeast Asia | 528 | 24 | 7 | 1.7 | 41 | 717 | 836 | 46 | 36 | 1,240 |
| Brunei | 0.3 | 25 | 3 | 2.2 | 32 | 0.5 | 0.7 | 24 | 67 | — |
| Cambodia | 12.1 | 38 | 12 | 2.6 | 27 | 21.2 | 29.0 | 80 | 16 | 260 |
| East Timor | 0.8 | 34 | 16 | 1.8 | 39 | 1.2 | 1.4 | 143 | — | — |
| Indonesia | 212.2 | 24 | 8 | 1.6 | 44 | 273.4 | 311.9 | 46 | 39 | 640 |
| Laos | 5.2 | 41 | 15 | 2.6 | 26 | 8.4 | 11.8 | 104 | 17 | 320 |
| Malaysia | 23.3 | 25 | 5 | 2.1 | 34 | 37.0 | 48.2 | 8 | 57 | 3,670 |
| Myanmar | 48.9 | 30 | 10 | 2.0 | 35 | 68.1 | 87.8 | 83 | 26 | — |
| Philippines | 80.3 | 29 | 7 | 2.3 | 31 | 117.3 | 139.6 | 35 | 47 | 1,050 |
| Singapore | 4.0 | 13 | 5 | 0.8 | 84 | 8.0 | 10.4 | 3.2 | 100 | 30,170 |
| Thailand | 62.0 | 16 | 7 | 1.0 | 70 | 72.1 | 71.9 | 22 | 31 | 2,160 |
| Vietnam | 78.7 | 20 | 6 | 1.4 | 48 | 109.9 | 123.7 | 37 | 24 | 350 |
| East Asia | 1,493 | 15 | 7 | 0.8 | 85 | 1,669 | 1,585 | 29 | 38 | 3,880 |
| China | 1,264.5 | 15 | 6 | 0.9 | 79 | 1,431.0 | 1,369.0 | 31 | 31 | 750 |
| China, Hong Kong SAR[e] | 7.0 | 7 | 5 | 0.3 | 256 | 8.6 | 7.6 | 3.2 | 95 | 23,660 |
| China, Macao SAR[e] | 0.4 | 10 | 3 | 0.7 | 96 | 0.6 | 0.8 | 6 | 99 | — |
| Japan | 126.9 | 9 | 8 | 0.2 | 462 | 120.9 | 100.5 | 3.5 | 78 | 32,350 |
| Korea, North | 21.7 | 21 | 7 | 1.5 | 48 | 25.7 | 26.4 | 26 | 59 | — |
| Korea, South | 47.3 | 14 | 5 | 0.9 | 82 | 53.3 | 51.1 | 11 | 79 | 8,600 |
| Mongolia | 2.5 | 20 | 7 | 1.4 | 50 | 3.4 | 4.1 | 34 | 52 | 380 |
| Taiwan | 22.3 | 13 | 6 | 0.7 | 97 | 25.3 | 25.2 | 6.6 | 77 | — |
| **EUROPE** | **728** | **10** | **11** | **-0.1** | **—** | **714** | **658** | **9** | **73** | **13,420** |
| Northern Europe | 96 | 12 | 11 | 0.1 | 653 | 101 | 100 | 6 | 83 | 21,640 |
| Denmark | 5.3 | 12 | 11 | 0.1 | 472 | 5.8 | 6.1 | 4.7 | 85 | 33,040 |
| Estonia | 1.4 | 8 | 13 | -0.5 | — | 1.3 | 1.0 | 9 | 69 | 3,360 |
| Finland | 5.2 | 11 | 10 | 0.2 | 433 | 5.3 | 4.8 | 4.2 | 60 | 24,280 |
| Iceland | 0.3 | 15 | 7 | 0.9 | 81 | 0.3 | 0.3 | 2.6 | 92 | 27,830 |
| Ireland | 3.8 | 15 | 9 | 0.6 | 116 | 4.5 | 4.5 | 6.2 | 58 | 18,710 |
| Latvia | 2.4 | 8 | 14 | -0.6 | — | 2.1 | 1.7 | 11 | 69 | 2,420 |
| Lithuania | 3.7 | 10 | 11 | -0.1 | — | 3.5 | 3.1 | 9 | 68 | 2,540 |

| | Population mid-2000 (millions) | Births per 1,000 pop. | Deaths per 1,000 pop. | Natural Increase (annual, %) | "Doubling Time" in Years at *Current* Rate | Projected Population (millions) 2025 | 2050 | Infant Mortality Rate[a] | Percent Urban | GNP Per Capita, 1998 (US$) |
|---|---|---|---|---|---|---|---|---|---|---|
| Norway | 4.5 | 13 | 10 | 0.3 | 217 | 4.9 | 5.1 | 4.0 | 74 | 34,310 |
| Sweden | 8.9 | 10 | 11 | -0.1 | — | 9.3 | 9.2 | 3.5 | 84 | 25,580 |
| United Kingdom | 59.8 | 12 | 11 | 0.1 | 546 | 64.1 | 64.2 | 5.7 | 89 | 21,410 |
| Western Europe | 183 | 11 | 10 | 0.1 | 612 | 188 | 181 | 5 | 79 | 26,160 |
| Austria | 8.1 | 10 | 10 | 0.0 | 2,310 | 8.1 | 7.7 | 4.9 | 65 | 26,830 |
| Belgium | 10.2 | 11 | 10 | 0.1 | 770 | 10.3 | 10.0 | 5.6 | 97 | 25,380 |
| France | 59.4 | 13 | 9 | 0.3 | 204 | 64.2 | 65.1 | 4.8 | 74 | 24,210 |
| Germany | 82.1 | 9 | 10 | -0.1 | — | 80.2 | 73.3 | 4.7 | 86 | 26,570 |
| Liechtenstein | 0.03 | 14 | 7 | 0.7 | 105 | 0.04 | 0.04 | 18.4 | — | — |
| Luxembourg | 0.4 | 13 | 9 | 0.4 | 198 | 0.6 | 0.6 | 5.0 | 88 | 45,100 |
| Monaco | 0.03 | 20 | 17 | 0.3 | 239 | 0.04 | 0.04 | — | 100 | — |
| Netherlands | 15.9 | 13 | 9 | 0.4 | 193 | 17.3 | 17.2 | 5.0 | 61 | 24,780 |
| Switzerland | 7.1 | 11 | 9 | 0.2 | 315 | 7.6 | 7.4 | 4.8 | 68 | 39,980 |
| Eastern Europe | 304 | 9 | 13 | -0.5 | — | 287 | 258 | 14 | 68 | 2,340 |
| Belarus | 10.0 | 9 | 14 | -0.5 | — | 9.4 | 8.5 | 11 | 70 | 2,180 |
| Bulgaria | 8.2 | 8 | 14 | -0.6 | — | 6.6 | 5.3 | 14.4 | 68 | 1,220 |
| Czech Republic | 10.3 | 9 | 11 | -0.2 | — | 10.2 | 9.3 | 4.6 | 77 | 5,150 |
| Hungary | 10.0 | 9 | 14 | -0.5 | — | 9.2 | 8.0 | 8.9 | 64 | 4,510 |
| Moldova | 4.3 | 11 | 11 | 0.0 | 1,733 | 4.5 | 4.2 | 18 | 46 | 380 |
| Poland | 38.6 | 10 | 10 | 0.0 | — | 38.6 | 33.9 | 9 | 62 | 3,910 |
| Romania | 22.4 | 11 | 12 | -0.2 | — | 20.6 | 17.8 | 20.5 | 55 | 1,360 |
| Russia | 145.2 | 8 | 15 | -0.6 | — | 136.9 | 127.7 | 17 | 73 | 2,260 |
| Slovakia | 5.4 | 11 | 10 | 0.1 | 866 | 5.4 | 4.7 | 8.8 | 57 | 3,700 |
| Ukraine | 49.5 | 8 | 14 | -0.6 | — | 45.1 | 38.4 | 13 | 68 | 980 |
| Southern Europe | 145 | 10 | 10 | 0.0 | 2,121 | 137 | 118 | 7 | 70 | 15,340 |
| Albania | 3.4 | 18 | 5 | 1.3 | 55 | 4.5 | 5.2 | 22 | 46 | 810 |
| Andorra | 0.1 | 11 | 3 | 0.8 | 92 | 0.1 | 0.2 | 6 | 95 | — |
| Bosnia-Herzegovina | 3.8 | 13 | 8 | 0.5 | 141 | 4.2 | 3.9 | 12 | 40 | — |
| Croatia | 4.6 | 11 | 12 | -0.1 | — | 4.4 | 3.9 | 8.2 | 54 | 4,620 |
| Greece | 10.6 | 10 | 10 | -0.0 | — | 10.4 | 9.7 | 6.7 | 59 | 11,740 |
| Italy | 57.8 | 9 | 10 | -0.1 | — | 52.4 | 41.9 | 5.5 | 90 | 20,090 |
| Macedonia | 2.0 | 15 | 8 | 0.6 | 112 | 2.2 | 2.1 | 16.3 | 59 | 1,290 |
| Malta | 0.4 | 12 | 8 | 0.4 | 182 | 0.4 | 0.4 | 5.3 | 89 | 10,100 |
| Portugal | 10.0 | 11 | 11 | 0.1 | 990 | 9.3 | 8.2 | 5.4 | 48 | 10,670 |
| San Marino | 0.03 | 11 | 7 | 0.4 | 193 | 0.04 | 0.04 | 9 | 88 | — |
| Slovenia | 2.0 | 9 | 10 | -0.1 | — | 1.9 | 1.6 | 5.2 | 50 | 9,780 |
| Spain | 39.5 | 9 | 9 | 0.0 | 6,931 | 36.7 | 30.8 | 5.7 | 64 | 14,100 |
| Yugoslavia | 10.7 | 11 | 11 | 0.1 | 866 | 10.7 | 10.2 | 10 | 52 | — |

[a]Infant deaths per 1,000 live births.
[e]Special Administrative Region

# Appendix B

## Additional Maps

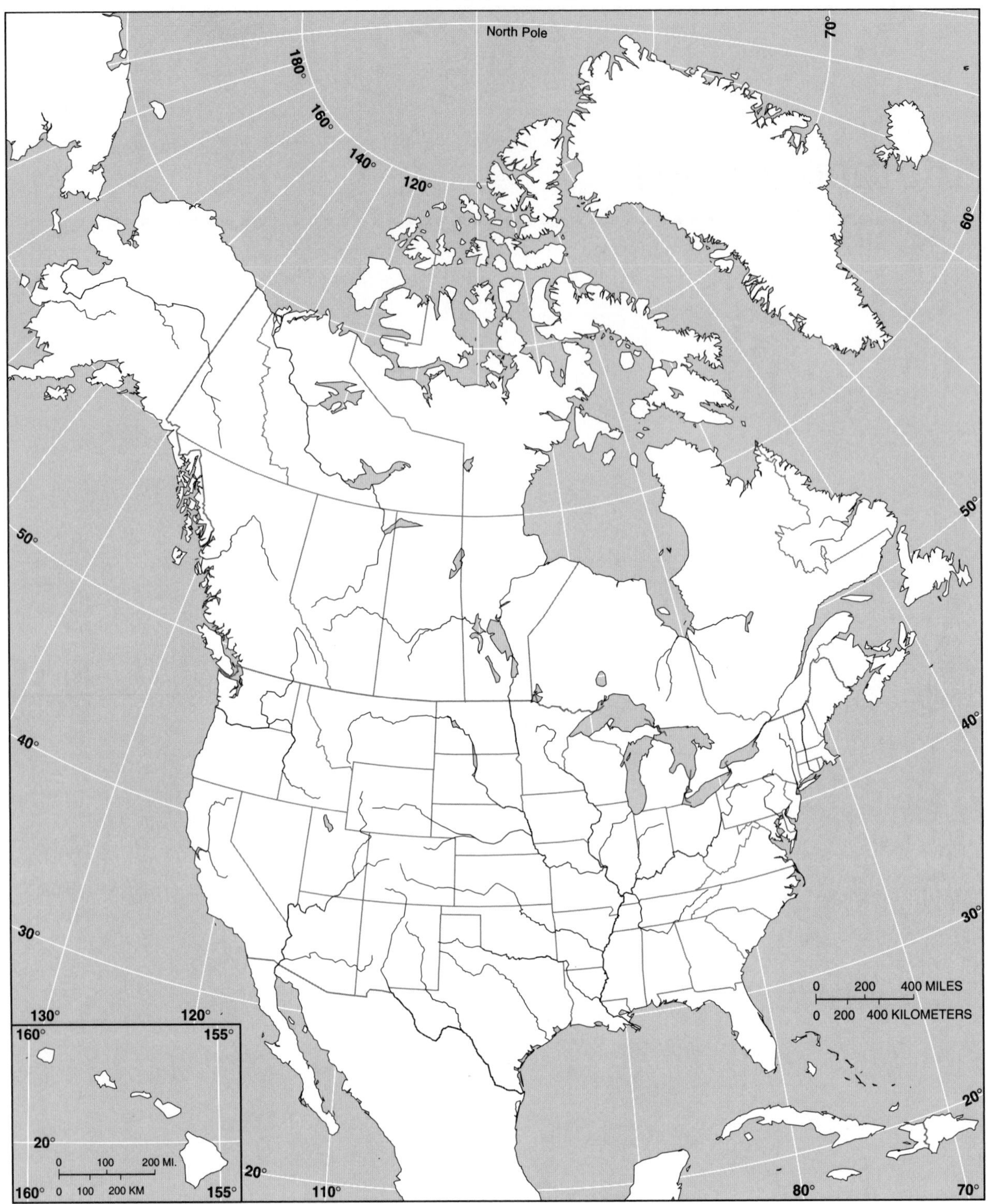


North America map.

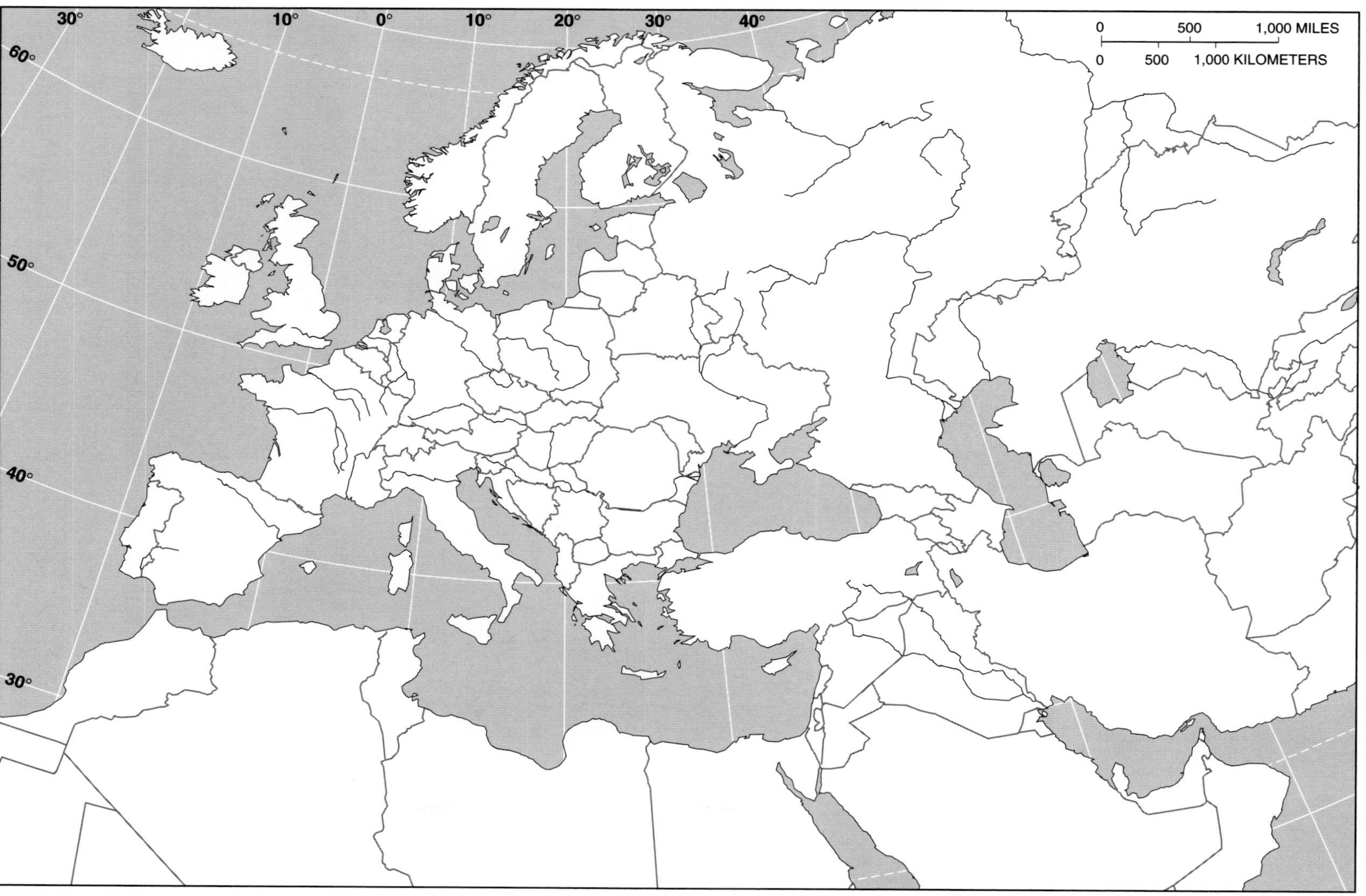
30°
10°
0°
10°
20°
30°
40°
0 500 1,000 MILES
0 500 1,000 KILOMETERS
60°
50°
40°
30°
Europe map.

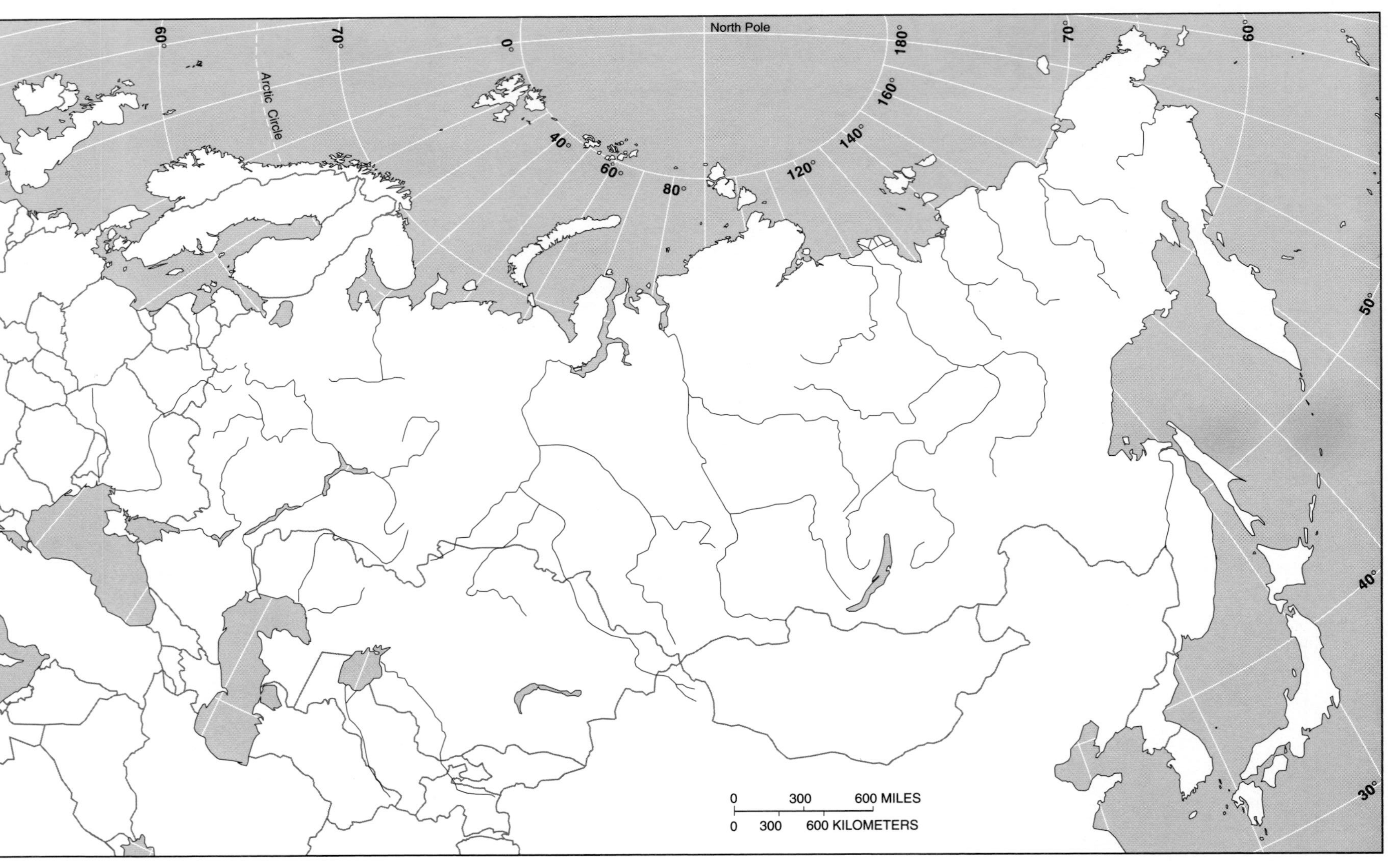


Russia, Transcaucasia, and Central Asia.

Australia and New Zealand.

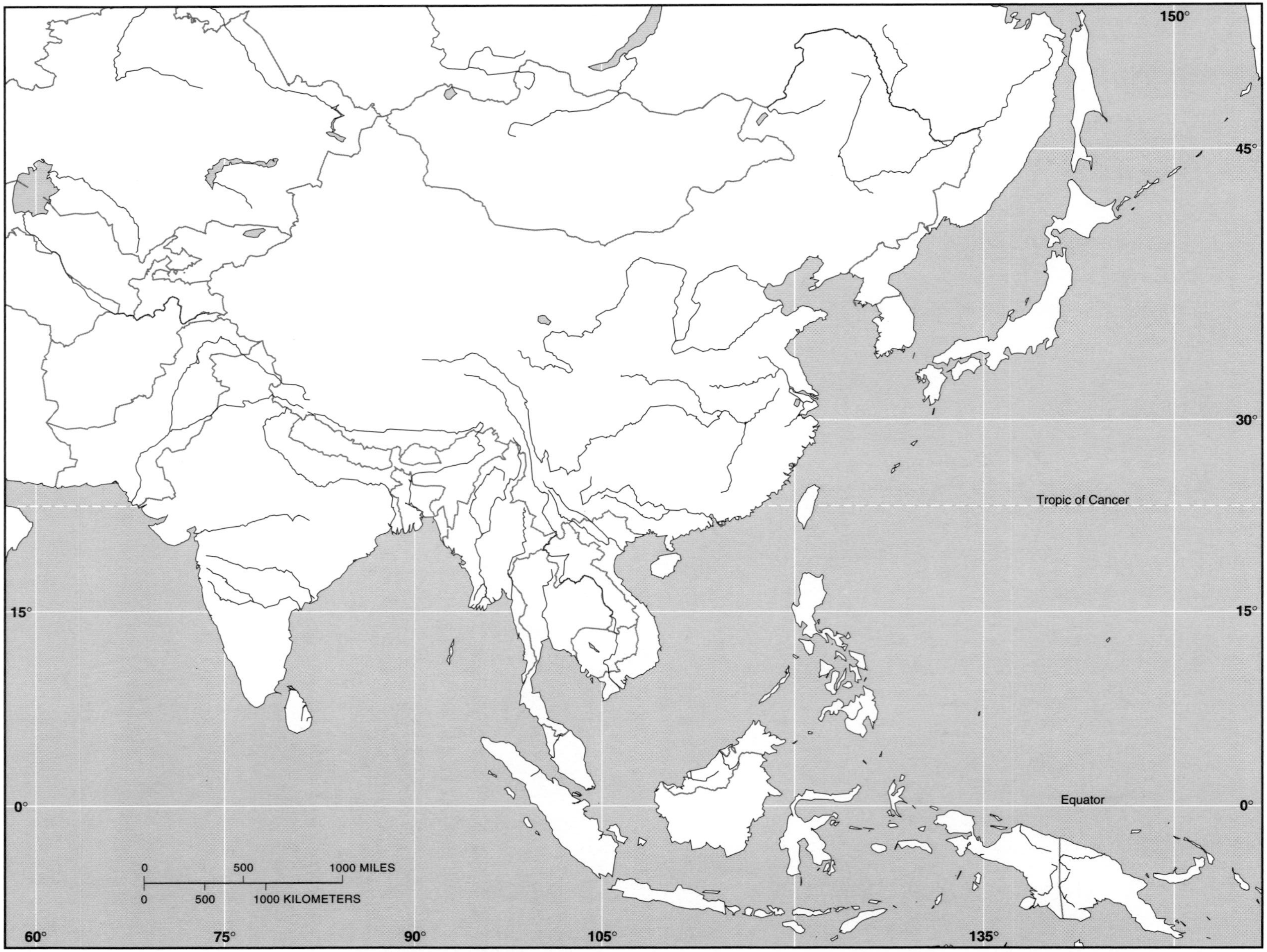

East Asian cultural region.

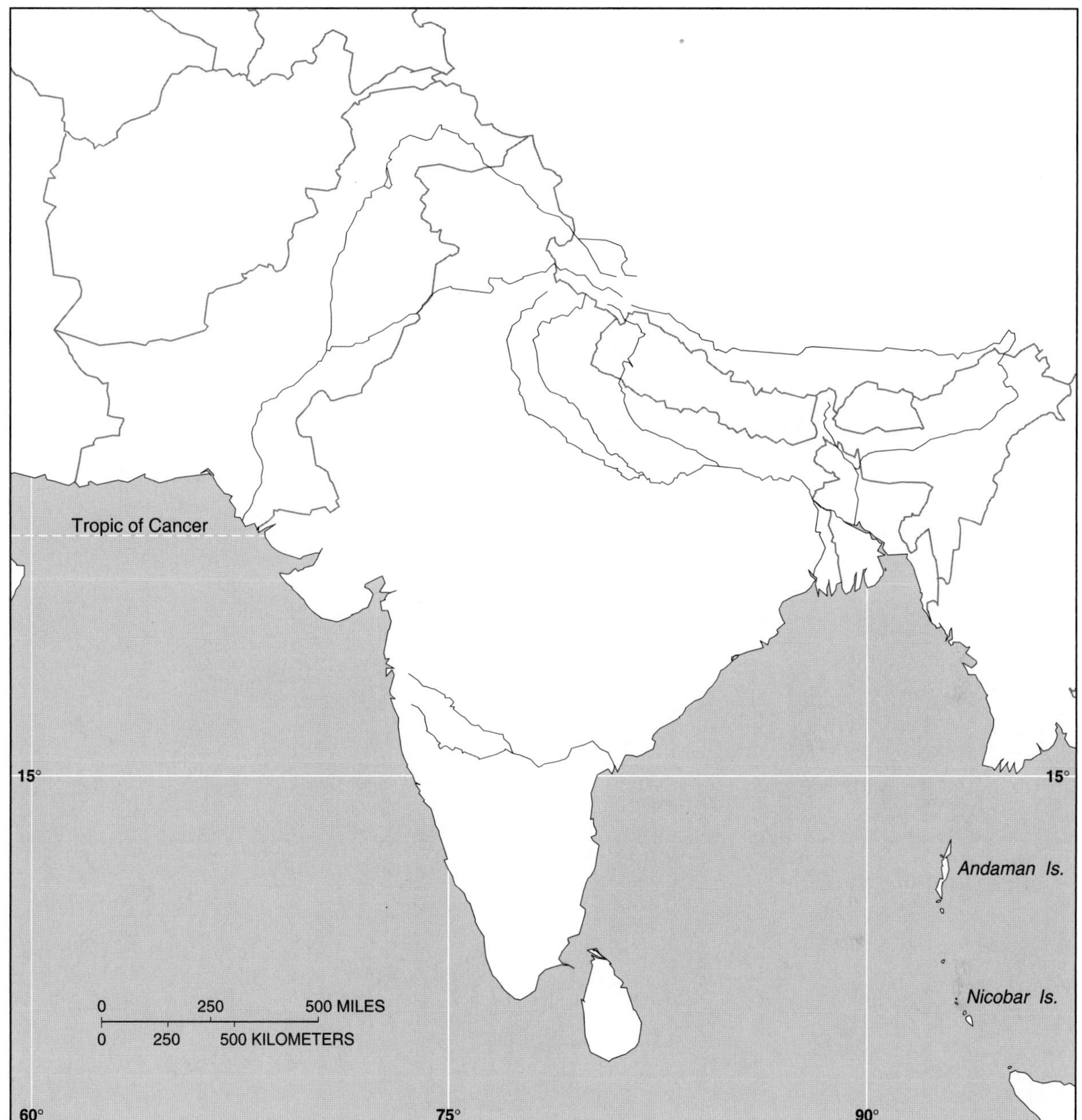


South Asian countries, capital, and major cities.

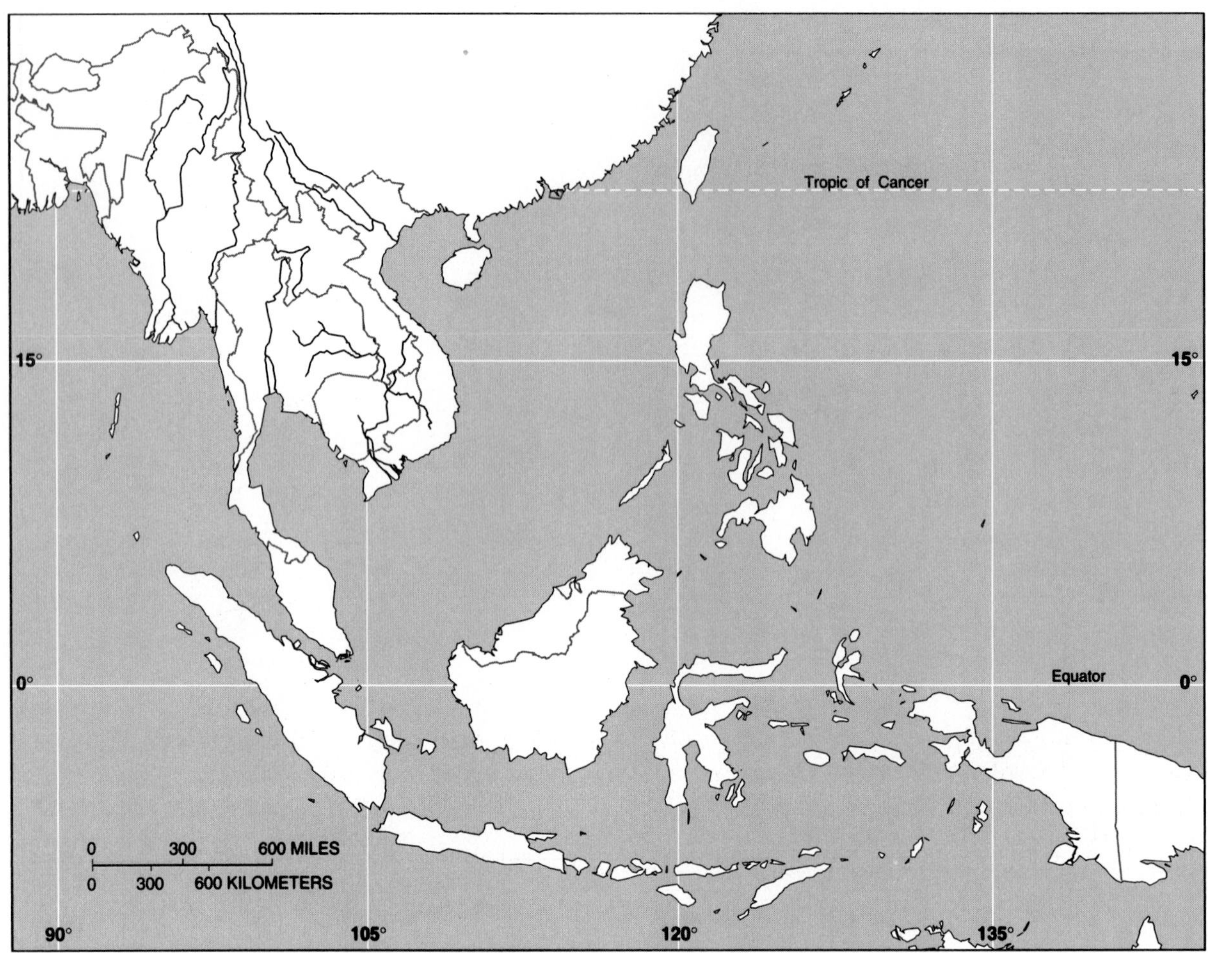


Southeast Asia map.

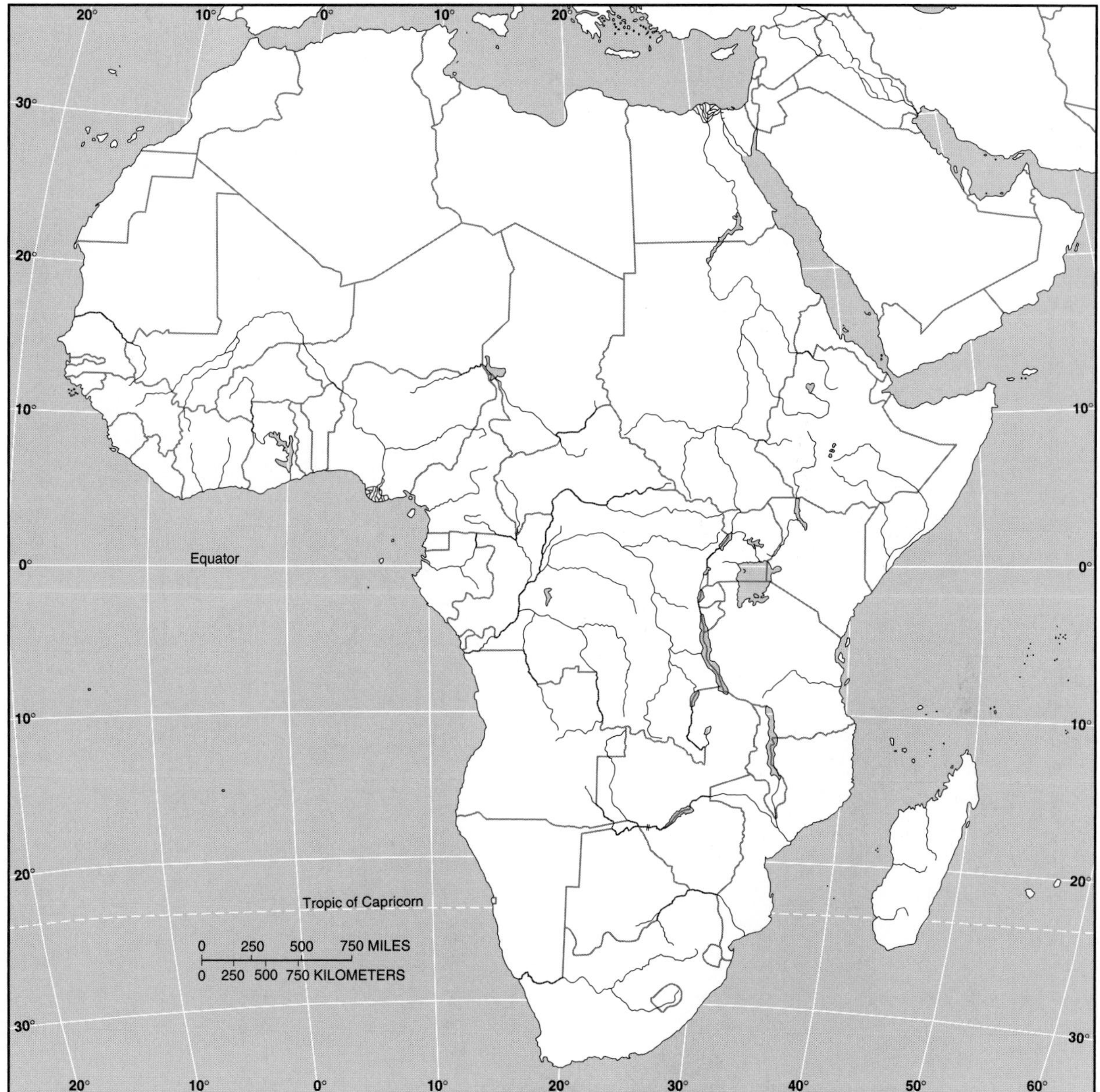
20°
10°
0°
10°
20°
30°
20°
10°
10°
0°
Equator
0°
10°
10°
20°
20°
Tropic of Capricorn
0 250 500 750 MILES
0 250 500 750 KILOMETERS
30°
30°
20°
10°
0°
10°
20°
30°
40°
50°
60°

Africa map.

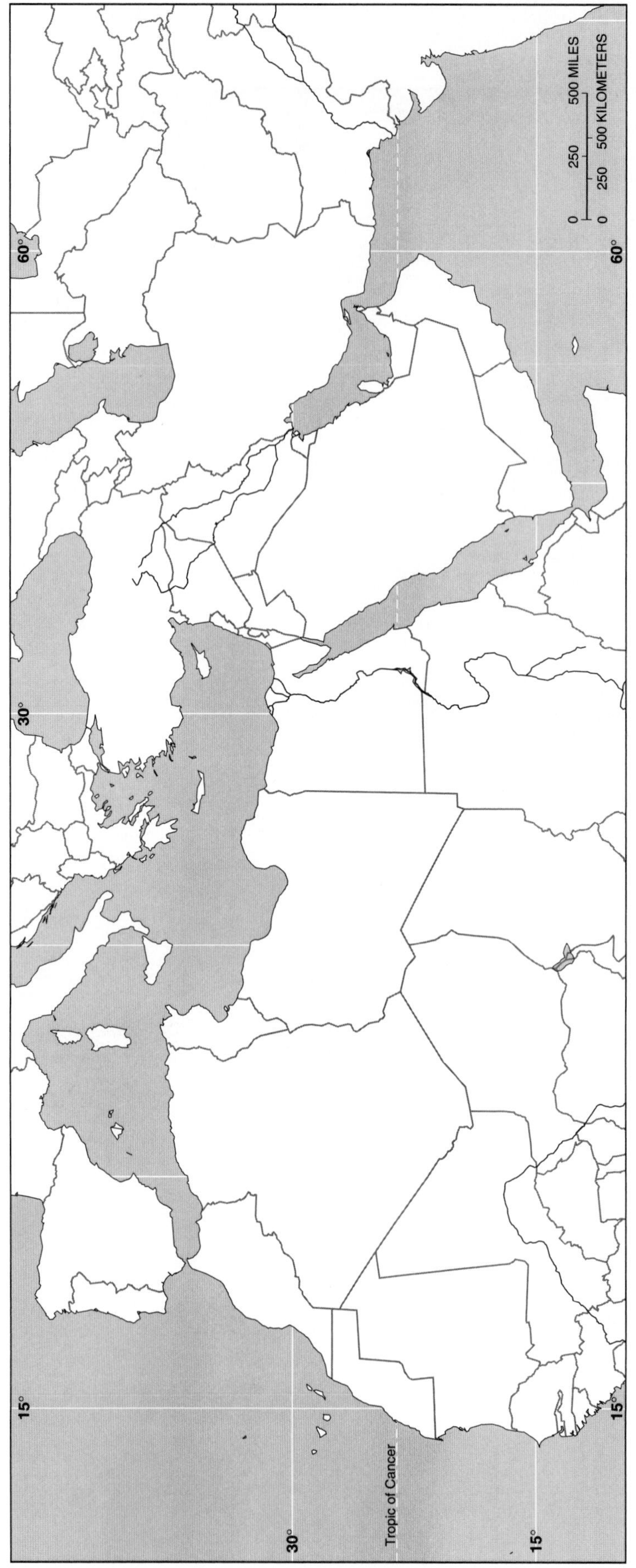


Middle East map.

South America map.

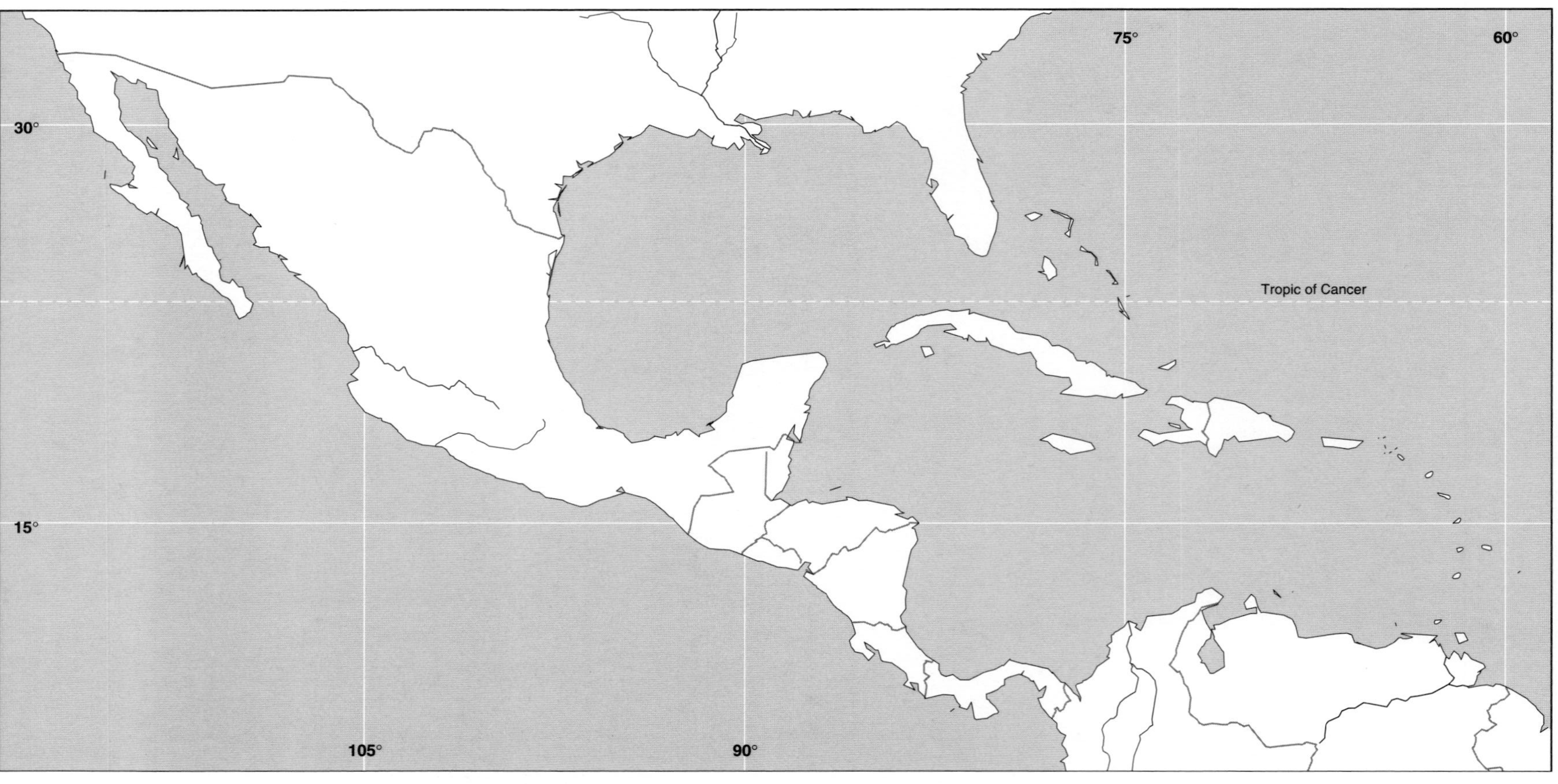


Middle American countries, capitals and other important cities.

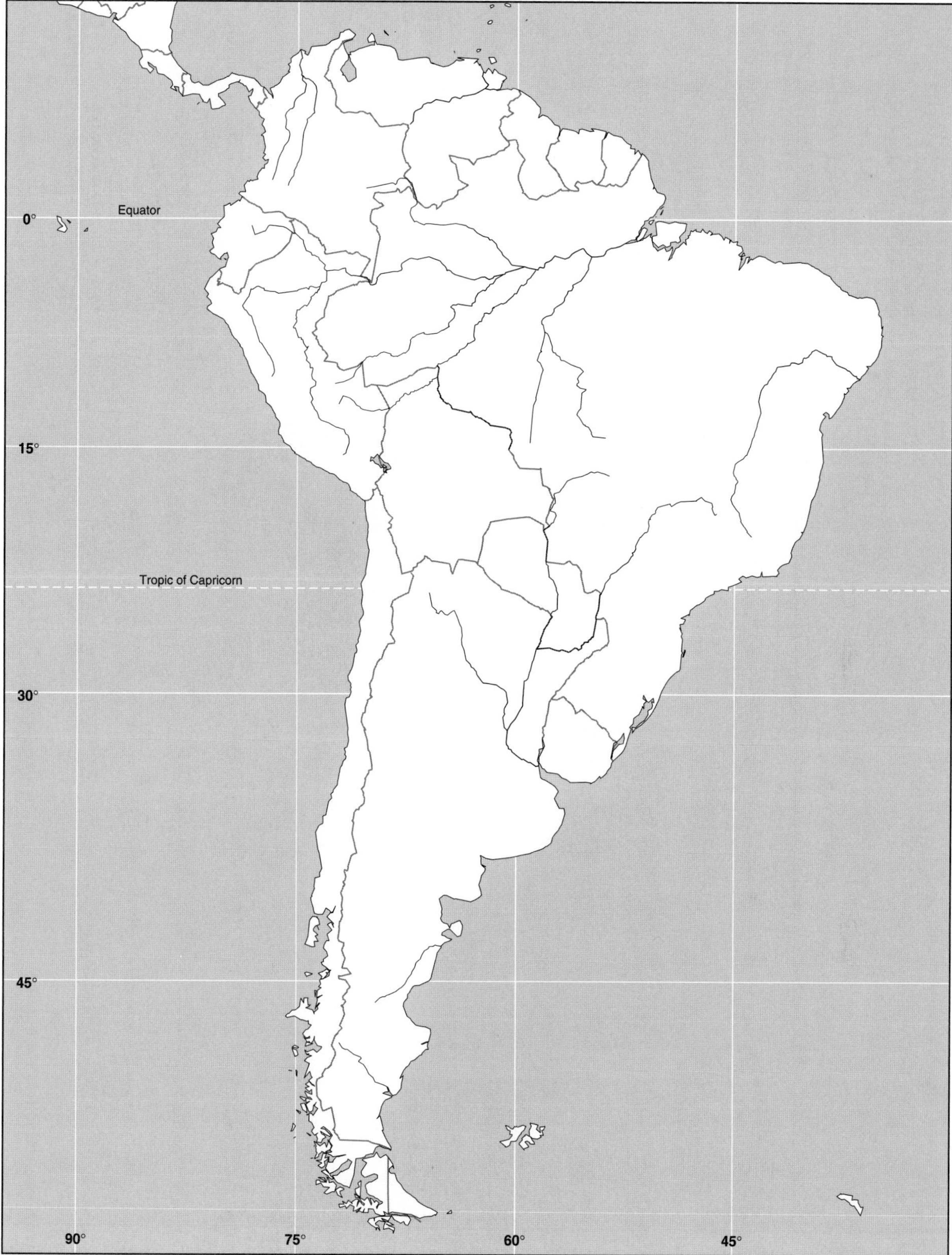


South America: Physical features.

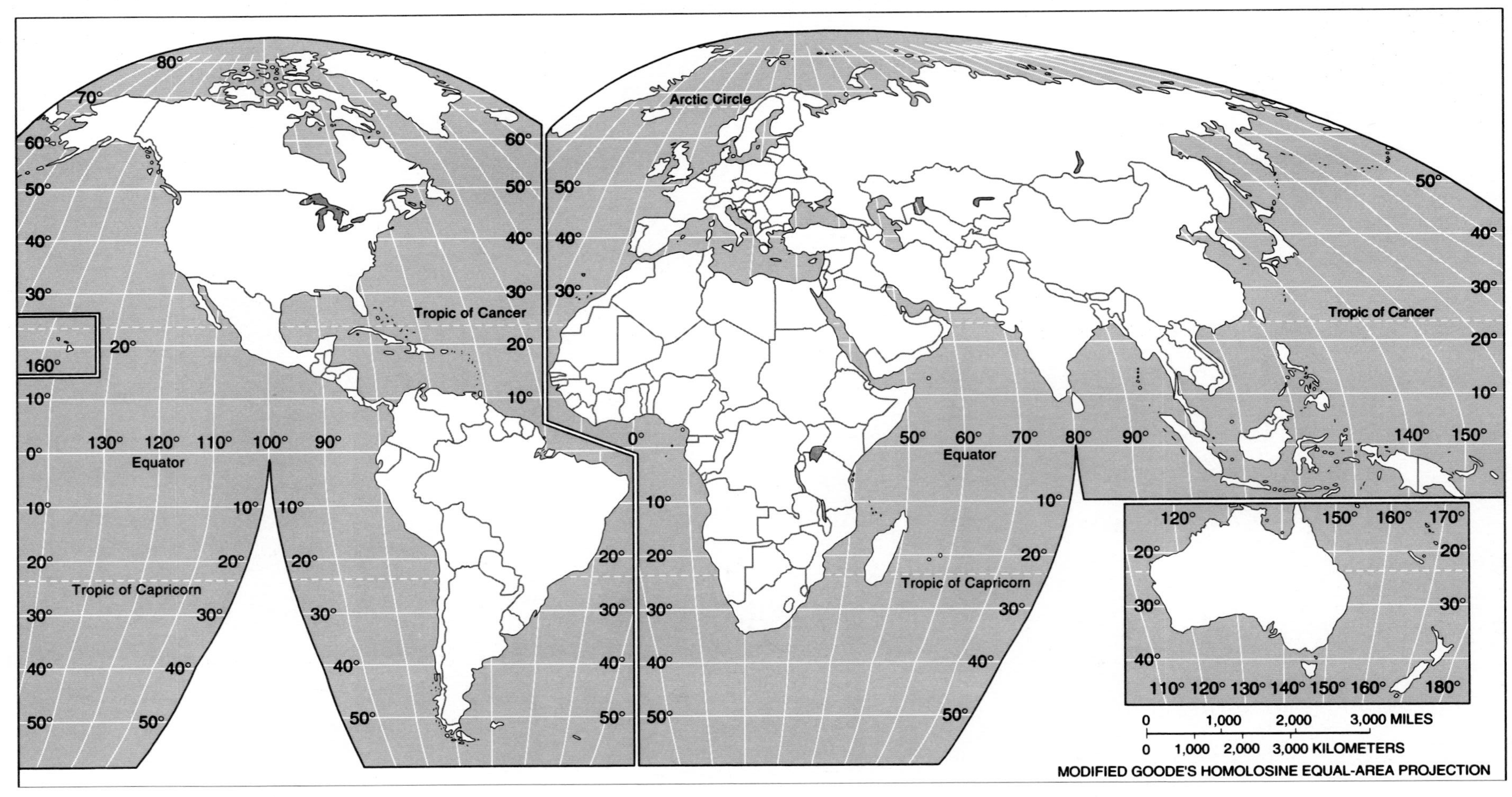


The 10 major world regions used in this workbook.

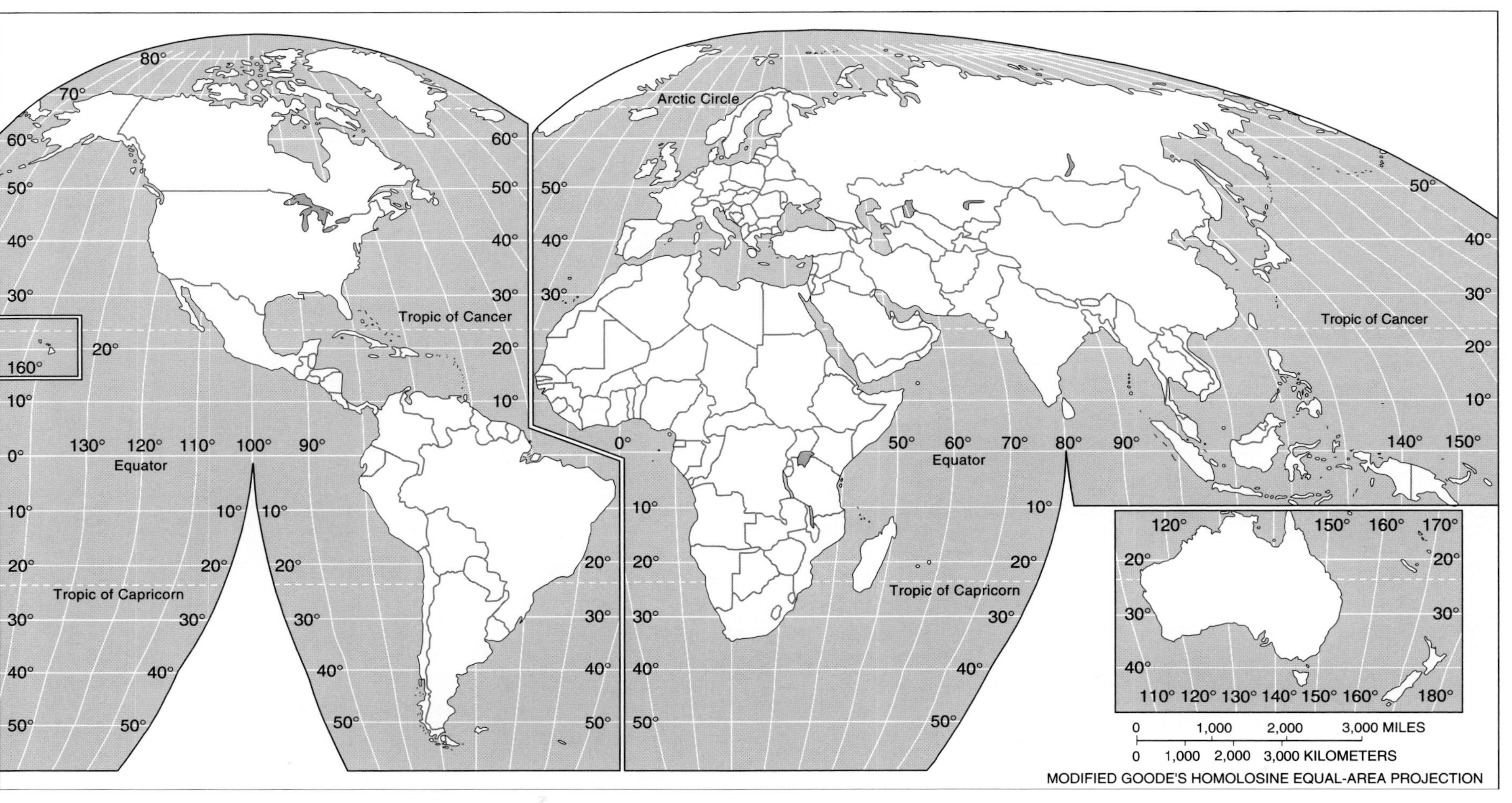


The 10 major world regions used in this workbook.

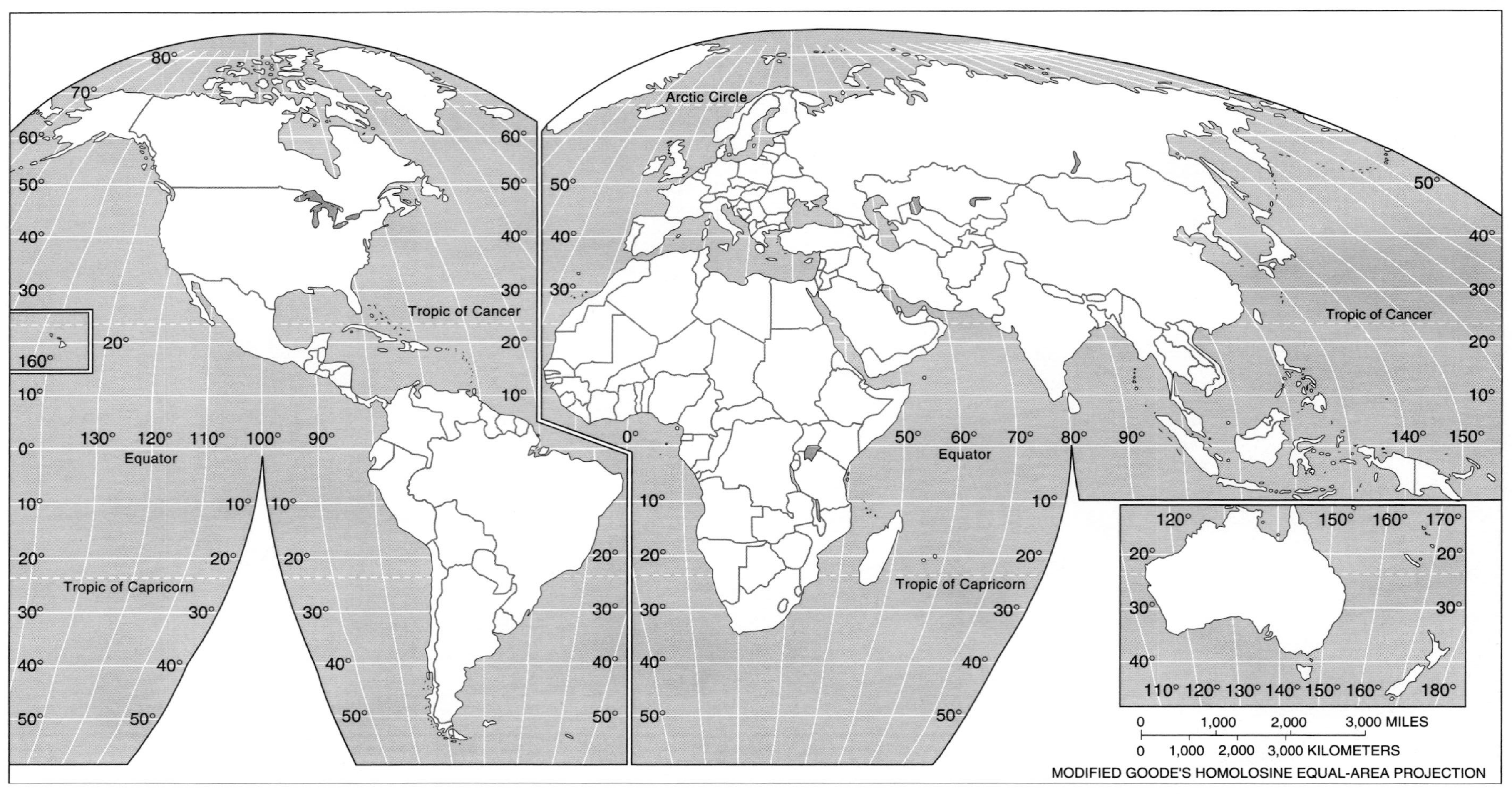


The 10 major world regions used in this workbook.

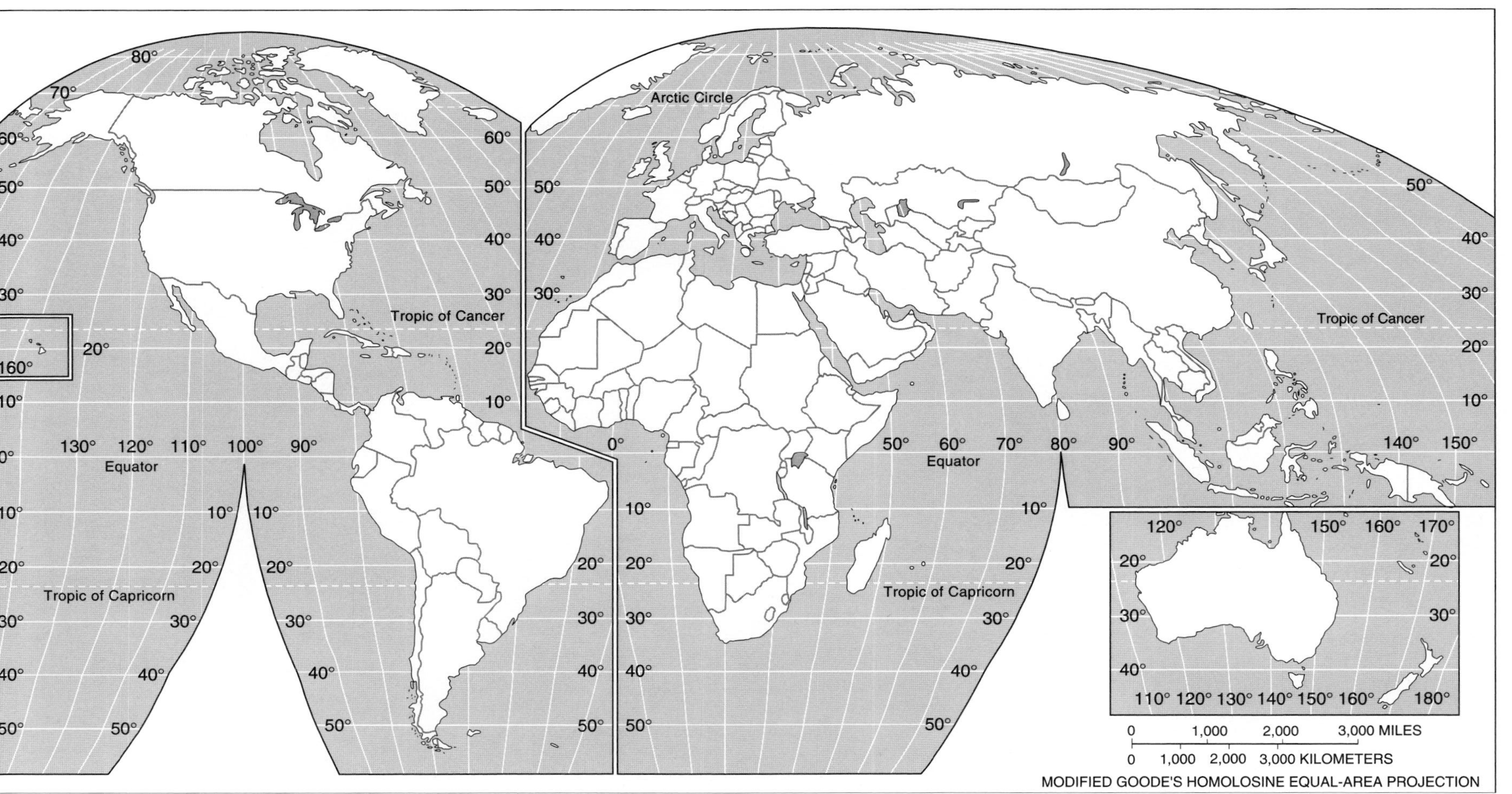


The 10 major world regions used in this workbook.

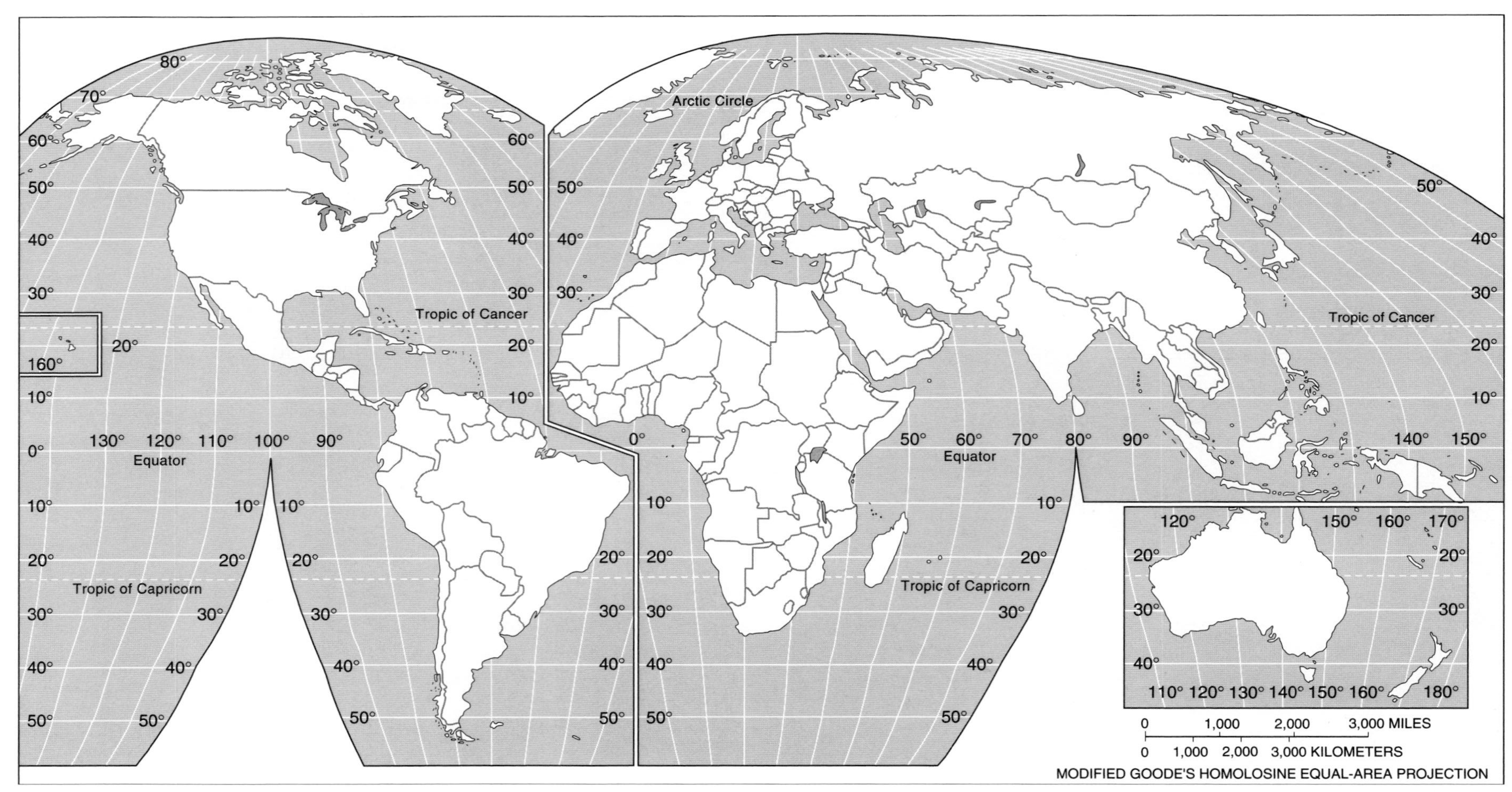


The 10 major world regions used in this workbook.

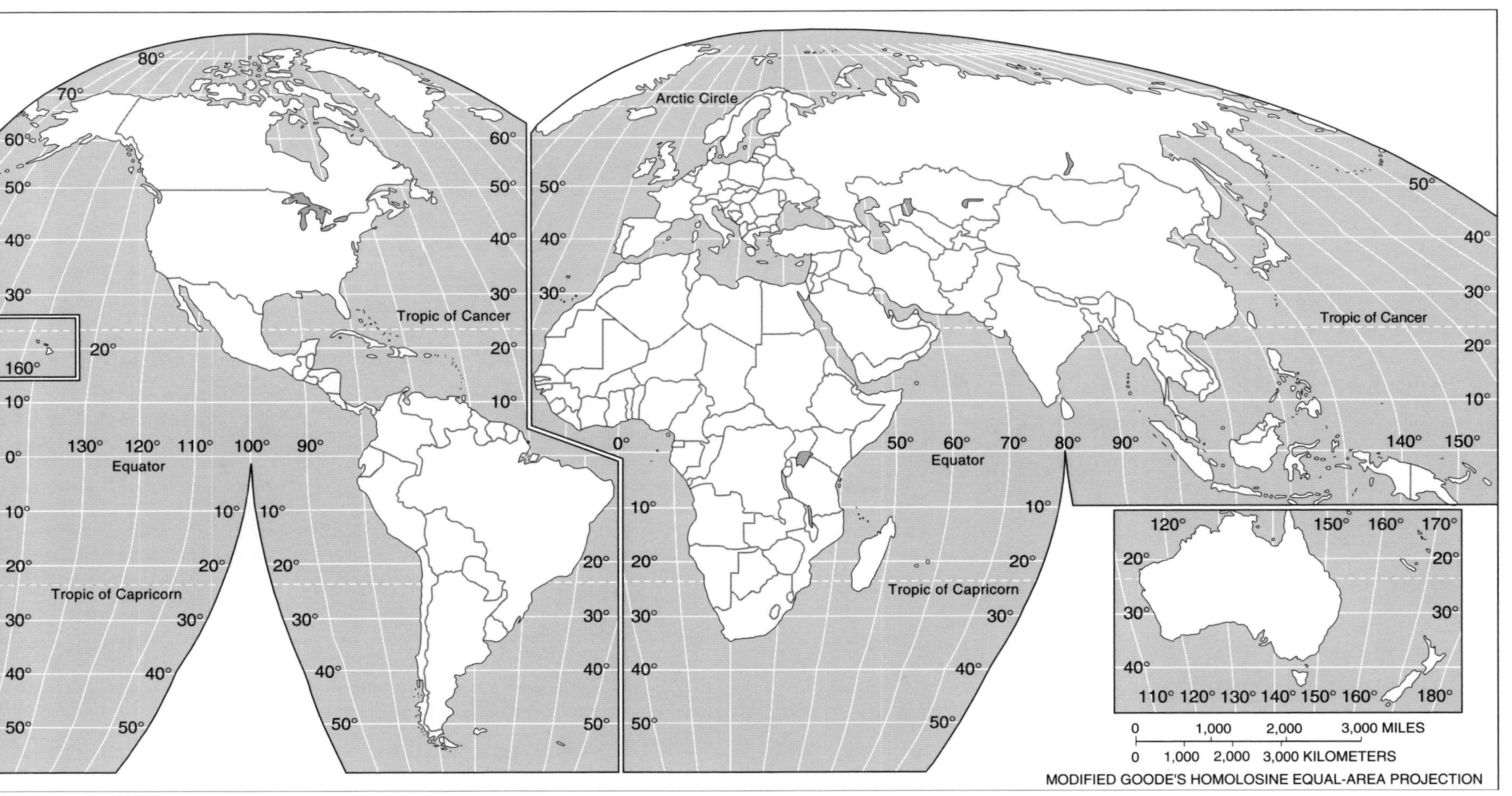


The 10 major world regions used in this workbook.

# Index